recent advances in phytochemistry

volume 24

Biochemistry of the Mevalonic Acid Pathway to Terpenoids

RECENT ADVANCES IN PHYTOCHEMISTRY

Proceedings of the Phytochemical Society of North America
General Editor: Helen A. Stafford, *Reed College, Portland, Oregon*

Recent Volumes in the Series

recent advances in phytochemistry

volume 24

Biochemistry of the Mevalonic Acid Pathway to Terpenoids

Edited by

G.H. Neil Towers

University of British Columbia
Vancouver, British Columbia, Canada

and

Helen A. Stafford

Reed College
Portland, Oregon

PLENUM PRESS • NEW YORK AND LONDON

Library of Congress Cataloging in Publication Data

Phytochemical Society of North America. Meeting (29th: 1989: Vancouver, B.C.)
Biochemistry of the Mevalonic Acid Pathway to Terpenoids / edited by G. H. Neil Towers and Helen A. Stafford.
p. cm.—(Recent advances in phytochemistry; v. 24)
"Proceedings of the twenty-ninth Annual meeting of the Phytochemical Society of North America, held June 16–20, 1989, in Vancouver, British Columbia, Canada"—T.p. verso.
Includes bibliographical references.
Includes index.
ISBN 978-1-4684-8791-6 ISBN 978-1-4684-8789-3 (eBook)
DOI 10.1007/978-1-4684-8789-3
1. Mevalonic acid—Derivatives—Synthesis—Congresses. 2. Terpenes—Synthesis—Congresses. 3. Botanical chemistry—Congresses. I. Towers, G. H. Neil. II. Stafford, Helen A., 1922– . III. Title. IV. Series.
QK861.R38 vol. 24
[QP801.M45]
581.19′2 s—dc20 90-7265
[581.19′2] CIP

Proceedings of the Twenty-ninth Annual Meeting of the Phytochemical Society of North America, held June 16–20, 1989, in Vancouver, British Columbia, Canada

A Division of Plenum Publishing Corporation
233 Spring Street, New York, N.Y. 10013

Softcover reprint of the hardcover 1st edition 1990

PREFACE

This series of lectures was delivered at the 29th meeting of the Phytochemical Society of North America, held at the University of British Columbia in Vancouver, B.C., Canada on June 16th-20th, 1989. Topics concerning terpenoids, consisting of isoprene units, are now so numerous that a judicious selection for a relatively limited symposium was difficult. We were able to assemble, however, a potpourri of reviews on topical areas of terpenoid chemistry, biochemistry and biology, by scientists who are making exciting contributions and whose work points the way to significant future research.

Because of the importance of terpenoids in the life of plants, and indeed in all living organisms, a periodical review of the mevalonic acid pathway and of the subsequent biochemical events leading to the biosynthesis of isoprenoids needs no justification. Life, as we know it, would not be possible without the ability of living organisms to employ this metabolic sequence which proceeds from condensations of three molecules of acetyl-CoA and terminates with the elaboration of the terpenoid precursors, isopentenyl pyrophosphate and dimethylallyl pyrophosphate. In addition to producing obviously essential compounds that are partially or completely of isoprenoid origin (Fig. 1), such as hormones, photosynthetic pigments, compounds involved in electron transport in respiration and in photosynthesis, oxidative enzymes and membrane components, plants elaborate thousands of novel terpenoids, many of which do not as yet have identifiable physiological, biochemical or even ecological roles, e.g. the cardenolides, ecdysones or saponins. We believe that the ecological importance for some of them, e.g., the volatile monoterpenes produced in plant trichomes serve as chemical cues for insects. Studies of the chemical signalling between plants and organisms, ranging from bacteria to mammals have expanded tremendously in recent years, and, in many cases terpenoids have been shown to be involved in these interactions.

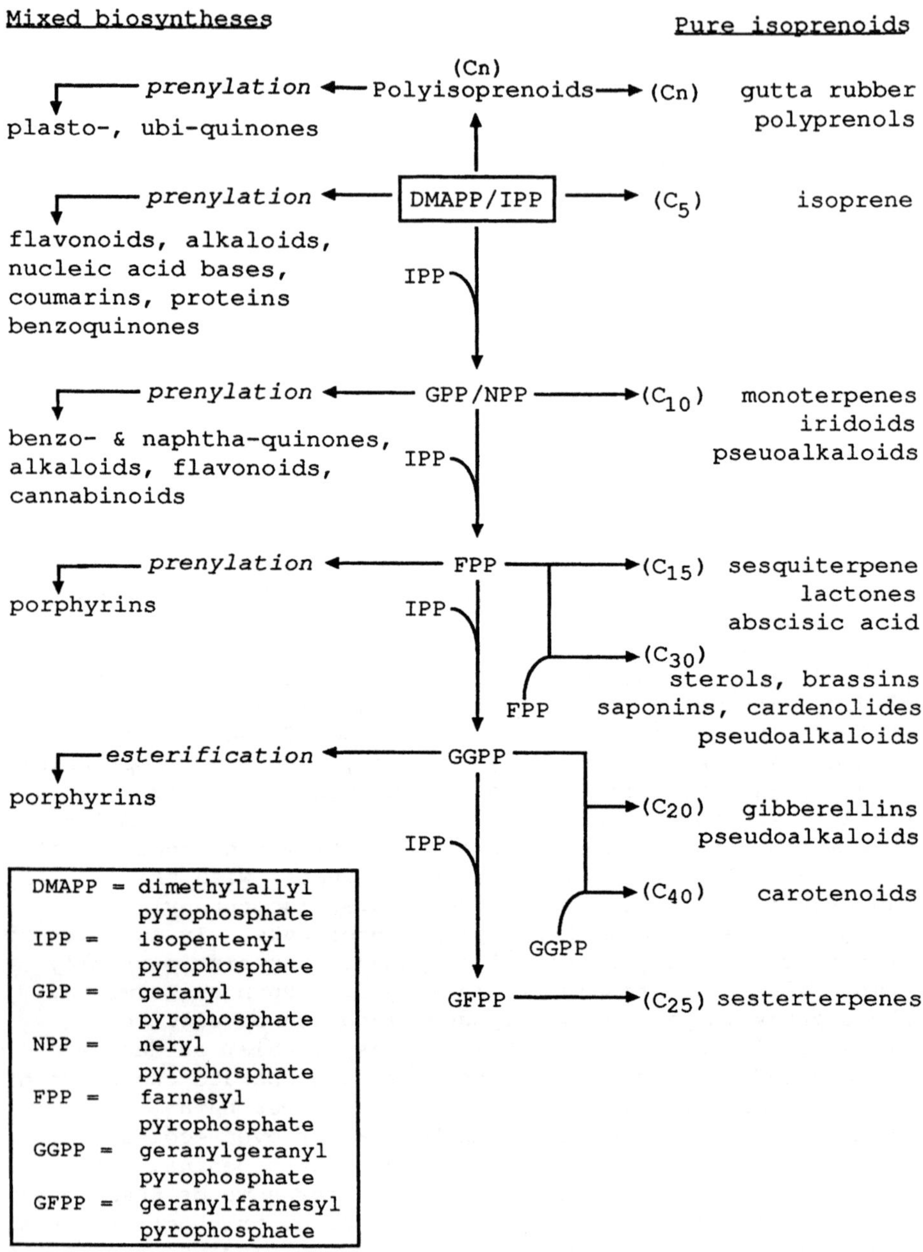

Fig. 1. Some Products of the Mevalonate Pathway in Plants

The sheer number of new terpenes and terpenoids, most of which of course bear trivial chemical names, in terrestrial and marine plants, presents an increasing problem for the phytochemist; surely it is time for the development of a new system that will allow for easy information retrieval in studies of the biochemical and biological relationships of these ever increasing families of phytochemicals. Professor Fischer points out (Chapter 4) that, in the ten years between 1979 and 1989, the number of identified sesquiterpene lactones alone rose from 500 to over 3200!

In Chapter 1 Bach and his colleagues present an up-to-date analysis of the biochemistry of the mevalonic acid (MVA) pathway, particularly the properties of hydroxymethylglutaryl CoA reductase (HMGR). Included in this chapter is a complete discussion of the currently conflicting views of compartmentalization of the enzymes involved and of regulation. They also discuss the potent fungal terpenoids, e.g. mevinolin, which are specific inhibitors of HMGR and which have been shown, by them, to act as growth inhibitors. These compounds are in current use in preventive medicine in reducing cholesterol biosynthesis. In Chapter 2 Boronat and his colleagues give us a close look at the structure of HMGR as well as aspects of its molecular biology.

In spite of the large number of monoterpenes in plants, many of which, e.g. camphor, have been known as plant constituents for over a hundred years, it was not until the pioneering investigations of Croteau's group, in the last decade or so, that an understanding of the nature of the cyclases, or enzymes yielding mono- and bicyclic terpenes from geranyl- or neryl- pryophosphates, was achieved. Gershenzon and Croteau provide the most complete account to date of the enzymology, regulation, compartmentalization, subcellular regulation and metabolism of monoterpenes. The results described should encourage plant physiologists to study the transport and further metabolism of this large class of terpenoids.

Fischer discusses the possible biochemical relationships of the fifteen or more classes of sesquiterpene lactones, inviting enzymologists to explore this vast area. It is a sad reflection on plant biochemists that not a single enzyme is known with regard to these compounds. In Chapter 5, Phinney and Spray pay particular attention to the pathway from GA_{12}-

aldehyde to the C_{19} gibberellins, with emphasis on the early-13-hydroxylation pathway leading to the hormone gibberellin (GA_1). Plant anti-microbial terpenoids are discussed by West and his colleagues, who provide a stimulating account of diterpene phytoalexins in soybeans and in rice tissue cultures with particular emphasis on regulation of their biosynthesis. The potential for studies of terpene biochemistry and molecular biology are clearly evident in these studies.

Isman, Proksch and Clark provide an opening to chemical ecology with a clear report on the insecticidal terpenoids of the American composites of the genus *Encelia*. These conspicuous plants of the Baja and other arid lands of Mexico and the U.S.A. excrete copious amounts of resins whose bioactivities were initially studied by Rodriguez's group in California. Many possibilities for plant/insect studies at a chemical level become evident in this chapter. In Chapter 8, by Andersen et al., we are introduced to the novel chemistry of a small selection of the 1700 or more new terpenoids identified in the 1980's in marine hydroids, sponges, nudibranchs and algae. Here again is an invitation for exciting work in chemical ecology.

The final chapter by Nes is a novel and perhaps highly controversial look at the world of sterols. It includes much food for thought about one of the most intriguing and still poorly understood groups of natural products.

We wish to thank all those who helped to arrange this meeting, all of the speakers and those who presented posters. The generous support of the Natural Sciences and Engineering Council of Canada was greatly appreciated.

We hope that the reader will enjoy this addition to our annual symposium series. We believe that it will provide much interesting reading about a very large group of natural products.

February, 1990

G.H.N. Towers
H.A. Stafford

CONTENTS

Chapter One

SOME PROPERTIES OF ENZYMES INVOLVED IN THE BIOSYNTHESIS AND METABOLISM OF 3-HYDROXY-3-METHYLGLUTARYL-CoA IN PLANTS

THOMAS J. BACH, THOMAS WEBER AND ANJA MOTEL

Botanisches Institut II
(Pflanzenphysiologie und Pflanzenbiochemie)
Universität Karlsruhe, Kaiserstr. 12
D-7500 Karlsruhe, F.R.G.

INTRODUCTION

There is hardly a single process in the living plant cell that does not require the biochemical and physiological involvement of isoprenoids. It is thus of great theoretical as well as practical interest to have a profound knowledge of how those various classes of compounds are synthesized, what the

Biochemistry of the Mevalonic Acid Pathway to Terpenoids
Edited by G.H.N. Towers and H. A. Stafford
Plenum Press, New York

properties of the enzymes participating are and how this synthesis is regulated by various factors depending on the demand of the developing and mature plant for various classes of isoprenoids and prenyllipids. Although a large body of information is available on the structures of the myriad of isoprenoids occurring in plants isolated so far, many of the details to be discussed in this contribution have frequently been learned through experiments first performed with animal cells or yeast, systems that, at least as far as the variety of isoprenoid end products is concerned, are not as complex as plants. Therefore, an important purpose of this article is to consider the feasibility of regulatory and of other models developed for non-plant systems that might be likely to explain some features of the enzymology of isoprenoid biosynthesis.

Much of our past as well as ongoing work has centered around gaining some knowledge of those enzymes at the entrance of the multi-branched isoprenoid pathway, where the site of coarse regulation of substrate flux was expected. Little data on plants are available in comparison to the rich data on mammalian cells; there are many reasons, however, to expect similar processes to occur. Rather surprisingly comparatively little attention has been focused in the literature on the complex interplay of enzymatic steps, e.g., those competing for central metabolic intermediates, such as acetyl-CoA, or serving as a substrate for several biosynthetic pathways in plants. In order to shed some additional light on the problem we include data concerning our recent attempts to characterize all enzymes positioned at the entry of the pathway from one or two plant species, a prerequisite for any deeper knowledge of how regulation of substrate flow to isoprenoids and prenyllipids is ensured. This discussion includes aspects of the differential utilization of substrates and the possible segregation of their synthesis by intracellular compartmentalization of enzymes, which is currently a matter of debate. The importance of an intact mevalonate biosynthetic pathway for the growth and development of plants will also be discussed.

BIOSYNTHESIS OF MEVALONIC ACID (MVA)

Early Research on the Biosynthesis of MVA

3-Hydroxy-3-methylglutaric acid (HMG), first isolated from flax seeds[1] as a branched-chain carbonic acid, was discussed early as a possible precursor of isoprenoids, synthesized through a pathway with dimethylacrylic acid as an intermediate.[2,3] Formation of an HMG-derivative (not further identified) from acetate and stimulated by exogenous ATP and HS-CoA in a cell-free flax system was interpreted as a condensation of acetyl-CoA with acetoacetate.[2] Rudney[4,5], Rudney & Ferguson[6] and Lynen et al.[7] demonstrated in homogenates from rat liver and in yeast, respectively, that the synthesis of HMG required the condensation of two molecules of acetyl-CoA, followed by a further condensation of acetyl-CoA to 3-hydroxy-3-methylglutaryl (HMG)-CoA with the release of free HS-CoA. Speculation on the "true" committed precursor molecule for isoprenoid biosynthesis terminated upon the isolation of mevalonic acid (MVA) from distillates of brewers yeast.[8] This was due to its ability as an acetate-replacing growth factor in the culture broth of *Lactobacillus acidophilus* and especially its excellent incorporation into cholesterol in rat-liver homogenates.[9,10] When *Mycobacterium sp.* was grown on synthetic (*R,S*)-MVA as a carbon source, (*S*)-MVA remained in the medium (biological selection of one enantiomer) and this was proof that (*R*)-MVA represents the natural, biosynthetically active enantiomer.[11] Soon thereafter, by monitoring the incorporation of labeled MVA into various plant isoprenoids, the principle that MVA serves as precursor of all isoprenoids could be verified and generalized (for literature see refs.[12-14]). The *in vitro* conversion of HMG-CoA to MVA in yeast-enzyme preparations with NADPH as the reducing coenzyme was independently shown by the research groups of H. Rudney[15-17] and of F. Lynen.[7,18,19] Brodie and Porter [20] demonstrated the conversion of ^{14}C-acetate into MVA in a cell-free system of rat liver; the reaction required the presence of microsomes, soluble proteins, ATP, HS-CoA, glutathione, Mg^{2+} and NADPH as cofactors. The enzyme HMG-CoA reductase (HMGR, EC 1.1.1.34), soon recognized and widely accepted as the key-regulatory enzyme in the cholesterol pathway, has since been the subject of intense studies in mammalian systems and man. For that reason, and in view of the fact that the enzyme's activity largely regulates endogenous cholesterol synthesis, with elevated blood cholesterol dramatically increasing the risk of atherosclerotic disease, health-risk factor number one in most civilized countries, HMGR has been the subject of several recent reviews and monographs.[21-25] As compared to the large body of information on the properties and regulation of the mammalian HMGR, the plant enzyme, at least until recently, was much less well studied (see Bach[26] and Gray[27] for review).

MVA Biosynthesis in Plants

The *in vitro* conversion of HMG-CoA to MVA by various membrane fractions has been achieved in an array of higher and lower plants (Table 1); this list will be certainly expanded in the future. From some of these studies it can be concluded that in plants the enzyme is closely controlled by phytochrome,[28-32] herbicides,[33] phytohormones,[29,32,34-36] by feedback mechanisms[34] and by endogeneous protein factors,[37] possibly through phosphorylation/dephosphorylation[32] or by action of calmodulin (Wititsuwannakul, personal communication). In the sweet potato (*Ipomoea batatas*),[38-40] and the potato[41] HMGR activity was induced after injuring the tissue and especially after infection with pathogenic fungi or by treatment with $HgCl_2$, followed by a rapid increase in the synthesis of isoprenoid phytoalexins. Similar observations have been made when cell cultures of *Solanum tuberosum* L.[42] and *Nicotiana tabacum*[43] were treated with arachidonic acid, an elicitor from *Phytophthora infestans*. In the case of cell and tissue cultures[44] of other plants such as roots of soybean infected with *P. megasperma* f. sp. *glycinea*, no clear induction was observed. As discussed later, HMGR might synthesize sufficient substrate for the formation of metabolic pools, later used for the synthesis of various isoprenoids such as phytoalexins (see ref.[45]), a situation which might also hold for certain stages of fruit ripening in tomato.[46]

HMGR is the rate limiting enzyme in rubber biosynthesis in *Hevea brasiliensis* [47-49] and is subjected to seasonal[48] and diurnal variation in activity.[49] The enzyme seems also to limit the synthesis of triterpenoids in the latex of *Euphorbia lathyris*,[65,66] as well as rubber synthesis in tissue of *Parthenium argentatum* Gray.[69] Many of the studies mentioned above are difficult to explain or are in disagreement with each other. The reason therefore might be that the variety of plant materials used were in quite different stages of development. This unsatisfactory situation prompted us and others to develop protocols for the solubilization and subsequent purification of this important enzyme.

Solubilization, Purification, and Characterization of Plant HMGR

As shown in Table 1 HMGR in plants is largely a membrane-bound enzyme, as in other eukaryotes. HMGR shares this membraneous localization with enzymes positioned later in the sterolic

Table 1. Assay of HMGR activity in plants.

A) Higher Plants	Membrane fraction	Ref.
Hevea brasiliensis (Latex)	600g pellet, 95% in the luteolid fraction, 5% with Frey-Wyssling-particles (49,000 x g pellet); 95% in 49,000 x g pellet	47,48 37,49-51
Pisum sativum (seedlings)	microsomes, mitochondria, plastids	28,29,31,32, 34,52
Ipomoea batatas (roots)	microsomes, mitochondria	38,39,53,54
(cell cultures)	microsomes	40
Nicotiana tabacum (seedlings)	microsomes	55
(suspension cultures)	10,000 + 100,000 x g pellets	42
(transformed cell cultures)	microsomes	26
Raphanus sativus (seedlings)	16,000 + 105,000 x g pellets	30,33,36,56, 57
Hordeum vulgare (seedlings)	microsomes, (plastids)	35
Capsicum annuum	chromoplasts	58
Daucus carota (cell cultures)	10,000 + 100,000 x g pellets	59
Glycine max L. (hypocotyls & cell cultures)	500 x g pellet	44
(roots of seedlings)	?	60
Nepetea cataria (leaf and callus tissue)	microsomes, plastids	61
Spinacea oleracea (leaf tissue)	microsomes	62

(continued next page)

(Table 1 continued)

Helianthus tuberosus (tissue explants)	15,000 + 105,000 x g pellets	63
Solanum tuberosum (tubers)	microsomes	41,64
(suspension cultures)	10,000 + 100,000 x g	43,45
Euphorbia lathyris (latex)	5,000 x g pellet	65,66
(leaf and stem tissue)	18,000 x g pellet	65
Sinapis alba L. (etiolated seedlings)	microsomes	67
Parthenium argentatum Gray (leaf tissue)	cytosol (S 45,000 x g), + plastids	68,69
Phaseolus radiatus (leaf tissue)	plastids	69
Lycopersicum exculentum (ripening fruits)	microsomes, plastids	46
Picea abies (seedlings, cell cultures)	microsomes, mitochondria	70,71
B) Lower plants		
Ochromonas malhamensis	microsomes	72,73
Dunaliella salina	plastids, mitochondria (?)	74

pathway that are involved in the synthesis of squalene and its conversion into the various phytosterols. For these enzymes it makes sense inasmuch as all substrates and products become increasingly hydrophobic and may interact with the lipid bilayer of the membrane. HMG-CoA, however, is freely water-soluble. Thus the "biological sense" behind this could be that for a close control of enzyme activity the cell has more options, for example, through changes in the lipid microenvironment of the enzyme, changes in membrane fluidity etc. Besides regulation through synthesis and degradation, lipid-induced changes in the kinetic properties of the native membrane-bound enzyme are conceivable. However, at this stage it has to be noted that enzymes preceding HMGR appear to be partially membrane-associated, as is shown later.

The purification of a membrane-bound enzyme requires its solubilization without denaturation. The well-established protocols for the solubilization of rat liver HMGR through repeated freezing and thawing of microsomes (cf. ref.[75] and literature cited therein) did not work successfully with membranes from radish.[57] Therefore we developed a new protocol using a detergent. Polyoxyethylene ether (Brij W-1) was found to be effective in activating and solubilizing radish HMGR. We continually use this material to solubilize enzymes other than HMGR because of its low costs and the property of not absorbing light at 280 nm, where proteins are usually monitored during column chromatography. A disadvantage is the insolubility of this detergent, which crystallizes below 18°C; however, in the case of radish HMGR, the enzyme is rather stable at room temperature. This stability in solution is greatly diminished upon passage of the enzyme through various materials such as DEAE-Sephadex or DEAE-Sepharose etc., where it can be largely inactivated. Such effects might be explained by induction of irreversible conformational changes which render the enzyme catalytically inactive. The methods for efficient solubilization and further purification of radish HMGR, described previously in sufficient detail,[76] could be immediately and successfully applied to the enzyme found in the latex of *Hevea brasiliensis* (R. Wititsuwannakul, personal communication, manuscript in preparation). Earlier attempts to solubilize HMGR from *Hevea* [50] or *Euphorbia lathyris*[65] with the aid of Triton X-100 were unsuccessful.

By applying detergent-solubilization we have also achieved a purification of HMGR from a microsomal fraction of etiolated maize seedlings (details to be reported elsewhere). This enzyme preparation was used for the kinetic studies reported below. To our knowledge this is the first characterization of HMGR from a monocot (see Table 1). In the course of experiments it was revealed that the repeated incubation at 37°C in the presence of Brij W-1 did not substantially increase the specific activity of the enzyme preparation. Therefore, the temparature was lowered to 30°C. In our maize system the specific activity associated with the heavy-membrane fraction

(P 16,000 x g) was not as high as that of the presumably microsome-bound enzyme (in contrast to the situation in radish). Therefore, we first concentrated on the latter membrane fraction. When detergent-solubilized membrane-enzymes were precipitated with ammonium sulfate (at 37% saturation in the case of preparations from radish, at 30% with maize) the subsequent centrifugation led to the formation of a floating layer consisting of proteins, lipids and carotenoids.[57,76,77] This procedure removed an interfering enzyme activity which cannot be inhibited by mevinolin (cf. refs.[36,57]), presumably HMG-CoA lyase (HMGL EC 4.1.3.4).[54] In maize this enzyme partially co-precipitated at 37% salt; thus the lower concentration was chosen, even risking uncomplete recovery of HMGR activity. However, in contrast to the radish enzyme, attempts to further purify the maize HMGR by chromatography on DEAE-Sepharose did not work out satisfactorily. Therefore, this step was replaced by adsorption chromatography on hydroxylapatite. Bound enzyme was washed off the column at phosphate concentrations between 200 and 400 mM, depending on the loading conditions. By far the major part of proteins as well as carotenoids and lipids were contained in the breakthrough fraction in the presence of about 0.5% detergent (Brij W-1). Although the further purification of the enzyme by Blue-Dextran agarose was successful, this step was omitted in favor of the more specific HMG-CoA agarose when sufficiently active enzyme had to be isolated for kinetic studies. The maize HMGR appears to be even more sensitive than the radish enzyme; for example, the loading and elution of the enzyme preparation from HMG-CoA agarose can destroy more than 90% of the initial activity. A microsome-bound form of HMGR was recently solubilized from potato tubers by trypsin treatment and subsequently purified.[64] In earlier studies we also used trypsin digestion of membrane preparations; however, the radish HMGR was very sensitive towards protease treatment, which was slightly diminished in the presence of substrates. The binding of HMGR from radish membranes to the various materials tested was not affected by the detergent. The binding properties to materials such as Blue-Dextran agarose and HMG-CoA agarose were quite similar to those reported for the rat-liver enzyme solubilized by freeze-thawing.[75] When detergent-solubilized and purified HMGR from radish was subjected to sucrose density centrifugation, activity could only be recovered when traces of Triton X-100 were present in the tube (since gradients were prepared by freeze-thawing, the use of Brij W-1 was impossible because of its insolubility). This can be interpreted to mean that under the influence of the strong shearing forces upon entering the gradient and under high-pressure conditions during further migration, the enzyme-bound detergent is stripped off, rendering the enzyme inactive. When the detergent is mainly removed from the enzyme by affinity chromatography on Extractigel D (Pierce), (it is not possible to remove it completely), the apparent HMGR activity is drastically reduced. It is therefore assumed that at least trace amounts of detergent are required to maintain the enzyme active once it has been detergent-solubilized.[57]

By sucrose-density centrifugation (in the presence of Triton X-100), the molecular mass of the purified radish enzyme was estimated to be 180 kD.[57] SDS-PAGE yielded a subunit molecular mass of 45 kD, therefore it was concluded that the enzyme was a tetramer. These findings are in contrast to the values reported for the potato enzyme; there a molecular mass of 110 kD for the active enzyme was estimated by gel filtration, consisting of two subunits of 55 kD.[64] However, our values have found some independent support from the data on purified *Hevea*-HMGR. By electrophoresis under non-denaturing conditions the apparent molecular weight was shown as 176 kD, with subunits of 44 kD (R. Wititsuwannakul, personal communication, manuscript in preparation). Proteins extracted from tobacco cultures supplied with [^{35}S]-methionine were subjected to affinity chromatography on HMG-CoA agarose. When those bound to the column were eluted and separated by means of SDS-PAGE, a strongly labeled band of 45 kD and some very weak bands appeared after fluorography (U. Vögeli & J. Chappell, personal communication). It cannot be excluded that proteolytic processes take place during detergent-solubilization, although addition of 5 mM phenylmethylsulfonyl fluoride (PMSF) or 10 μM leupeptin as inhibitors of protease did not result in any decrease in the solubilization efficiency.[56] However, as yet it has not been possible to keep the enzyme active during purification in the presence of rather high concentrations of PMSF. This does not tell us too much since a full loss of enzyme activity can also happen in the absence of the inhibitor, depending more on the time required to load and to elute enzyme preparations from various materials during conventional column chromatography. Meanwhile a cDNA encoding an entire radish HMGR-protein has been cloned and sequenced (manuscript in preparation). From this result it appears that the "true" molecular mass for the monomeric enzyme comprising 583 amino acids is about 63 kD. It is attractive to localize the site of proteolytic cleavage in the proline-rich "PEST" region linking the two aminoterminal membrane-spanning domains and the larger carboxy-terminal enzyme part where the active site of the radish enzyme resides. If so, a molecular mass of about 45 kD for the catalytically active fragment would be the result. The additional band below 66 kD appearing in SDS-gels upon analyzing the peak fraction from the HMG-CoA agarose column may not be due to contamination by bovine serum albumin, as interpreted earlier,[57] but represents instead some intact non-degraded HMGR-protein. Since this cDNA for radish HMGR sequenced so far bears very strong homologies (up to 95%) to the corresponding gene in *Arabidopsis*, the reader is referred to the article of Dr. Boronat's group concerning the molecular cloning of HMGR, elsewhere in this volume. There is some strong evidence for the existence of at least one additional gene, as revealed by restriction analysis of another cDNA clone and by the Southern blot technique.

The HMGR purified from yeast has an apparent molecular mass of about 270 kD.[35,78] The native enzyme is also membrane-bound and contains a higher number of membrane-spanning domains,[79] as was shown for the mammalian HMGR, with the monomer having a molecular mass of 97 kD,[80-84] for HMGR from sea urchin,[85] or from insects.[86] The mammalian enzyme can be cleaved into two domains, the membrane-spanning part and the larger carboxyterminal fragment of 52 kD containing the active site, by the action of endogeneous lysosomal proteases during the conventional freeze-thawing procedure (see ref.[87] for a review of early controversies concerning the molecular mass of HMGR).

The elegant work by Brown & Simoni[82] and of Chin et al.[80] has shown that at a site of the aminoterminal domain the protein is co-translationally glycosylated and oriented towards the lumen of the endoplasmic reticulum; the glycosyl residue is of the high-mannose type.[82,83] A small part of HMGR activity in crude solubilized enzyme extracts from radish membranes was retained on concanavalin A agarose. In contrast to α-L(-)-fucose, α-methyl-D-glucoside and α-methyl-D-galactoside, only α-methyl-D-mannoside eluted this portion of the activity. The data available on the sequence of *Arabidopsis* HMGR favors the assumption that a potential glycosylation site exists in the N-terminal region (see Monfar et al., in this volume). However, with radish enzyme purified to a higher degree, there was no binding to observe. The full-length cDNA sequenced so far encoding a radish HMGR protein does not contain glycosylation sites in the N-terminal region (unpublished observations). Thus, the question of whether the plant enzyme is a

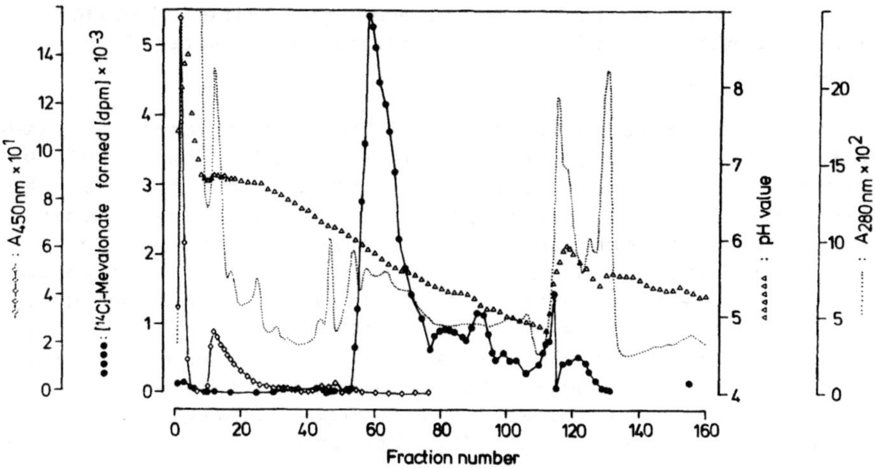

Fig. 1. Chromatofocusing of a crude solubilized enzyme preparation from a heavy-membrane pellet (P 16,000 x g) isolated from a cell-free homogenate of 4-day-old etiolated radish seedlings. The protein fraction loaded (60 mg protein) containing HMGR activity had been partially purified through 37.5% ammonium sulfate precipitation. The PBE column (Pharmacia) was equilibrated with 25 mM imidazole-HCl (pH 7.4), containing 0.33% (w/v) polyoxyethylene ether (Brij W-1) and 3 mM dithioerythritol (DTE). The column was developed with 1:10 PB 74, pH 4 (HCl), containing 0.6% (w/v) Brij W-1 and 3 mM DTE. Proteins that remained bound to the material were eluted with 100 mM KCl in the latter buffer system. HMGR activity is expressed as ^{14}C dpm MVA synthesized by 35 µl fraction during incubation for 40 min at 37°C (from ref. 26, with permission).

glycoprotein remains open at present. This is also true of the functional size of plant HMGR *in situ*. In the case of rat liver HMGR, a size of about 200 kD was determined by the aid of radiation inactivation,[88] apparently a dimeric enzyme. The value of 210 kD, independently determined by the same technique,[89] was reduced to 120 kD upon feeding the animals with mevinolin (an inhibitor of HMGR, see below) and colestipol, a material that is able to absorb bile acids in the intestine. Since under those dietary conditions, the half life of the enzyme was strongly increased, in addition to a dramatic increase in apparent in HMGR activity,[90] it might be possible that the monomeric form of the enzyme is more stable against proteolytic turnover. Whether this dimeric enzyme consists of non-covalently bound monomers[88] or was synthesized through a covalent disulfide bridge,[89,91] is currently a matter of discussion.

A considerable purification of HMGR detergent-solubilized from radish membranes (P 16,000) was achieved by chromatofocusing (Fig. 1). The enzyme eluted at a pH of about 5.9, resulting in a remarkably stable preparation. Theoretically, this value should represent the pI of the enzyme. However, in recent experiments to establish protocols for the purification of other enzymes besides HMGR, a new free-solution isoelectric focusing (IEF) method revealed that HMGR prepared from radish and maize membranes was an even more acidic protein (see Figs. 9-11). Enzyme occurring in the 30% ammonium precipitate solubilized from maize microsomes (P 140,000), although more sensitive during the free solution IEF, behaved in a less acidic way with an apparent pI of 6.45 (Fig. 11) than in the case where the precipitation step was omitted (apparent pI about 4.5). We suspect these differences to be caused by formation of phospholipid-protein complexes that shift the apparent pI towards the acidic range. Whether the differences found between chromatofocusing and free solution IEF lie in non-specific interaction with the column material during chromatofocusing or are caused by complex formation of proteins during separation remains to be investigated and is a matter of current studies in our lab.

Heat treatment was a part of the purification protocol developed for yeast[17] and rat-liver HMGR.[75] This step alone removed up to 90% of contaminating proteins, whereas HMGR remained in solution. When crude detergent-solubilized extracts of the P 16,000 x g fraction from radish seedlings were kept at 65°C for 5 min (in the absence of glycerol), a considerable portion of HMGR activity was found in the centrifugation pellet. This was possibly due to the formation of inclusion aggregates between HMGR and denatured proteins rather than heat-denaturation of the protein *per se*, as was described for the detergent-solubilized rat-liver enzyme.[92] When HMGR purified from radish membranes was incubated at 67.5°C in the presence of Brij W-1, 30% glycerol and > 40 mM KCl and > 40 mM DTE, and aliquots were withdrawn every minute, the enzyme remained stable after a slight activation during the first 5 min (Fig. 2). A similar, but much more dramatic increase of apparent enzyme activity could be demonstrated with the membrane-bound enzyme. A similar increase, but in the time range of hours, has been observed during incubation of a membrane preparation at 37°C in the presence of DTE, whereas

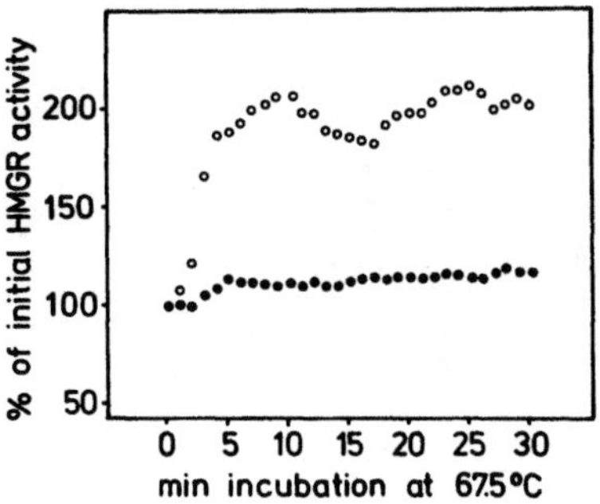

Fig. 2. Heat-stability of purified radish HMGR and of membrane-bound enzyme. Closed symbols: purified; open symbols: membrane-bound. Note the activation in the latter case, which is largely due to the presence of high concentrations of DTE. The purified enzyme (of P 16,000 x g) was kept in 50 mM K-phosphate pH 7.5 containing 400 mM KCl, 30% glycerol, 2% Brij W-1, and > 40 mM DTE. The membranes had been resuspended in 100 mM K-phoshate pH 7.5, 5 mM EDTA, 100 mM KCl, and > 40 mM DTE. The enzyme preparations were incubated at 67.5°C and aliquots were withdrawn every minute and assayed for HMGR activity.[57]

preincubation of the suspension at 0°C only stabilized the activity at the initial value.[30] The interpretation developed was that plant HMGR activity might be regulated *in vivo* by oxidation/reduction of thiol groups.[30] Similar findings have been reported for an enzyme preparation from *Parthenium argentatum*.[68] A large body of evidence of mechanisms like this also exists for the mammalian enzyme (see refs.[91,93-96] and literature cited therein). Loss of apparent enzyme activity during the isolation of membranes (by oxidation?) might explain a lack of correlation of the *in vitro* activity of HMGR with the rate of phytoalexin formation in elicitor-challenged tobacco cultures *in vivo;* [43] this might be eliminated if the enzyme could be similarly activated by DTE. However, when purified radish HMGR was diluted into the assay solution, stability was diminished with a temperature optimum of < 50°C; from the Arrhenius plot an activation energy of 92 kJ/mol was calculated.[57]

HMGR purified from radish, maize and yeast has a broad pH optimum around 6.5 to 7.5, as determined by parallel assays (Fig. 3).

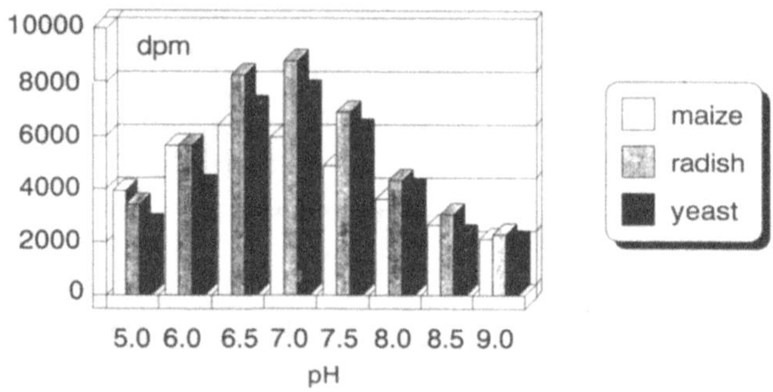

Fig. 3. pH optima of HMGRs purified from radish (P 16,000 x g), maize (P, 140,000 x g), and yeast. Enzyme activity was radiochemically assayed as described in detail,[57] but the buffer concentration (potassium phosphate) was increased by a factor of 4.

Kinetic Properties of Plant HMGR

The purified enzyme from etiolated radish seedlings was used for kinetic studies.[56,57,77] Similar studies have been undertaken with an enzyme purified from a microsomal pellet from etiolated maize seedlings (Figs. 4 a,b, 5 a,b, 6 a,b,c, 7 a,b,c). These studies were somewhat hampered by the fact that the activity of plant HMGR, even if purified, can only be exactly determined by a radiochemical method (for details see [57]). Anything else, such as optical tests, especially when using crude membrane suspensions (cf. refs. [28,31]), will not give reliable results. The radiochemical assay creates the dilemma that there are no exact v_o conditions (see discussion elsewhere).[35,36] However, the problem can be partially solved by using correction terms derived from the integrated Michaelis-Menten equation. In any case, all values obtained from the plots shown in Figures 4-7 are meant to be apparent activities and apparent kinetic constants. The NADPH used up by the HMGR reaction was regenerated by a system comprising glucose-6-phosphate and glucose-6-phosphate dehydrogenase.[57] In the case of product inhibition studies only NADPH was added, but this was without significant changes in the result. With the purified maize enzyme we could not detect any use of NADPH or HMG-CoA by an activity possibly interfering with the assay.

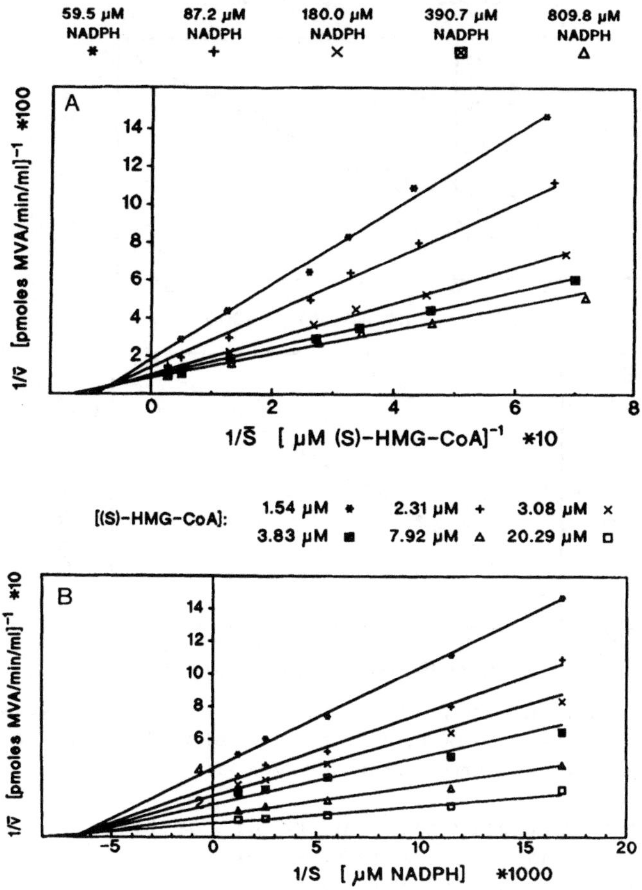

Fig. 4. Kinetic characterization of purified maize HMGR. A) Determination of K_m with respect to (*S*)-HMG-CoA in presence of variable concentrations of NADPH as indicated in the figure. The concentrations of NADPH were kept constant *via* a regenerating system.[57] B) Determination of K_m with respect to NADPH at variable concentrations of (*S*)-HMG-CoA.

In the double reciprocal plot (Fig. 4 A,B) when the concentration of substrate was varied with a constant concentration of cosubstrate and *vice versa*, a converging pattern was observed with lines intersecting above the X-axis. From this it can be concluded a) that the mechanism of substrate binding is sequential, and b) that the process of substrate binding is

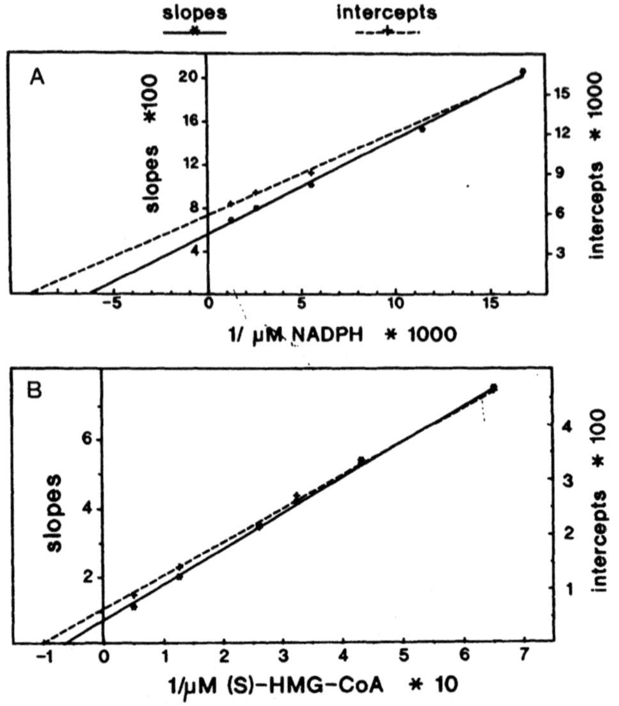

Fig. 5 A,B. Secondary plots derived from the Lineweaver-Burk plots in Figure 4. A best fit to the experimental data was achieved by linear regression analysis.

ordered since obviously the binding of the first substrate affects the affinity of the enzyme towards the second one. From the secondary plots (Fig. 5 a,b), the following apparent kinetic constants could be derived:

K_A = 10 µM
K_{iA} = 14 µM
K_B = 107 µM
K_{iB} = 172 µM;

A refers to (*S*)-HMG-CoA, B to NADPH. K_A means the K_m with respect to (*S*)-HMG-CoA, K_{iA} is the dissociation constant between enzyme and (*S*)-HMG-CoA etc. The sequence of substrate binding as indicated by the indices A and B,[97] was chosen in analogy to the findings with purified radish HMGR, as determined through substrate inhibition by NADPH at low concentrations of HMG-CoA.[57] At concentrations below 50 µM NADPH the measurements

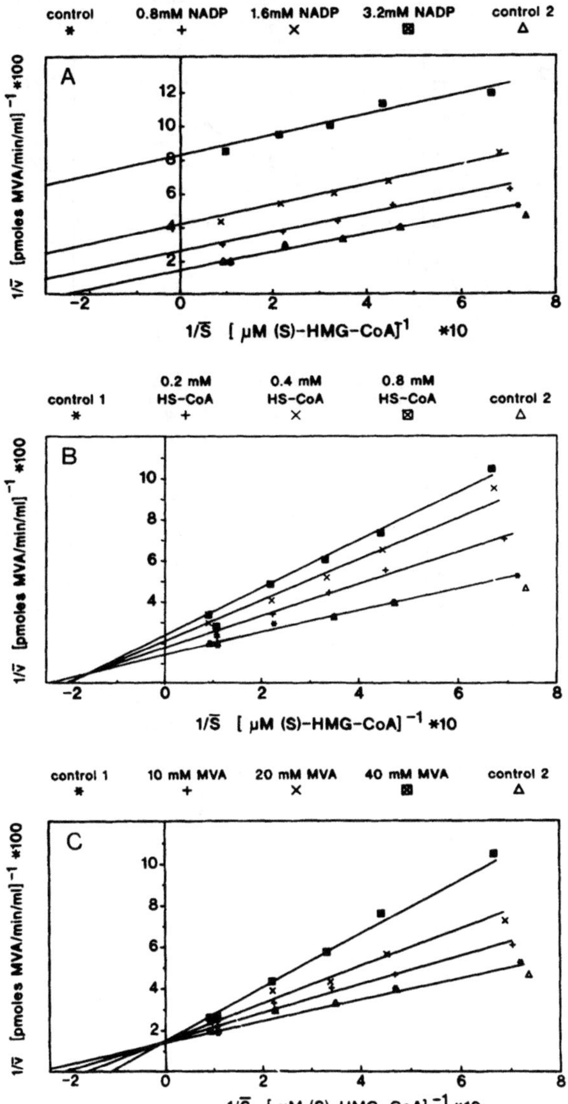

Fig. 6 A, B, C. Kinetic properties of purified maize HMGR: Product inhibition study. Variable substrate: HMG-CoA; NADPH: 600 μM.

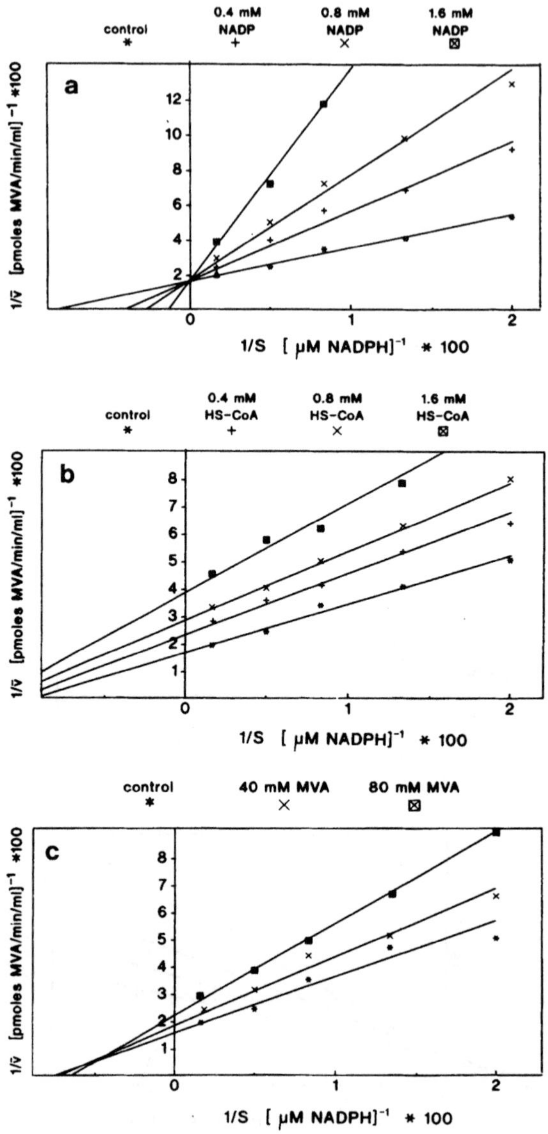

Fig. 7 a, b, c. Kinetic properties of purified maize HMGR: Product inhibition study. Variable substrate: NADPH; (*R,S*)-HMG-CoA: 22.8 μM.

became unreliable at the limited activity of enzyme available. Thus a possibly sigmoidal dependence of enzyme acticity on NADPH cannot be excluded at present. There does exist some evidence of this for the microsome-bound enzyme from radish.[30,98] From the product inhibition studies as shown in the double reciprocal plots (Figs. 6, 7), secondary plots were constructed that allowed for the calculation of inhibition constants that are summarized in Table 2.

The K_m of about 10 μM against (*S*)-HMG-CoA for purified microsomal HMGR from maize seedlings is clearly higher than the K_m of 1.5 μM reported for the radish enzyme,[57] but lies within the range of values determined in crude membrane preparations from other plant species (Table 3). The difference between the values reported for the microsomal and putatively plastidic HMGR(s) from *Pisum sativum*, points to the existence of isozymic forms.[31] A break in the double-reciprocal plot when a membrane fraction from *Ipomoea batatas* (containing organelles and microsomes) was used, might also indicate the presence of isozymes with different substrate affinities. The K_m value of the maize enzyme of 107 μM with respect to NADPH is about four times higher than for the purified radish enzyme (see Table 3). In view of the metabolic function of enzymes, their substrate affinities should roughly reflect the range of intracellular substrate concentrations. Therefore, in maize we would expect a rather high synthetic capacity for HMG-CoA, and indeed we have some evidence of this. A comparably high K_m against NADPH (cf. cytosolic and plastidic enzyme in *Parthenium argentatum*, see Table 3) could indicate an easy access to reduction equivalents in this photosynthetically active tissue. A contrasting example is the microsomal HMGR in non-photosynthetic transformed tobacco cells (strain LA-6) (Table 3). Under those conditions the supply of NADPH could limit the pathway leading to sterols.

Hitherto, mainly yeast HMGR has been the subject of extensive kinetic studies (see Qureshi & Porter[22] for literature) and the reaction sequence was analyzed in the forward and reverse direction. Tanzawa & Endo[99] used enzyme, solubilized (and fragmented) by the usual freeze-thawing method from rat-liver microsomes . They also obtained converging patterns in the Lineweaver-Burk plots when both substrates were varied. The product inhibition pattern was similar to that found for the maize enzyme (Table 2) but the inhibition constants appear to be different in the case of MVA by about three orders of

Table 2. Product inhibition study with maize HMGR.

Substrate	Inhibitor	Inhibition type	K_{is} [mM]	K_{ii} [mM]
HMG-CoA[a]	NADP	uncompetitive[c] (positively paraboloid)[d]		0.766
HMG-CoA	HS-CoA	non-competitive (negatively paraboloid)	0.478	0.796
HMG-CoA	MVA	competitive (linearly)	24.7	
NADPH[b]	NADP	competitive (positively paraboloid)	0.663	
NADPH	HS-CoA	non-competitive (slightly negatively paraboloid)	1.98	1.12
NADPH	MVA	non-competitive (linearly?)	121	178

K_{ii}: Inhibition or dissociation constant calculated from the changes in the intercepts on the y-axis in primary plots as a function of inhibitor concentration; K_{is}: dissociation constant calculated from the change in the slopes. HMGR was purified from microsomes (P 140,000 x g) isolated from 7-day-old etiolated maize seedlings. The enzyme was solubilized by incubation in the presence of Brij W-1 and of glycerol, by a slight modification of the method described for the radish HMGR.[57] Further purification steps included ultracentrifugation at > 100,000 x g, precipitation with 30% ammonium-sulfate, followed by column chromatography on hydroxylapatite and HMG-CoA agarose. The enzyme activity was assayed for 60 min at 37^{0}C as described,[57] but in the absence of a NADPH regenerating system and at pH 7.0.

[a] At 600 µM NADPH.
[b] At 22.8 µM (*R,S*)-HMG-CoA.
[c] Determined from primary plots.
[d] Determined from secondary plots; where required the constants were determined by linear regression under exclusion of the values at the highest inhibitor concentration.

Table 3. K_m values of plant HMGR preparations.

Plant	Membrane fraction	K_A [μM]	K_B [μM]	Ref
Pisum sativum	Microsomes	80		31
Pisum sativum	Plastids	0.385		31
Ipomoea batatas	P 105,000 x g	6.5[a]		38
		21.0[b]		38
Raphanus sativus	Microsomes	2.4	75[c]	30,36
Raphanus sativus	P 16,000	2.2-7.3[d]		36
Raphanus sativus	P 16,000(purified)	1.5	27	57,77
Hevea brasiliensis	P 103,000	28		50
Nicotiana tabacum (transformed cells)	Microsomes	19	15	26
Solanum tuberosum	Microsomes	30[e], 20[f]		
				41
Solanum tuberosum	Microsomes[g]	6.4	25	64
Parthenium	Cytosol(?)	125	310	68
argentatum		9	420	68
Parthenium argentatum	Plastids			

K_A: K_m against (*S*)-HMG-COA; K_B: K_m against NADPH.

[a] Below 50 μM (*S*)-HMG-CoA.

[b] Above 50 μM (*S*)-HMG-CoA.

[c] Value at half-maximal saturation, sigmoidal dependence.

[d] Dependent on the treatment of seedlings with light and phytohormones, etc.

[e] Before treatment with $HgCl_2$.

[f] After treatment with $HgCl_2$.

[g] purified.

magnitude. This might be due to a wrong calculation, since the work of Rogers & Rudney[100] provided convincing evidence that MVA does not affect immunotitration curves of microsome-bound as well as purified rat-liver HMGR. This was in contrast to HS-CoA, HMG, HS-CoA together with HMG, HMG-CoA or mevinolin, which all reacted with the enzyme by inducing conformational changes, rendering the enzyme more insensitive towards the binding of monospecific antibodies. A comparison with the data on purified radish HMGR reveals that in that case MVA as well as HS-CoA are competitive inhibitors of the enzyme with respect to HMG-CoA. This might indicate that both products are randomly released from the enzyme, whereas the maize enzyme is inhibited by HS-CoA in a non-competitive manner. If we also assume with the maize enzyme that HMG-CoA is the first substrate to bind, a competitive inhibitor (MVA) must compete for the same free enzyme. Thus the second NADP must be released before MVA. The interpretation is somewhat difficult, because the CoA-moiety of HMG-CoA possibly reacts with a further binding site in addition to that reacting with either MVA or the HMG-part of HMG-CoA. For a partially purified enzyme from *Parthenium argentatum,* MVA was shown to be an uncompetitive inhibitor against NADPH (wrongly interpreted as being non-competitive);[68] a K_i was not determined.

Because NADPH is bound twice during the course of reaction, linear patterns in the secondary plot (Fig. 5B) in the case of varied NADPH concentration and constant HMG-CoA, indicates that an irreversible step occurs between the points of binding,[101] with NADP being released from the enzyme before the second NADPH can bind. The uncompetitive inhibition by NADP against HMG-CoA, both for the radish and the maize HMGR is further proof that HMG-CoA is the first substrate to react with free enzyme, since apparently NADPH can only bind to the enzyme-HMG-CoA complex. The transition to an apparently non-competitive inhibition at high concentrations of NADP, as was observed with the microsome-bound[35] and purified radish HMGR,[57] could indicate that NADP reacts with the enzyme-mevaldyl-thiohemiacetal to form a dead-end complex. To elucidate the exact mechanism and to prove that the complete rate equation for the ordered sequential reaction we derived[102] is right, it will be necessary to measure the course of partial reactions in both directions including *in vitro* NMR techniques. However, HMGR purified from plants in the conventional way is not abundant and active enough for this purpose. Modern cloning techniques can be expected to resolve the problem. Although the HMGR reaction largely favors the formation of the products, in accordance with the early findings of Lynen's group,[19] in the presence of "cleaving enzyme" (HMG-CoA lyase, HMGL), the catalysis of the reverse reaction could be forced. This is one of the reasons that prompted us to study HMGL in greater detail and to develop methods for its purification from plants.

Immunological Studies Concerning Plant HMGR

Despite the sequence homology between all eukaryotic HMGRs, at least around the active site, the radish enzyme (to our surprise) did not cross-react with monospecific antibodies against the rat-liver enzyme (from the labs of Dr. Harry Rudney, Cincinnati and Dr. Gene Ness, Tampa, respectively) and yeast HMGR (from Dr. J.W. Porter, Madison).[57] However, even in view of the strong sequence homologies between the insect and mammalian HMGRs,[86] similar results have been reported when antibodies against rat-liver enzyme were tested with HMGR from the insect *Diploptera punctata*[103] or from the sea urchin *Strongylocentrotus purpuratus*,[104] which also share clear homologies around the active site.[85] Polyclonal monospecific antibodies raised against HMGR from *Pseudomonas* did not cross-react with the rat-liver enzyme;[105] this observation is much less surprising since the bacterial enzyme catalyzes the NAD-dependent oxidation of MVA, and from this alone the evolutionary relationship to the eukaryotic enzyme(s) can be expected to be more distant.[106]

Unfortunately, detergent-solubilized and purified radish HMGR appeared only weakly immunogenic, and it took considerable effort to produce two rabbit sera that react with the enzyme. In double diffusion tests there was only one band visible, indicating monospecific recognition. One serum clearly cross-reacts with solubilized enzyme preparations from maize. With the same antibodies we could observe some *in vitro* inhibition of HMGR isolated from yeast. In order to detect rockets in immunolelectrophoresis it is necessary to delipidate the membrane preparations by precipitation with cold acetone and to replace Brij W-1 by Triton X-100 in agarose gels. The antibodies currently being tested, together with cDNA probes now available (see elsewhere in this volume), will serve as powerful tools for the detailed study of the regulation of this important enzyme at the transcriptional and post-transcriptional level. Antibodies have also been raised against the potato[64] and *Hevea* HMGRs (R. Wititsuwannakul, personal information).

Enzymatic Synthesis and Metabolism of HMG-CoA

The synthesis of HMG-CoA from acetyl-CoA requires the action of two enzymes, a) acetoacetyl-CoA thiolase (AACT, EC 2.3.1.9) and b) of HMG-CoA synthase (HMGS, EC 2.3.1.9) (Scheme 1). The equilibrium of the thiolase reaction is far on the side of formation of acetyl-CoA (Ac-CoA) from acetoacetyl-CoA (AcAc-CoA) and HS-CoA. The reason for this lies in the chemical mechanism, which includes a Claisen-type condensation: one Ac-CoA serves as an electrophilic, the other one as nucleophilic reactant. The abstraction of a proton from the α-methyl group of the Ac-CoA entering the reaction is energically unfavorable. In contrast to this, the reaction as catalyzed by HMGS favors the further aldol condensation of an Ac-CoA under release of HS-CoA. Coupling both reactions enables the cell to synthesize HMG-CoA in a thermodynamically favorable manner. This mechanism for the synthesis of HMG-CoA including acylated enzyme intermediates has been reported for the yeast system as well as for vertebrate cells and tissues (for literature see refs.[22,107-109]). The corresponding enzymes have been fairly well characterized and even the genes have partially been cloned and characterized.[110-113] In hamster cells the promoter sequence of the HMGS gene contains two sterol regulatory elements[112] homologous to the regulatory octamers identified in the 5' flanking regions of the HMGR gene[114] and of the LDL receptor gene,[115] which are responsible for sterol-dependent regulation. In the HMGR gene this sequence is located within a cluster of binding sites for proteins that resemble nuclear factor 1 (NF-1), a positive transcriptional activator.[116]

Our rather limited knowledge of the situation in plants is best characterized by a citation from J.C. Gray's recent review:[27]

"The enzymes involved in the formation of HMG-CoA from acetyl-CoA in plants have not been characterized. Acetoacetyl-CoA thiolase (EC 2.3.1.9) has not been purified from any plant source and its properties are essentially unkown. . . . HMG-CoA synthase (EC 4.1.3.5) has similarly not been purified from any plant source and its presence is largely inferred from the formation of HMG-CoA from acetyl-CoA in plant extracts. As HMG-CoA synthesis in animal cells has been reported to contribute to the control of cholesterol synthesis in animal cells (White and Rudney, 1970), further information on the nature of this enzyme in plants is urgently needed."

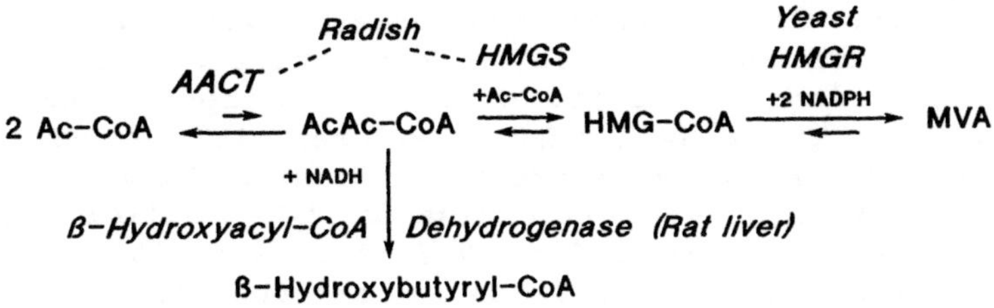

Scheme 1. The pathway to mevalonic acid (MVA) from acetyl-CoA (Ac-CoA). (See Table 6.)

As mentioned before, the synthesis of an uncharacterized HMG-derivative was reported for a cell-free extract of flax seedlings, the same plant tissue from which HMG had first been isolated.[1] Lynen[47] determined the activity of several enzymes of the rubber biosynthetic pathway in the latex of *Hevea brasiliensis*, among them AACT and HMGS. Of all enzymes assayed HMGR had the lowest specific activity; from that Lynen[47] concluded that this enzyme is a bottle-neck in the flow of acetate units to rubber. As was revealed shortly later,[48] this extremely low activity was due to the freezing of the enzyme source, since HMGR is apparently cold-sensitive. However, even then the activity remained sufficiently low to regulate the whole pathway.[49,50] The properties of both AACT and HMGS in *Hevea* have not been investigated further. In a cell-free system of sweet potato (*Ipomoea batatas*) the synthesis of a HMG-derivative, most probably HMG-CoA, from Ac-CoA was reported. This biosynthetic capacity was found in cytosolic and microsomal preparations; the enzymes involved, which have not been further defined, did not require cofactors such as ATP, Mg^{2+}, HCO_3^-, or NADPH for activity.[117] An iodoacetamide-sensitive AACT activity has been identified in vesicles from mature orange fruits;[118] the attempts to assay HMGS and HMGR activities failed. Quite recently AACT activity has been found in plastid preparations from *Parthenium argentatum* and *Phaseolus radiatus*.[69]

Once synthesized, HMG-CoA, instead of entering the isoprenoid pathway, can be cleaved by the activity of HMG-CoA lyase (HMGL, EC 4.1.3.4) in a stereochemically controlled Claisen-type retrocondensation reaction to yield acetoacetate and Ac-CoA,[119] an important reaction that contributes to the formation of ketone bodies in mammalian cells (mitochondria) (see Scheme 2). Indeed Hepper and Audley[48] provided some evidence for the existence of this enzymic activity in the latex of *Hevea brasiliensis*. Yu-Ito et al.[54] described an enzyme activity, most likely HMGL, that interferes with the HMGR assay in (mitochondrial?) membrane preparations from *Ipomoea batatas*. Recently, Skrukrud et al.[66] have assayed HMGL in the soluble fraction of latex collected from *Euphorbia lathyris*. The presence of HMGL in latex explains in part the failure to detect any significant incorporation of Ac-CoA, HMG-acid or of HMG-CoA into triterpenoids.[120]

We have determined the activity of HMGS and of HMGL in cell-free systems of radish[121] and of maize seedlings by the aid of a radiochemical method for the determination of enzyme activity in birds, which was introduced by Clinkenbeard et al.[122-124]. The underlying principle is rather simple: HMGS activity makes use of the incorporation of [2-^{14}C]-Ac-CoA into (*S*)-[4-^{14}C]-HMG-CoA in the presence (or absence, see below) of unlabeled AcAc-CoA. After stopping the reaction by addition of acid, subsequent heating and evaporation of the sample to dryness leads to the cleavage of CoA-esters; however, in contrast to acetate, HMG acid is not volatile and radioactivity incorporated can be conveniently estimated. The assay of HMGL activity is based on the same principle; synthetic (*R,S*)-[3-^{14}C]-HMG-CoA is incubated with enzyme, and the sample is processed as described above. HMG-CoA that was not metabolized remains, whereas [3-^{14}C]-acetoacetate is decarboxylated to volatile acetone; a maximum of 50% of the synthetic substrate can be used up since HMGL stereospecifically and exclusively reacts with (*S*)-HMG-CoA.

During the course of experiments concerning HMGS activity, we concluded that we could not distinguish between HMGS and AACT activity; [2-^{14}C]-Ac-CoA was incorporated into acid-stable HMG-CoA even in the absence of AcAc-CoA. In fact our assay system measures both enzymes. Similar findings have been reported when rather crude enzyme preparations from yeast[125] or rat liver have been used.[126] However, as we will see later, both enzymes must closely co-operate in the plant systems tested so far. First we checked for the intracellular distribution of AACT/HMGS (and of HMGL) activities (Table 4). We found considerable activity associated with a heavy-membrane pellet, which also exhibits the highest activity of HMGR in radish. For that reason and in view of the salt-sensitivity of cytosolic AACT/HMGS, we first concentrated on the further characterization of activity in this particulate fraction.

Table 4. Intracellular distribution of AACT/HMGL activities.

Enzyme system	Fraction	spec. activity [pmol/mg/min]	total activity [nmol/min]
AACT/HMGS[a]	Homogenate	747	1872
	P 16,000 x g	944	242
	S 16,000 x g	654	1195
	P 140,00 x g	221	28
	59% $(NH_4)_2SO_4$	165	99
	95% $(NH_4)_2SO_4$	104	43
HMGL[b]	Homogenate	143	358
	P 16,000 x g	255	66
	S 16,000 x g	33	60
	P 140,000 x g	97	12
	S 140,000 x g		
	59% $(NH_4)_2SO_4$	13	8

Four-Day old etiolated *Raphanus* seedlings were homogenized and membranes were isolated as described (Bach & Lichtenthaler 1984). "P": (membrane) pellet; "S": supernatant after centrifugation.

a) Assay: 20 µl enzyme solution + 20 µl 200 mM Tris/HCL + 10 µl "start mix" (1.25 µl [2-^{14}C]-acetyl-CoA = 27,460 dpm dissolved in 50 mM KH_2PO_4 pH 4.5, + 1.25 µl 2 mM acetyl-CoA (end concentration of acetyl-CoA: 52 µM), 1.25 µl 2 mM AcAc-CoA (end concentration 50 µM) + 6.30 µl Tris/HCL pH 7.5, 10-30 min at 30°C, preincubation over 5 min before addition of start mix. The reaction was stopped by addition of 175 µl 6 N HCl followed by heating to 100-105°C for > 4 h. The acid-stable radioactivity of HMG and of HMG-CoA, respectively required the addition of 200 µl water, 30 min incubation under continuous shaking, followed by addition of 4 ml Quickszint 2000 (Zinsser) in minivials, repeat of shaking for 30 min and measuring of the radioactivity by the aid of a LSC (Packard 2000-CA).

b) Assay system: 25 µl 100 mM Tris/HCl pH 8.0, 1.0 µl *(R,S)*-[3-^{14}C]-HMG-CoA (0.01 µCi, specific activity 52 mCi/mmol), 0.75 µl 4 mM cold *(R,S)*-HMG-CoA, each in 40 mM KH_2PO_4 pH 4.5; with enzyme solution and buffer in a final volume of 50 µl. Incubation for 10 min at 30°C; start of reaction by addition of a substrate-buffer mixture in 10 µl to the enzyme solution after a preincubation of 2 min at 30°C. The reaction was terminated after 10 min by addition of 175 µl 6 N HCl, transfer of the samples into minivals, incubation at 115°C for 3-4 h. Acetoacetate formed is decarboxylated to volatile acetone, whereas non-metabolized HMG-CoA remains. After addition of 100 µl H_2O, and after 30 min incubation under shaking the radioactivity was assayed as described above.

The AACT/HMGS activity could be easily released from the membrane by incubation with Brij W-1 by slightly varying the protocol developed for the solubilization of HMGR. Resuspensions containing enzyme were incubated with 2% Brij W-1 in buffer system A (see [57]) for 30 min at 30°C in the absence of glycerol and the material was repeatedly homogenized by a tightly fitting teflon pestle. After centrifugation at 100,000 x g for 45 min. practically all activity was found in the supernatant. Among many different methods to purify the enzyme system, including chromatography on ion exchange materials, affinity resins, and by chromatofocusing under various start and end conditions, only gel filtration (Amicon Cellufine GC-200 and GC-700) resulted in a reasonable recovery and purification of AACT/HMGS activity (Fig. 8). The enzyme eluted at an apparent molecular mass of about 52 to 56 kD. The sensitivity of AACT/HMGS towards KCl at > 100 mM hampered the attempts to develop further purification protocols. By introducing a negative precipitation step through addition of 0.3% polyethylenimine (at this concentration AACT/HMGS remains in solution whereas the predominant portion of contaminating proteins is precipitated), a considerable purification factor was achieved (Table 5).

Very recently we have started to separate crude solubilized enzyme preparations by the aid of free solution IEF (see Figs. 9-11). The enzyme system AACT/HMGS proved to be rather sensitive and lost most of its activity (in contrast to HMGR and HMGL), which could partially be reconstituted by addition of Fe^{2+} or Sn^{2+} ions. The enzyme activity, still able to synthesize HMG-CoA from Ac-CoA, had an acidic pI of about pH 4.5, similar to HMGR, separated and assayed in parallel. Usually, each fraction from column eluates or from IEF was twice assayed for HMG-CoA synthesis, once only in the presence of ^{14}C-Ac-CoA, and once after being additionally supplied with unlabeled AcAc-CoA. If we had a fraction exclusively containing HMGS, we should observe formation of ^{14}C-HMG-CoA only in the second case. There were some slight variations in the corresponding assays (Table 5) in the presence or absence of AcAc-CoA, which by itself was revealed to be a strong inhibitor of the enzyme reaction (data not shown). This was certainly not due to any substrate dilution effect. With this partially purified enzyme a K_m of 15 μM for Ac-CoA was determined.[121]

The observation that these two enzyme activities somehow act together was further confirmed through the experiment out-

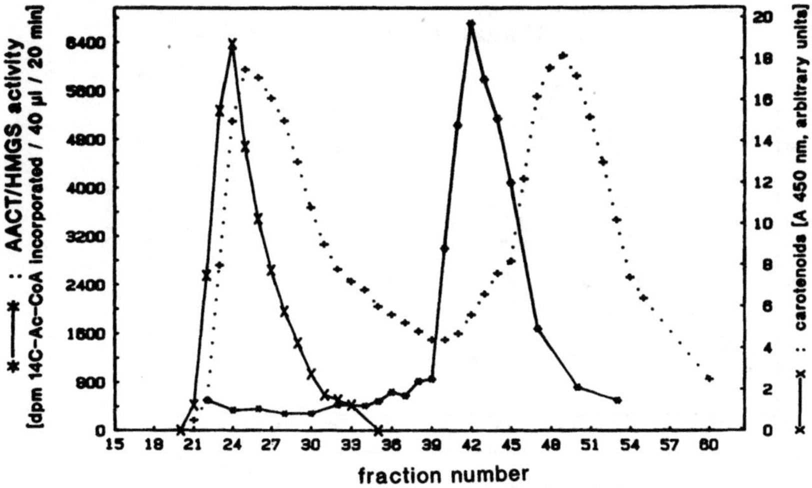

Fig. 8. Gel filtration on Cellufine GC-200 of the enzyme system AACT/HMGS assayed in a crude solubilized extract of a heavy-membrane pellet isolated from etiolated radish. The column (Pharmacia C 16/70, bed volume 131 mL, void volume 51 mL) was pre-equilibrated with a buffer system consisting of 0.2 M K-phosphate pH 7.5, 0.35 M sorbitol, 3 mM DTE, 5 mM $MgCl_2$, 7 mM EDTA (= buffer "A"), and 0.2% Brij W-1. Enzyme was solubilized by incubation of resuspended membranes (in buffer A) in the presence of 2% Brij W-1 for 30 min at 30°C, followed by centrifugation for 45 min at 100,000 x g. From this solubilisate 7.5 mL were loaded to the column. The void volume of the column was determined by Blue Dextran. AACT/HMGS activity eluted at an apparent molecular mass of 56 kD as determined by comparison with known molecular-weight standards.

lined in Table 6 and in Scheme 1. Here we could also prove that the enzyme system produced ^{14}C-HMG-CoA, because highly purified yeast HMGR completely converted the product into ^{14}C-MVA. This reaction was not affected by the addition of NADH, the cosubstrate for mammalian β-hydroxyacyl-CoA dehydrogenase (β-HAC-DH). High levels of this enzyme in the assay system should have been capable of trapping any intermediate AcAc-CoA, converting it into β-hydroxybutyryl-CoA, and thereby preventing any formation of ^{14}C-HMG-CoA. However, the apparent inability of this enzyme to affect HMG-CoA (or MVA) accumulation led us

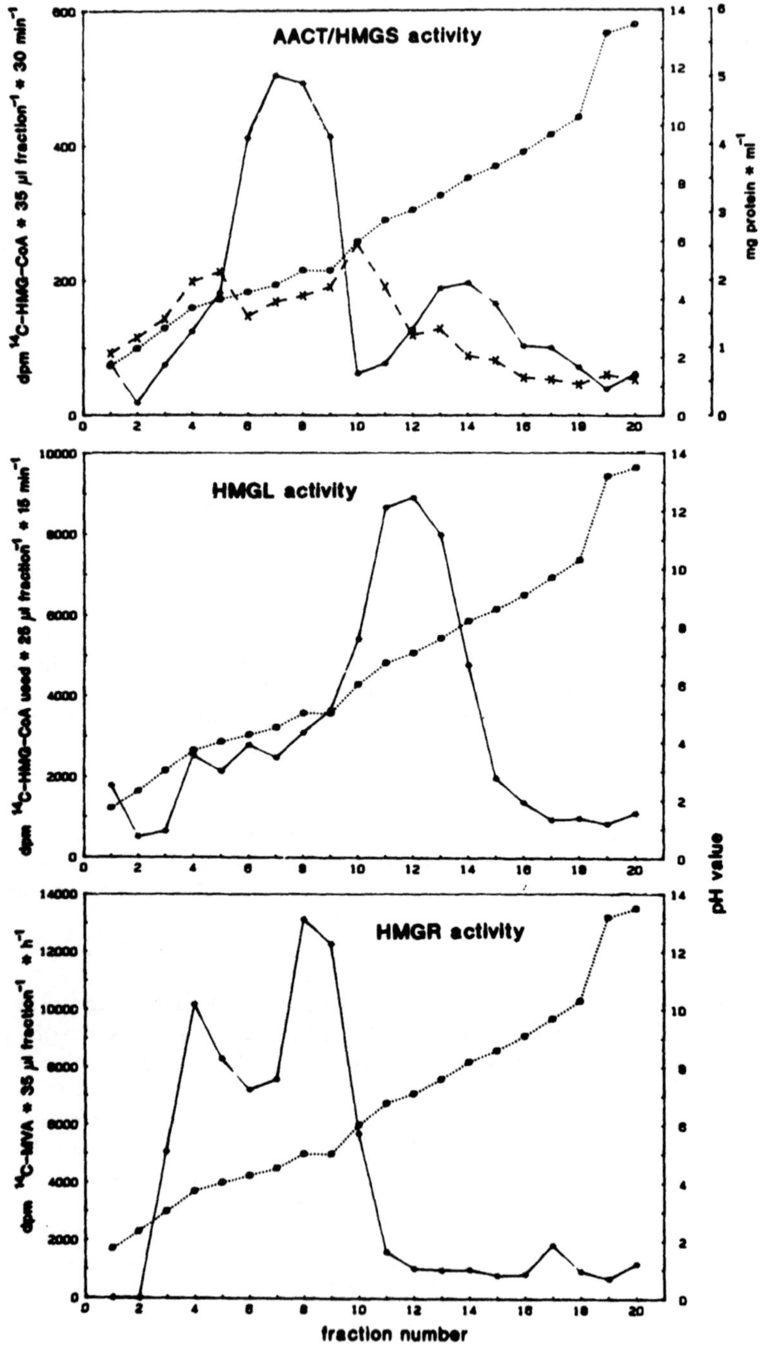
AACT/HMGS activity
dpm ¹⁴C-HMG-CoA * 35 µl fraction⁻¹ * 30 min⁻¹
mg protein * ml⁻¹
HMGL activity
dpm ¹⁴C-HMG-CoA used * 25 µl fraction⁻¹ * 15 min⁻¹
HMGR activity
dpm ¹⁴C-MVA * 35 µl fraction⁻¹ * h⁻¹
pH value
fraction number

to conclude that there might exist a very close co-operation through formation of a tightly fitting complex. The product arising from the AACT reaction could be immediately transferred to the active site of the consecutive enzyme HMGS without any substantial diffusion into and exchange with the medium. We have some further evidence for this hypothesis inasmuch as the antibiotic F-244 recently described as a highly specific inhibitor of yeast[125] and of rat liver HMGS[126] with K_i values in the range of 0.1 µM, is completely inactive in our system when tested at concentrations up to 0.1 mM. It is reasonable to assume that through formation of this enzyme complex the active site of HMGS is somehow shielded, preventing the binding of the inhibitor, but otherwise allowing for the direct transfer of AcAc-CoA to enter the HMGS-catalyzed reaction. Such a consecutive arrangement of both enzymes would also solve the problem of the thermodynamically unfavorable equilibrium of AcAc-CoA synthesis. In view of the numerous enzymes competing for the central intermediate Ac-CoA, a directed flux of carbon units into the isoprenoid pathway makes sense. Such short distances between enzymes create microdomains of comparably high substrate concentrations. The final proof of the formation of tightly bound enzyme complexes (perhaps including HMGR), even if they occur only temporarily *in vivo*, requires careful *in vitro* NMR studies, a matter of future investigations.

At this stage we cannot exclude the possibility that the enzyme system does not consist of two singular proteins when

---

Fig. 9. Free solution IEF of a crude solubilized enzyme preparation from a heavy-membrane pellet (P 16,000 x g) of etiolated radish seedlings. Enzymes were solubilized as described in Fig. 8, however the membranes were resuspended in 1:4 diluted buffer A in order to decrease the salt concentration. The start mixture of the IEF consisted of a 30 mL solution of > 5 mM DTE, 0.2% Brij W-1, 4% betaine (monohydrate, Sigma), 2 mL Servalyt 4-9 T (Serva) plus 8 mL of solubilisate; running time 3.5 h at 12.5°C; after 3 h the voltage was 700 V, 13 mA, 10 W. Enzyme activity (AACT/HMGS) was assayed as described in the legend to table 4; however, the concentration of Tris buffer was increased by a factor of 4 and 10 mM Fe^{2+} and 20 mM EDTA were added. HMGL was assayed as described in Table 4, however at a four-fold concentration of Tris and in the presence of 20 mM $MgCl_2$. HMGR activity was assayed as described,[57] but at a buffer concentration three times higher and at pH 7.0.

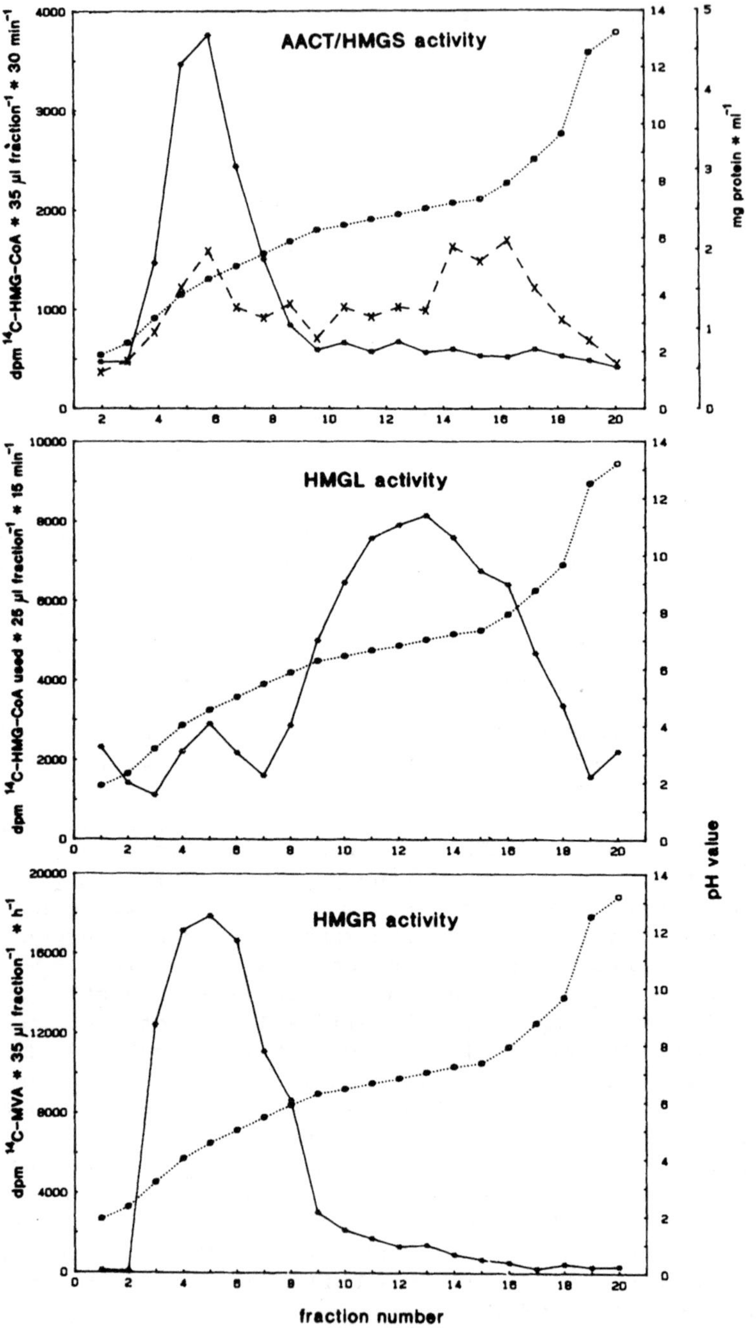
AACT/HMGS activity
dpm ¹⁴C-HMG-CoA * 35 µl fraction⁻¹ * 30 min⁻¹
mg protein * ml⁻¹
HMGL activity
dpm ¹⁴C-HMG-CoA used * 25 µl fraction⁻¹ * 15 min⁻¹
pH value
HMGR activity
dpm ¹⁴C-MVA * 35 µl fraction⁻¹ * h⁻¹
fraction number

Table 5. Solubilization and partial purification of the enzyme system AACT/HMGS from the P 16,000 x g of etiolated radish seedlings.

Fraction	Spec. Activity (nmol/min/mg protein)		Purification Factor	Yield
P 16,000	0.89[a]	0.62[b]	1	100
P 16,000 + Brij	1.6	0.79	1.8	179
S 100,000	4.5	1.4	5.1	222
P 100,000	0.15	0.02	0.17	17
S 0.3% Imin	26.3	7.6	29.4	55
Gel filtration (Peak fraction)	111.4	31.4	129	5.4

The solubilization was achieved by incubation of membranes with 2% Brij W-1 in a phosphate buffer system (0.13 M K-phosphate pH 7.5, 0.21 M sorbitol, 6.7 mM EDTA, 3.3 mM Mg^{2+}, > 20 mM DTE) for 30 min at $30^{0}C$ and by repeated vigorous homogenisation (teflon-homogenizer). Non-solubilized proteins were separated by centrifugation at 100,000 x g. "P": Pellet or sediment; "S": Supernatant after centrifugation.

[a] Substrate: 82 μM [2-^{14}C]-acetate (27,470 dpm)
[b] Substrate: 82 μM [2-^{14}C]-acetate (27,470 dpm) in presence of 50 μM AcAc-CoA.

Fig. 10. Free solution IEF of a crude solubilized enzyme preparation from a heavy-membrane pellet (P 16,000 x g) of etiolated maize seedlings. Proteins were detergent-solubilized as described in Figure 9 with the exception of increasing the final concentration of sorbitol to 300 mM. The start mixture (30 mL) contained 0.2% Brij, > 5 mM DTE, 2% glycerol, 5% betaine, 300 mM sorbitol, 2 mL Biolyt 5-8 (Bio-Rad), and 8 mL of solubilized proteins. The separation was performed over 3.5 h at $12.5^{0}C$. After 3 h a voltage of 800 V was reached (12 mA, 10 W). Enzymes were assayed as described in the legend to the previous figure.

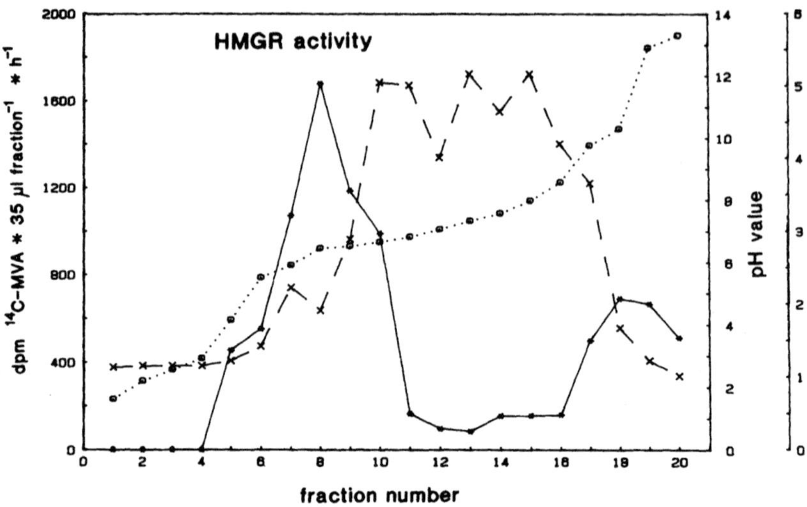

Fig. 11. Free solution IEF of a partially purified solubilized fraction from microsomal membranes of etiolated maize seedlings. HMGR activity was solubilized from the membrane as described in the text. The enzyme was partially purified by precipitation with ammonium sulfate (30% saturation). 5.5 mL of a resuspension of this precipitate were added to 36 mL of start mix having the same composition as described in the legend to Figure 10. The other running conditions were also identical.

considering the apparent molecular mass of 56 kD. A retarding reaction of the material used for gel filtration seems rather unlikely. Western blot analysis using polyclonal antibodies raised against two synthetic oligopeptides that comprise parts of the hamster HMGS protein failed to detect any signal after transfer of partially purified plant enzyme from SDS-gels to nitrocellulose (Bach & Boronat, unpublished observations). Attempts to select the HMGS gene from a genomic DNA library of *Arabidopsis* by the use of a cDNA coding for the hamster enzyme (a strategy that has been successfully applied in the case of HMGR) have failed so far (A. Boronat, personal communication and this volume). It is possible that AACT and HMGS, because of a less stringent selection pressure, have evolved a greater divergence in eukaryotes than HMGR.

Table 6. Co-operation between AACT and HMGS, solubilized from membranes (P 16,000 x g) of radish seedlings.

Treatment	Product formation (dpm)	% of control
control	10,789[a]	100
+ NADPH		
+ HMGR[c]	10.790[b]	100
+ NADPH	10,606	98.3
+ NADPH	10,680[b]	99.0
+ HMGR		
+ NADPH		
+ β-hydroxyacl-CoA-DH		

HMGR was assayed as described.[57] ^{14}C-Mevalolactone formed was separated from unreacted substrate by TLC. The values were corrected for recovery by the aid of an internal standard of [5-^{3}H] MVA. The enzymatic synthesis of ^{14}C-HMG-CoA was monitored as described in the legend to the previous table.

[a] Determined as ^{14}C-HMG-CoA.

[b] Determined as ^{14}-C-MVA.

[c] HMGR was purified from yeast.

Clinkenbeard et al.[123,124] demonstrated that feeding of animals with cholesterol resulted in down regulation of cytoplasmic AACT and HMGS activity, whereas the mitochondrial isozymes remained unaffected. This is a further proof of the different functions of the enzyme systems. In mitochondria they contribute to the formation of ketone bodies in the so-called HMG-CoA cycle, in which the HMG-CoA synthesized is cleaved by the activity of HMGL to aceto-acetate and Ac-CoA,[7,123,127] whereas in the cytosol they are involved in cholesterogenesis. The interference of both pathways is then precluded by the different location of enzymes. The inhibition of mitochondrial HMGS (from bovine liver) by physiological concentrations of succinyl-CoA is based on an auto-catalyzed succinylation of the active site,[107] which can be reversed in the presence of Ac-CoA with a $t_{1/2}$ of 17 min. The ketogenic substrate flow could then be increased, e.g. by glucagon, by decrease in the intramitochondrial concentration of succinyl-

CoA and thereby through the degree of succinylation.[128] When we tested AACT/HMGS solubilized from radish and maize membranes, we could not detect any time-dependent inhibition at concentrations of succinyl-CoA up to 500 µM (data not shown).

As reported for HMGR, the HMGS in the cytosol of rat liver and adrenal cortex (there only one form of HMGS was found) exhibits a diurnal variation in activity,[129] the first indication of co-ordinated control. The common regulation of AACT, HMGS and HMGR by sterols in somatic mutants of Chinese hamster ovary (CHO-K1) cells[130,131] has also been reported. In addition, in man the genes of HMGS,[132] HMGR,[133,134] and of the LDL-receptor[134] are localized in chromosome 5, which would facilitate co-ordinated control.

In yeast there are two forms of AACT,[135] one being cytosolic (thiolase 1, having a pI of 5.3, a molecular mass of 140 kD, a K_m against AcAc-CoA of 0.35 mM and against HS-CoA of 0.16 mM), the other being mitochondrial (thiolase II, with a pI of 7.8, 65 kD, a K_m of 20 µM and 0.16 mM, respectively). Both isozymes reach their highest activity at different times within the growth cycle, thiolase II at the beginning of the logarithmic phase ("early enzyme"), thiolase II at the end of this phase ("peak enzyme").[136] Recent studies of the regulation of ergosterol synthesis in yeast revealed a feedback inhibition of AACT and HMGS, whereas HMGR was less affected;[137] this was also demonstrated in several mutant (ergosterol-auxotroph) strains of yeast[138] where HMGS activity appeared to limit substrate flow. Other authors still ascribe this role to HMGR.[139-141] There are some indications that lanosterol, rather than ergosterol is the real feedback regulator.[142] Amplification of AACT by transformation of yeasts (*S. uvarum* and *S. cerevisiae*) did not result in higher ergosterol synthesis; the authors hypothesized that in such strains the activity of MVA kinase might be limiting.[143] From the molecular weight as deduced from the amino-acid sequence, the isolated AACT gene from yeast codes for the mitochondrial isozyme.[113]

The physiological role the enzyme HMGL plays in plants remains to be elucidated (see below). Enzyme activity is found in the cytosolic fraction as well as being associated with membrane pellets, indicating variable intracellular location. For the same reasons as with the system AACT/HMGS, we have initially studied the enzyme associated with the heavy-membrane fraction (P 16,000 x g) of etiolated seedlings of radish, and more recently of maize. The enzyme can be released from the

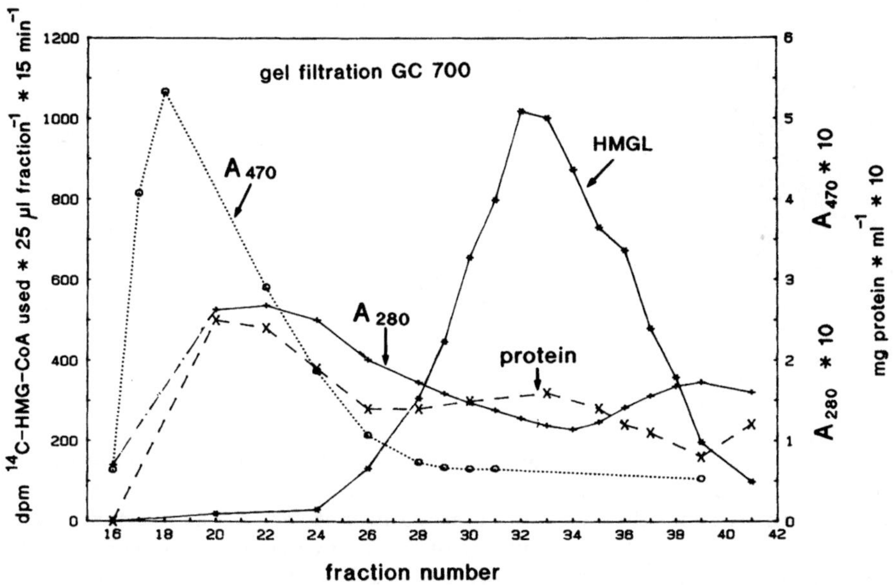

Fig. 12. Determination of the apparent molecular mass of HMGL solubilized from a 16,000 g-membrane pellet of etiolated radish seedlings. The column (Cellufine GC-700, C16/70, gel volume 130 mL, void volume 50 mL) was pre-equilibrated with a buffer system A (see fig. 8) diluted 1:2 with water and containing 0.2% Brij W-1. The peak activity of the enzyme eluted at an apparent molecular mass of 70 kD.

membrane by mild treatment with detergent and accompanies HMGR found in the same particulate fraction. However, in contrast to AACT/HMGS and especially to HMGR, repeated washing and centrifugation of membrane pellets results in considerable enrichment of activity in the supernatant, which might indicate its being soluble but included in vesicles or organelles. The enzyme was more sensitive than the HMGR; it lost most of its activity when kept over-night at 4°C, but it was sufficiently stable when stored at -20°C. A one-step purification by a factor of 20 to 40 of solubilized enzyme can be achieved by affinity chromatography on HMG-CoA agarose. However, the enzyme destroys the material which renders the method unacceptable. We are currently trying to synthesize agarose derivatives with covalently bound acyl-CoA, but in forms that cannot be cleaved by the enzyme. The enzyme can be partially purified by gel flitration on Cellufine GC 200 and 700, eluting at an apparent molecular mass of 70 kD (Fig. 12).

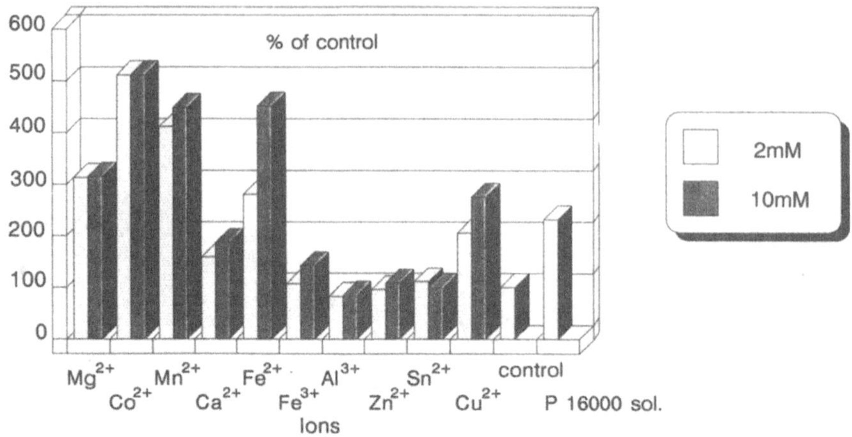

Fig. 13. Effect of bivalent cations on gel-filtrated HMGL activity solubilized from radish membranes (P 16,000 x g). Enzyme was released from the membrane pellet by treatment with detergent as described in the previous figures. An aliquot was subjected to gel filtration on a PD-10 column (Pharmacia) in order to remove low molecular-weight compounds.

The pI of the the maize and the radish HMGL is around 7, as shown by free solution IEF (see Figs. 9, 10). The enzyme survives this latter treatment well and can easily be separated from HMGR and from AACT/HMGS. The mass of lipid and carotenoid material migrates towards the acidic side, thus the fractions containing HMGL activity are fully transparent. By a combination of gel filtration and repeated free solution IEF, first with wide and subsequently with narrow pH ranges, we hope to get plant HMGL completely purified within the near future. Nevertheless, the preliminary data presented here give the first example of any partial purification and characterization of this enzyme from a plant source.

During the course of experiments we observed that the composition of the isolation buffer exerted some effect on the activities of HMGL as well as of AACT/HMGS, possibly indicating the presence of low molecular-weight molecules having a regulatory function. A lower concentration of EDTA in the isolation buffer increased apparent HMGL activity, but decreased that of AACT/HMGS. We tried to evaluate the effects of certain cations on the activity of gel-filtrated enzyme preparations. Whereas

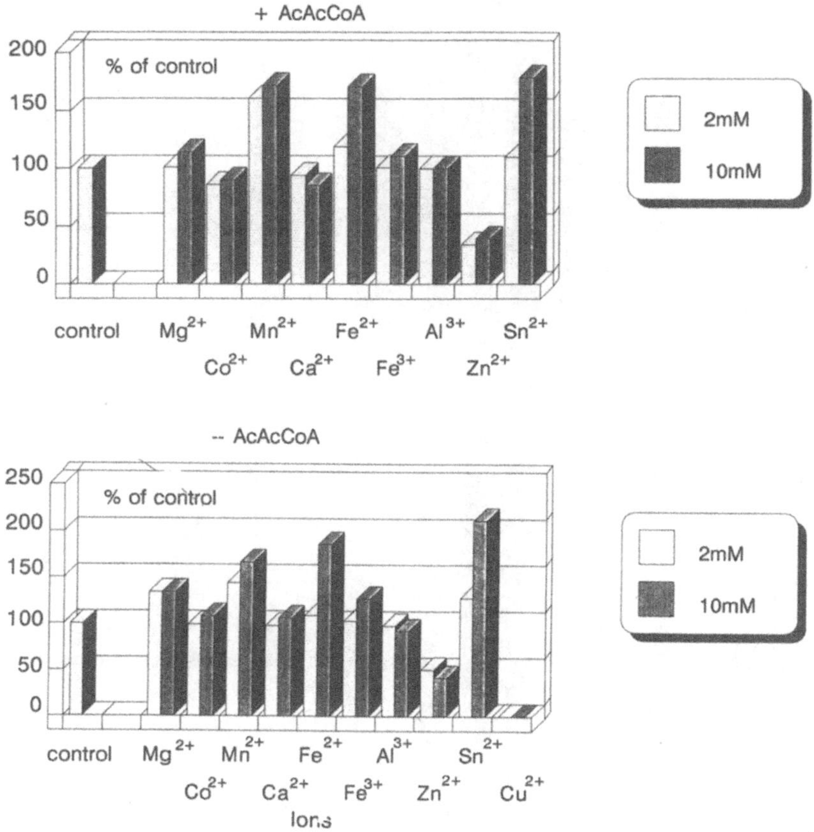

Fig. 14. Effect of various bivalent cations on the in vitro activity of the enzyme system AACT/HMGS solubilized from radish membranes (P 16,000 x g). The solubilized enzyme was gel-filtrated as described in the previous Figure.

HMGR retained its activity after gel filtration and was only inhibited by ions able to form sulfides (e.g. Co^{2+}, Mn^{2+}, Fe^{2+}, Zn^{2+}, data not shown), resulting in inactivation of the enzyme, HMGL and AACT/HMGS lost a large part of their activity after passage through Sephadex G-25 or G-10. Thus, the effects on the latter enzyme systems (see Figs. 13-15) involve a reconstitution of activity, rather than an activation *per se*. However, even then the effects exerted by either Fe^{2+} or Sn^{2+} on apparent *in vitro* activity of AACT/HMGS containing preparations were striking (see Fig. 15). This activation could even be enhanced

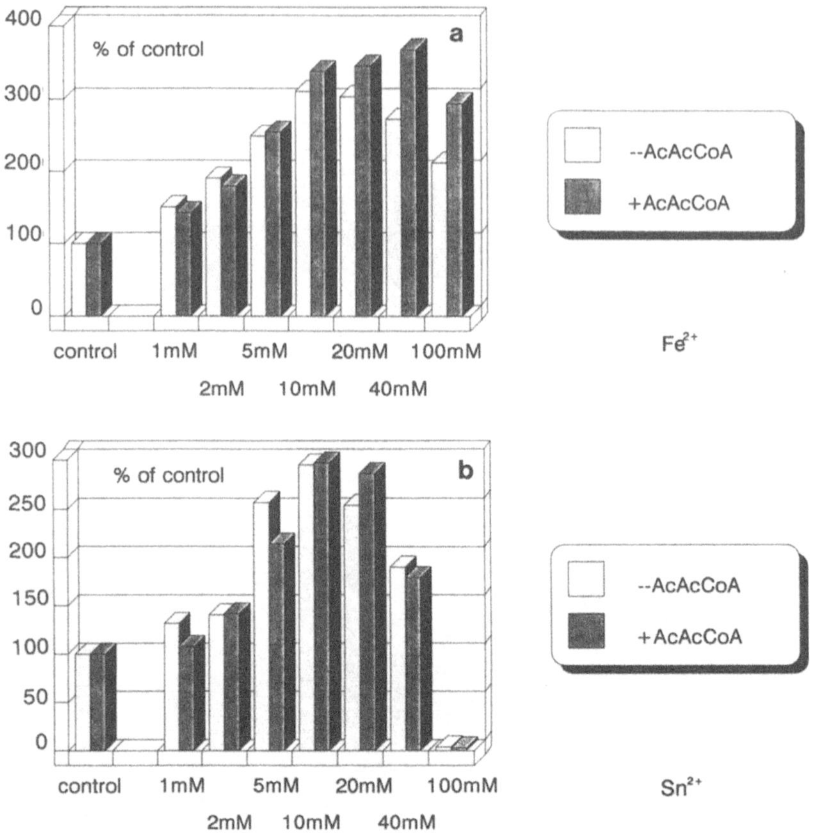

Fig. 15. In vitro effect of Fe^{2+} and Sn^{2+} on gel-filtrated AACT/HMGS activity. The enzyme solution was preincubated for 5 min with the ions, followed by addition of a substrate mix and incubation for 10 min at 30°C. (a) Influence of Fe^{2+} on radish AACT/HMGS activity; (b) Influence of Sn^{2+} on radish AACT/HMGS activity.

by addition of the ions in form of an EDTA-complex. We first thought of these ions as possibly shielding reactive thiol groups from becoming oxidized. However, their existence remains somewhat obscure, since AACT/HMGS solubilized from the P 16,000 x g both from radish and maize is not sensitive to iodoacetamide up to 100 mM, in contrast to HMGL, which is inactivated. Whether a regulation *in vivo* includes formation of complexes with certain ions yet to be identified remains to be investigated.

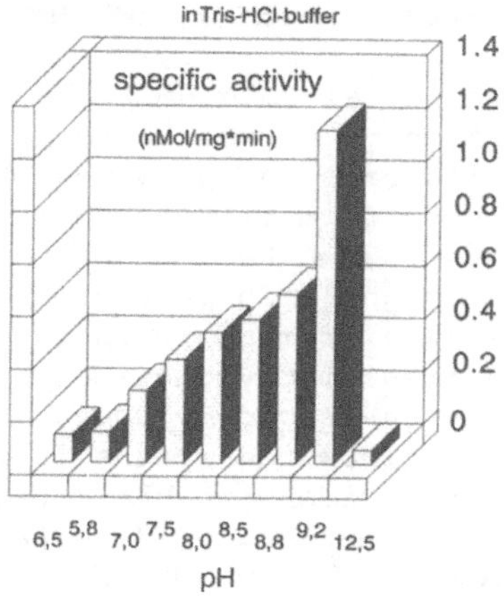

Fig. 16. pH optimum of HMGL activity solubilized from radish membranes (P 16,000 x g). The assay was performed in the presence of Tris buffer. Essentially the same result was obtained when this system was replaced by phosphate or Theorell-Stenhagen buffer.

Some dependence of HMGL activity from duck liver on the presence of Mg^{2+} has been reported.[144] The stability of chicken liver HMGL was enhanced in the presence of Mg^{2+} and Ca^{2+}.[123] The bovine liver enzyme was found to have a K_m against Mg^{2+} of 8 μM.[145] This enzyme had an apparent molecular mass of about 48 kD, with a pH optimum of 9.5, similar to the values earlier reported by Bachhawat et al.[146] In our system the radish HMGL had a pH optimum in the alkaline region (Fig. 16); however, since under such conditions the stability of CoA thioesters cannot be warranted, the influence of artifact formation cannot be excluded. For that reason the activity of radish or maize HMGL was routinely assayed at pH 8.0. Higgins et al.[147] compared the data obtained with several mammals and tissues as an enzyme source and assumed the enzyme to contain bound Ca^{2+}, which was not affected by chelators above pH 8. Our preliminary observations as to the influence of the homogenizing buffer composition (especially of EDTA concentration) indicate a requirement of bivalent ions that is not absolute.

The Possible Metabolic Function of HMGL

The exact physiological role HMGL plays in the plant cell remains obscure at present. As yet there is no clear data available that would indicate the presence of a so-called HMG-CoA cycle and ketone body formation as described for vertebrate systems.[7,123,127] However, besides being involved in leucine catabolism, the enzyme might also participate in the process of shunting carbon units away from their inclusion into the isoprenoid pathway *via* a trans-methylglutaconyl-CoA or MVA shunt mechanism as originally envisaged by Popjak.[148] The existence of this shunt, which combines the synthetic pathways leading to fatty acids and to isoprenoids with the catabolism of leucine, has been demonstrated in vertebrates (see Landau & Brunengraber[149] for review of literature), in insects[150] and in wheat seedlings, as demonstrated by incorporation of [2-^{3}H]-, but not of [5-^{3}H]-MVA into long-chain fatty alcohols (LCFA).[151] Tritium in the chromatographically pure LCFA was found associated primarily with C_{22}, C_{24}, and C_{26} (components isolated from subcellular membranes), while ^{14}C from acetate fed simultaneously, was present additionally in C_{28}, the major LCFA isolated from the epicuticular wax. Therefore, it was hypothesized that acetate formed by the shunt and presumably compartmentalized, is preferentially used for the synthesis of LCFA having a chain-length distribution more suitable for membrane than wax construction.[151] In the literature some evidence has been presented for the existence of different acetate pools used for the synthesis of monoterpenes,[152] for sterols and fatty acids.[153,154] In the meantime the existence of the shunt in plants was also demonstrated by incorporation of [2-^{14}C]-MVA into docosanol (C_{22}) in leaves of *Sorghum bicolor*.[155]

It will be much more difficult to answer the question of whether the exact reaction sequence as proven for animal systems (see Landau & Brunengraber[149]) is also true of plants. Baisted & Nes[156] showed that dimethylacrylic acid was not efficiently incorporated into sterols of pea seedlings, possibly indicating that this intermediate of the shunt cannot directly enter the synthesis of IPP but is first degraded to acetate. Essential reactions of the MVA shunt (see Scheme 2) such as methylcrotonyl-CoA carboxylase (MCC) have not yet been measured in plants. Our first attempts to assay this enzyme in preparations from radish by acid-stable incorporation of $^{14}CO_2$ into methylcrotonyl-CoA have not been successful. However, there are some indications that further biotin-containing polypeptides occur besides the predominating acetyl-CoA car-

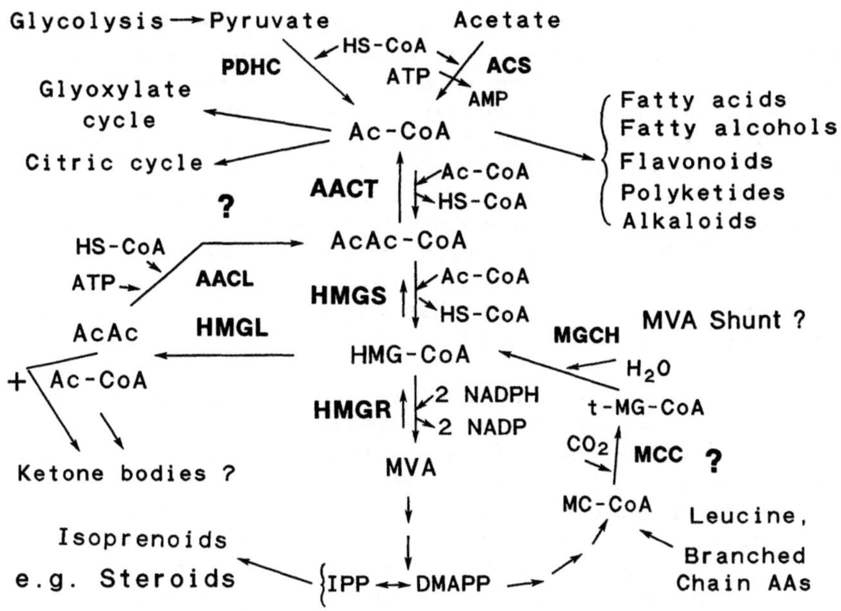

Scheme 2. A speculative scheme concerning the enzymology of HMG-CoA synthesis, its metabolism to acetoacetate, which might serve as a substrate for lipid synthesis and ketogenesis. The scheme is largely based on findings that have been made with animal systems.

AACL: acetoacetyl-CoA ligase; AACT: acetoacetyl-CoA thiolase; ACS: acetyl-CoA synthase; HMGR: 3-hydroxy-3-methylglutaryl-CoA reductase; HMGS: 3-hydroxy-3-methylglutaryl-CoA synthase; HMGL: 3-hydroxy-3-methylglutaryl-CoA lyase; MCC: methylcrotonyl-CoA carboxylase; MGCH: methylglutaconyl-CoA hydratase; PDHC: pyruvate dehydrogenase complex.

boxylase (cf. refs.[157-160]). Also no clear data exists on the presence of methylglutaconyl-CoA hydratase (MGCH) (see Yu-Ito et al.[54]). In addition, what happens to the acetoacetate produced by the HMG reaction ?

In animal cells a considerable activity of cytosolic acetoacetyl-CoA ligase (AACL), also known as acetoacetyl-CoA synthetase (cf. refs.[161-163] and literature cited therein) activates acetoacetate for lipogenesis. The enzyme is also controlled by low density lipoprotein (LDL) and by 25-hydroxycholesterol as HMGR and HMGS[162] and is considered to be negatively regulated by HS-CoA, fatty acyl-CoA and especially by acetoacetyl-CoA.[161] As was shown for HMGR and HMGS, mevinolin or compactin induce AACL activity.[163] In any case, this enzyme considerably contributes to cholesteroneogenesis.[162] The existence of this enzyme in plants has not yet been proved; however its presence might be expected.

Tritium from [2-^{3}H]-MVA, if routed through the MVA shunt, must appear in the acetoacetate fragment of the HMGL-catalyzed cleavage of HMG-CoA. Because the label appeared in the LCFA, this acetoacetate had to be used as a precursor of fatty-acid biosynthesis; whether through conversion to acetate or not remains uncertain. Acetoacetate could also arise from β-oxidation of fatty acids (cf. Harwood[164]) by the action of slightly related, multi-functional enzymes with enoyl hydratase, 3-hydroxyacyl-CoA epimerase and 3-hydroxyacyl-CoA dehydrogenase activity.[165] In this regard it is interesting to note that a purified NADH-dependent acetoacetyl-CoA, reductase from *Euglena gracilis*, was inhibited by acyl-carrier protein;[166] this enzyme could contribute to the *de novo* biosynthesis of fatty acids if the supply by the classical way with NADPH is limited. The HS-CoA-dependent AACT would convert AcAc-CoA into two molecules of Ac-CoA, favored by the thermodynamic equilibrium. In peroxisomes of the yeast *Candida tropicalis,* both a thiolase exclusively reacting with AcAc-CoA and the more unspecific 3-ketoacyl-CoA thiolase could co-operate in the complete degradation of fatty acids to acetyl-CoA.[167] Acetone produced by decarboxylation of acetoacetate *via* an enolized intermediate was shown to participate in the synthesis of tropane alkaloids in tobacco.[168] Finally, acetone as a solvent for sterols or other lipophilic compounds at moderate concentrations, is not at all toxic to cell cultures of *Apium gravedens* [169] or to intact radish seedlings. However, more information on the possible existence of ketone body formation in plants is urgently awaited. Since the apparent activity of HMGL far exceeds that of HMGR in the plant cell (see also Skrukrud et al.[66]), the organism must have mechanisms that preclude their interference, possibly through compartmentation or by hitherto unknown regulatory mechanisms, a matter of our future research.

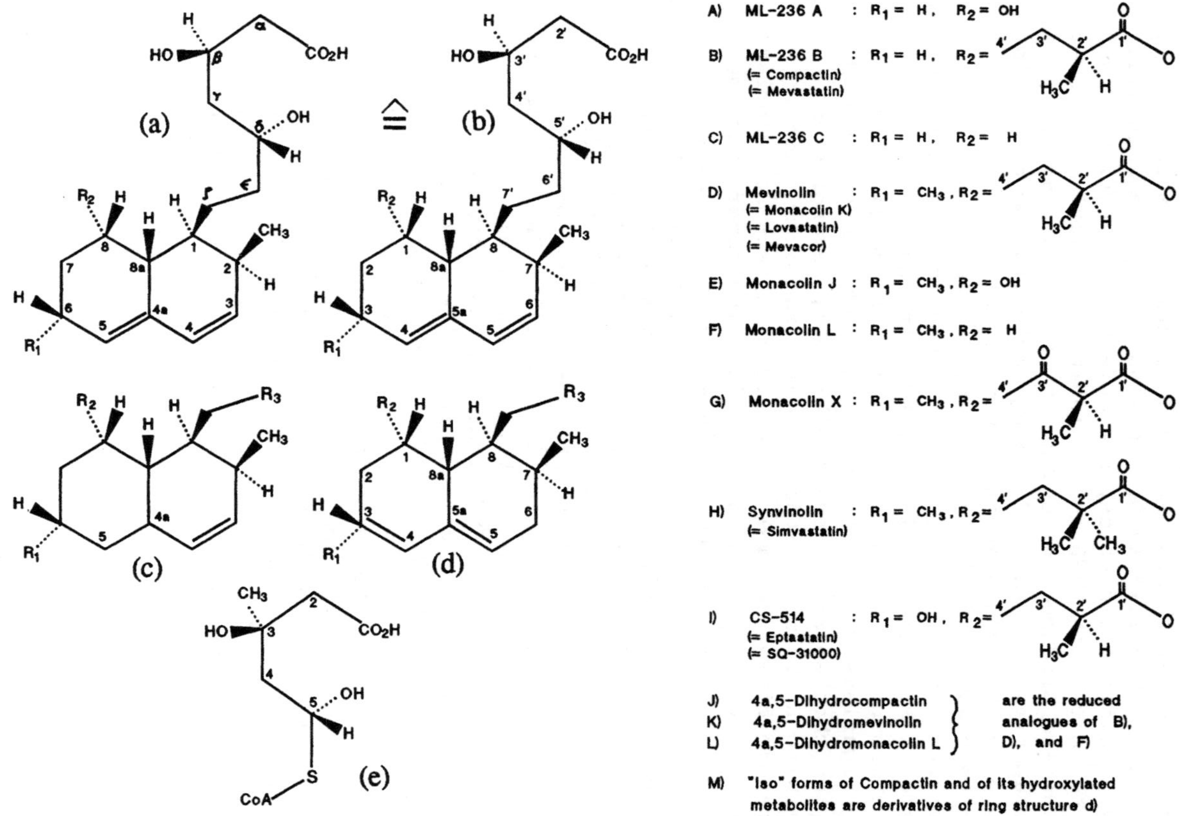

Fig. 17. Structures of mevinolin and its analogues. The numbering system a) was used by the group of Merck Sharp & Dohme, and b) by the Sankyo group; e) structure of mevaldyl thiohemiacetal, the enzyme-bound intermediate in the HMGR-catalyzed reduction of HMG-CoA to MVA. For literature see ref. 171.

Mevinolin, a Highly Specific Inhibitor of MVA Synthesis in Plants

That MVA biosynthesis plays an essential role in the growth of plant and microbial cells is shown by the existence of highly specific antibiotics, the only known function of which is to react with HMGR and to inhibit its activity. These metabolites are produced by several strains of ascomycetes, the most studied examples (Fig. 17) being mevinolin and compactin, which are of great pharmaceutical interest because of their hypocholesterolemic activity (see Grundy[170] and literature cited therein). From the natural occurrence of the microorganisms in the rhizosphere, it is assumed that the biological role of those compounds is the blockage of the isoprenoid pathway in competing organisms.[171] As yet all eukarytotic and prokaryotic HMGR enzymes checked so far are inhibited by mevinolin and its analogues with K_i-values in the nanomolar range (see refs.[98,171] and literature cited therein). Therefore, it was no surprise when mevinolin was revealed as a specific inhibitor of plant growth.[172-174] As has been noted elsewhere (cf. [Bach & Lichtenthaler[171] and Bach[102] and the literature cited therein), the ability of certain microorganisms to metabolize compounds like mevinolin or compactin and their analogues might indicate the existence of natural detoxification mechanisms, this reflects a part of the complex interplay between producers of antibiotics and target organisms in the soil. In addition, the observation of plant growth inhibition by mevinolin strongly supports the view that the HMGR-reaction has a bottle-neck function in the regulation of substrate flow from acetyl-CoA, e.g. to sterols. The basic idea was that inhibition of an enzyme capable of controlling the rate of a pathway *in vivo* should result in clear morphological responses. The radish root-growth test was also applied to test other compounds affecting later steps in the steroid pathway[171,175] and appears to be quite sensitive. When synthetic HMGR inhibitors (Hoechst HR 780 and S 87 4592 A) were tested *in vitro*, using purified HMGR from radish, the inhibitory activity was reflected by the effect on root growth *in vivo* (Figs. 18, 19). The potency of HMGR-inhibitors as plant-growth regulators has been confirmed by treating seedlings of *Medicago sativa* with compactin.[176] Exogeneous MVA, the immediate product of the inhibited reaction at concentrations of > 1 mM could overcome the growth inhibition.[174,176]

When seedlings are cultivated on water, mevinolin, as its freely soluble sodium salt is taken up by the developing roots

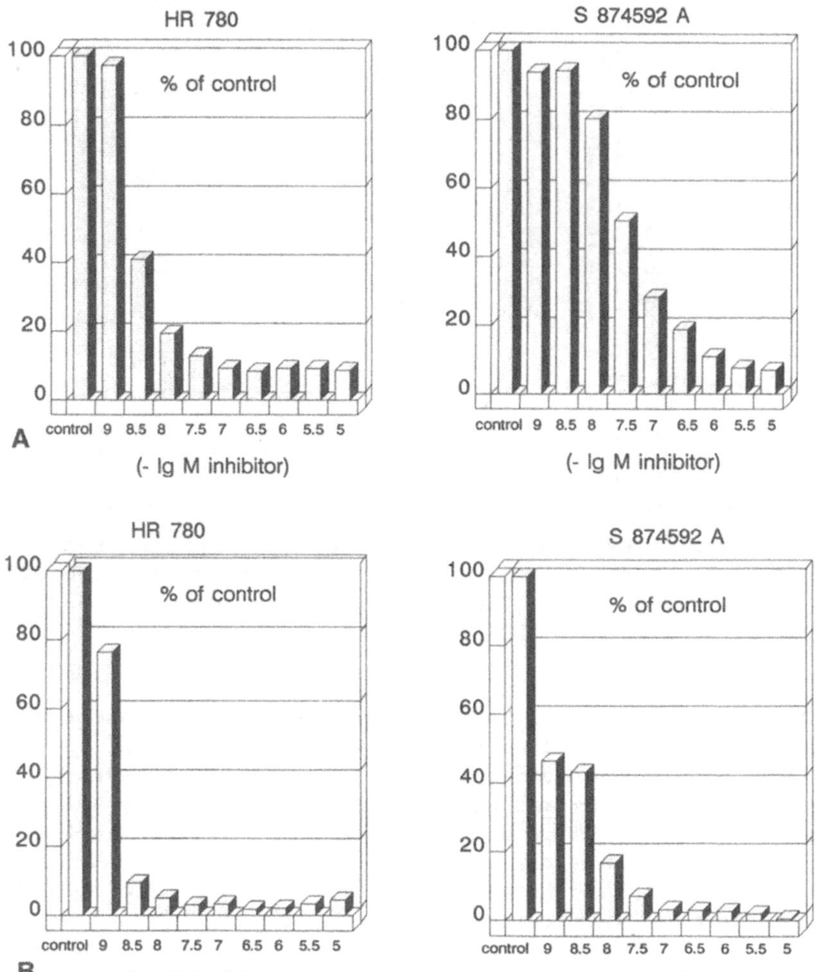

Fig. 18. In vitro efficacy of synthetic HMGR-inhibitors on enzyme purified from yeast and radish. (A) Radish HMGR; (B) Yeast HMGR

and is translocated to other parts of the plant. If this transport is blocked, shows some barriers, a gradient in growth inhibition or other morphological and biochemical responses should be the result. At least a significant inhibition of

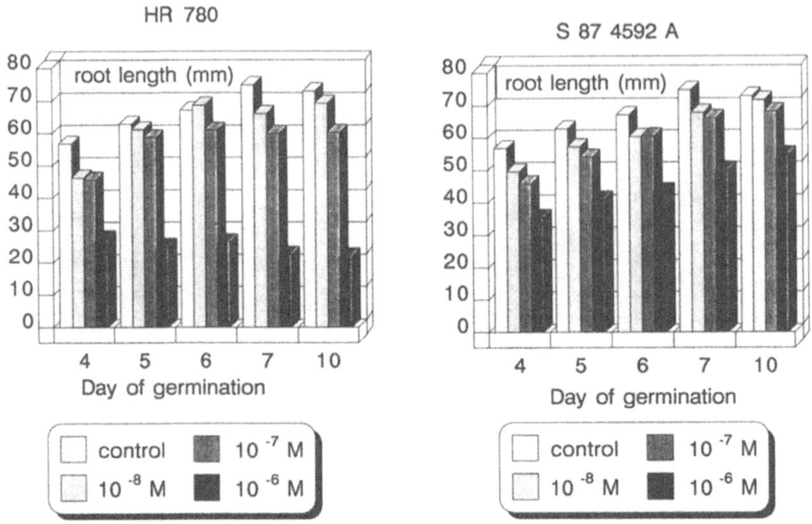

Fig. 19. In vivo efficacy of synthetic HMGR-inhibitors on the growth of radish roots. The data is based on the average of 25 plants per single value. The standard deviations were as reported earlier.[171-175].

hypocotyl growth needed a two to four-fold higher concentration of mevinolin as compared to roots. The primary effect on growth was due to the blockage of *de novo* sterol biosynthesis. When roots, hypocotyls and cotyledons were separately analyzed for prenyllipids, this graded response on mevinolin treatment of sterol accumulation was obvious.[177] Intact as well as excised radish seedlings treated with the inhibitor did not exhibit significant differences in chlorophyll and carotenoid content.[174,177] At low inhibitor concentrations, the concentration of chlorophyll a and b per cotyledon was even higher as compared to the controls, as was the concentration of other plastidic prenyllipids such as plastoquinone and phylloquinone (vitamin K1); only with mevinolin at > 5 µM was there a non-significant 10 to 15% inhibition in their accumulation. α-Tocopherol behaves like plastoquinone and phylloquinone, an independent indication of its exclusively plastidic location of synthesis (see discussion in paragraphs below); previously cytoplasmic as well as plastidic sites of synthesis were assumed for this compound (cf. [Pennock & Threlfall[178]). The exact location of its synthesis is the plastid envelope.[179-181]

The accumulation of mitochondrial ubiquinone was clearly suppressed by mevinolin treatment of radish seedlings. However, the inhibition at 5 μM was identical in all plant parts analyzed;[177] obviously there is a threshold beyond which mevinolin cannot exert any further inhibitory effect. However, the relative portion of the ubiquinone homologue Q-9 increased at the expense of Q-10, the latter being the predominant Q-homologue in radish.[182] Treatment of leaf sections of 9-day-old etiolated wheat seedlings (*Triticum aestivum* L.) with extremely high doses of mevinolin (125 and 500 μM) during an 18 h greening period resulted in some inhibition of plastidic prenyllipid biosynthesis.[174] The accumulation of chlorophyll b was inhibited more strongly than that of chlorophyll a. Besides the possibility of partially anaerobic conditions during the incubation period (leaf segments floating on the inhibitor solution), this effect might be due to the fact that during the greening phase the formation of chlorophyll-protein complexes of the photosynthetic reaction centers I and II, mainly containing Chl a, precedes the assembly of rudimentary light-harvesting chlorophyll protein complexes. The light-induced change in the carotenoid pattern is suppressed by mevinolin, largely at the expense of β-carotene.[174] The relatively low sensitivity of plastidic prenyllipid synthesis in primary leaves from wheat to mevinolin is even diminished in younger seedlings. Under comparable conditions, however mevinolin, when incubated in the presence of ^{14}C-acetate and ^{3}H-MVA, completely blocked the incorporation of acetate but not of MVA into phytosterols.[151,183] The predominant effect of mevinolin on steroid synthesis also becomes evident through the inhibition of saponin accumulation in seedlings of *Avena sativa* L.[184] Incorporation experiments using protoplasts from spinach, incubated in the presence of mevinolin, support our observations.[185] Recently it was shown[186,187] that exogeneous MVA (2 mM) increases the content of sterols in roots of *Medicago sativa* L., especially of Δ^7-sitosterol, 24-methylene-cycloartenol and squalene. The conversion of those biosynthetic precursors of 4,4-desmethylsterols requires aerobic conditions, which might not exist when the seedlings are cultivated on water. A gradient in the inhibition of sterol accumulation by the mevinolin-analogue compactin from roots to hypocotyls and cotyledons was also observed with *M. sativa*. An increase of sterol content over the controls in the presence of exogeneous MVA[186,187] provides further evidence of the putative role of HMGR-activity as being rate-limiting for sterol biosynthesis [Bach 1986, 1987].[26,56] Secondary physiological responses of radish seedlings treated with mevinolin, such as senescence retardation, increased accumulation at later stages of development. The complete lack of side root formation at concentrations > 1 µM, have been interpreted as reflecting a change in the balance between newly synthesized steroidal brassinolide(s) and of the cytokinin isopentenyladenine. The synthesis of brassinosteroids as derivatives of phytosterols could require an undisturbed *de novo* biosynthesis when the storage cotyledons have been depleted of sterols as precursors.

In mammalian cells it was shown that saturation of the pathway leading to cholesterol requires much higher concentrations of IPP-units in the cytoplasm than is needed for the synthesis of ubiquinone,[188] dolichol,[189,190] or of isopentenyl-t-RNA.[191] This might be due to large differences in the substrate affinities of prenyltransferases competing for IPP or farnesyl-PP. Thus it was hypothesized (cf.[171]) that for the synthesis of isopentenyladenine, also a MVA-derivative, only low concentrations of IPP are required, whereas the synthesis of steroids is more drastically influenced by inhibition of HMGR activity. The result would then be a dominating effect of endogenous cytokinins, mainly synthesized in the roots,[192,193] over the brassinosteroids. The morphological appearance of mevinolin-treated radish seedlings closely resembles that observed in the presence of exogenous cytokinins.[194,195]

Narita & Gruissem[46] used ripening tomato fruits as a model system to study the role of an intact MVA biosynthesis in plants. When HMGR activiy was inhibited *in vivo* by mevinolin at an early stage, further development was prevented. Inhibition at a later stage, however, did not affect fruit ripening and lycopene synthesis. The authors concluded that a pool of MVA needed for the synthesis of phytosterols was accumulated mainly during the first half of fruit ripening.

Mevinolin: Growth Inhibition in Plant Cell Cultures

The use of cell cultures permits studies with less of a problem of apoplastic and symplastic transport of the inhibitor to the site of action. For these experiments we used suspension cultures of *Silybum marianum*[171,196] that were grown in the presence of increasing concentrations of mevinolin (0-10 µM). Inhibition of cell propagation, as determined by recording changes in turbidity, by packed cell volume and other parameters such as dry-weight increase, protein content etc., was more prominent at the end of the logarithmic growth period (day 6) than at the beginning (day 3 after application of mevinolin). As in intact and excised seedlings of radish, mevinolin gradually affected the accumulation of prenyllipids: sterols > ubiquinone(s) > plastoquinone ~ carotenoids ~ chlorophylls. Besides the reduction of ubiquinone content by a maximum of 50% per matter dry weight (if related to fresh weight the inhibition would appear even more drastic), at the expense of Q-10, mevinolin induced a shift in the pattern of Q-homologues towards Q-8, with Q-9 being the by far dominating homologue in *Silybum marianum*.[197] The growth of transformed tobacco cells (strain LA-6) was inhibited within the same range of concentrations, whereas that of habituated cells (strain NW) was less affected (Roth & Bach, unpublished observations). In analogy to the findings with mammalian cell systems, secondary responses of plant cell cultures upon treatment with mevinolin, such as the reduction of fresh and dry weight, protein content, and prevention of cell division, have been interpreted as indicating some interference of the inhibitor with the cell cycle.[26]

The first plant system where an inhibitor of HMGR was tested was cell cultures of *Acer pseudoplantanus*.[198] Compactin at 5 mg/L (about 12.5 µM) blocked the incorporation of ^{14}C-leucine and ^{14}C-acetate into 4-desmethylsterols by 95 and 99%, respectively. Within the time range of at least 6 h after application of compactin, sterol biosynthesis continued at a reduced rate, whilst the cells used up a pool of non-identified sterol precursors, as determined by monitoring the incorporation of [Me-^{14}C]-methionine into the sterol sidechain. This pool, which was gradually depleted by the process of sterol biosynthesis, could not be replenished in the presence of the inhibitor.[198] The most likely candidate for building up a precursor pool is squalene, according to the labeling studies recently reported.[153] By analogy it might be suggested that phytoene acts as the intermediate ensuring a sufficient supply of substrate during those later stages of ripening and lycopene formation in tomato fruits which are not affected by mevinolin.[46]

Compactin was reported to inhibit growth of tobacco callus cultures.[199] At 5 µM the average inhibition of fresh weight was between 45 and 80%, at 10 µM up to 95%. Cytokinin (isopentenyladenine and kinetin) at concentrations up to 1 µM and 5 µM, respectively, could not overcome the inhibition by compactin; however this is what can be expected, since at such concentrations the hormones alone might exert some growth-inhibitory effects. Mevinolin as well as compactin at 25 μM completely suppressed the development of tissue explants of *Helianthus tuberosus*.[63] Only MVA (2 mM), but not farnesol or squalene alone or in combination with abscisic acid, dolichol monophosphate or ubiquinone, could completely compensate for this effect. However, in the presence of 0.1 mM farnesol, even 0.2 mM MVA was fully effective,[63] thereby indicating that a non-steroidal factor derived from the branched isoprenoid pathway was essential for cell growth and division. In this regard it has to be noted that MVA itself, in the presence of optimal concentrations of cytokinin and auxin, was reported to stimulate the growth of callus cultures.[200,201]

The accumulation of sesquiterpenoid phytoalexins in cultures of *Solanum tuberosum*, challenged with the elicitor arachidonic acid, was inhibited by 60% at 40 nM mevinolin.[42] In particular lubimin was affected much more than rishitin. This observation was interpreted to mean that a pool of lubimin as a possible intermediate in the synthetic pathway leading to rishitin could not be replenished fast enough during blockage of HMGR activity, and thus an inhibitory effect on lubimin synthesis should be apparent earlier.[42] Another explanation, however, could be that different and compartmentalized sites of synthesis for both phytoalexins exist within the plant cell, in analogy to our findings on plastidic prenyllipids, which are not affected by mevinolin. The synthesis of capsidiol, the dominant sesquiterpenoid phytoalexin in solanaceous species, *Nicotiana tabacum*, was diminished by 80% at 10 μM mevinolin.[43] Recent findings indicate that the elicitor-induced increase in HMGR activity is a transient process that obviously ensures the supply of intermediates, positioned in the pathway behind MVA, for sesquiterpenoid biosynthesis.[45] Fine regulation of substrate flow from MVA and its immediate derivatives is mediated through suppression of squalene synthetase activity and accompanied by the rapid antiparallel increase in farnesylpyrophosphate sesquiterpene cyclase.[45] One μM mevinolin, isolated from a strain of *Aspergillus* that was freshly isolated from soil in Taiwan (thereby pointing to the ubiquitous ecological role of antibiotics of the mevinolin-type) efficiently blocked the somatic embryogenesis in cultures of *Daucus carota*, an effect which was partially reverted by 1 μM MVA.[202]

If one considers the cultivation of algae as a system comparable to cell-suspension cultures of higher plants, experiments with mevinolin added to the cell wall-free chrysophyte *Ochromonas malhamensis* are relevant. Mevinolin at 10 µM initially inhibited 90% of the synthesis of poriferasterol, which in these algae can be up to 1% of dry matter.[73] Under long-term conditions mevinolin at 10 µM induced a 10 to 15-fold increase in HMGR activity, which rapidly dropped to control levels after removal of the inhibitor.[73] This increase in HMGR activity is comparable to the situation found in mammalian cells[130,203-209] and is hitherto the only such report available on plant material. When radish seedlings were grown on 2 mg/L mevinolin to increase HMGR activity, the apparent enzyme activity was only 20% of the control, even after several steps of purification which should have washed away any inhibitor (Bach 1983, unpublished observations). However, HMGR purified in this way, for the first time yielded clearly visible bands in SDS-gels after silver staining.[57] There are two explanations: first, the type of inhibition by mevinolin is not solely competitive, as determined *in vitro,* and the inhibitor sticks firmly to parts of the enzyme not necessarily identical to the active site, and second, mevinolin induces the synthesis of (inactive?) HMGR protein. Immunological techniques should provide help to resolve these questions in future studies.

Mevinolin, a Molecular Probe to Study the Intracellular Distribution of MVA Biosynthesis

From incorporation studies using $^{14}CO_2$ and ^{14}C-MVA,[212-213] the so-called segregation model was developed, which postulated separate pathways for the synthesis of MVA and isoprenoid derivatives in the organelles and the cytoplasm, respectively (Fig. 20a). Although this assumption was supported by independent experimental data,[214,215] this model was also challenged by *in vitro* incorporation studies as well as direct determination of enzyme reactions using isolated oganelles.[62,67,216-221] Kleinig's group (cf. Fig. 20b) ascribed to IPP a pivotal role as a common and exclusive precursor molecule of all prenyllipids, independent of their intracellular localization. Accordingly, IPP, exclusively synthesized in the cytoplasm, should then be translocated into the organelles, which thereby are completely dependent upon the cytoplasmic capacity for MVA and IPP synthesis.

A main argument in favor of this assumption was the purity of isolated organelles used: preparations in which HMGR activity had been assayed[31,35,52,61,68,69], however certainly do

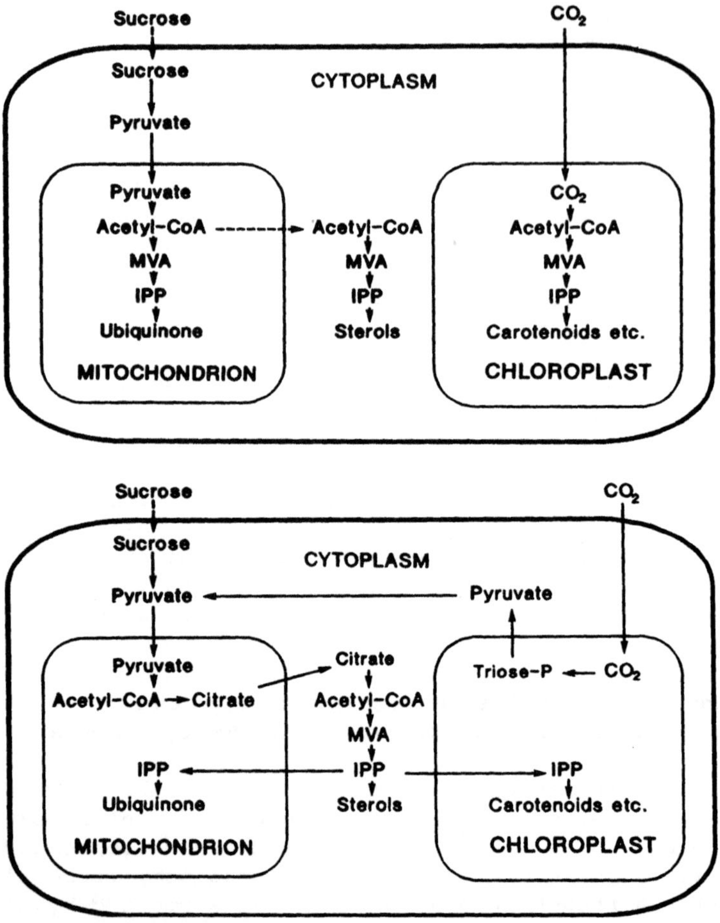

Fig. 20. Models concerning the compartmentalization of isoprenoid biosynthesis. a) according to Goodwin and his associates: existence of three separate pathways; b) according to Kleinig and his colleagues: IPP is exclusively synthesized in the cytoplasm and then transported into the organelles. For literature see text.

not fulfill the criteria of purity, *viz.* lack of contamination by cytoplasmic membranes. "Pure" chloroplasts, for example, are usually isolated from mature leaf tissue (e.g. spinach), and rates of synthesis of isoprenoids, are expected to be rather low after the build-up of a functioning photosynthetic apparatus. Thus the small quantity of MVA synthesized might be sufficient to supply the plastid with the substrate required to maintain the turnover of carotenoids (mainly of β-carotene) and side-chains of chlorophylls and quinones. This leads one to expect only little enzyme activity, possibly too low to be assayed. To be unable to detect an enzyme activity does not provide proof of its non-existence.

The predominant effect of mevinolin on the accumulation of phytosterols was interpreted as indicating that HMGR plays a limiting role in their synthesis, as in the case of mammalian cells. However, this only partially applies to ubiquinone(s), but not at all to plastid prenyllipids. The interpretation is simple: mevinolin is taken up by the plant cell and is able to completely block the cytoplasmic synthesis of sterols, but the mitochondrial envelope is a barrier for the inhibitor of mitochondrial HMGR. There is a body of evidence indicating that HMGR occurs in mitochondria.[30,39,52] The slight inhibition of ubiquinone accumulation possibly indicates some partial coupling of mitochondrial supply of IPP with the cytoplasmic synthetic capacity and is therefore not in complete disagreement with the hypotheses of Kleinig and his colleagues.[67,219] The mevinolin-induced shift in the Q-pattern towards shorter side chains[171,177,196] can be readily explained by an adaptation of the cells to a limited supply of IPP-units.

The almost total inability of mevinolin to prevent the synthesis of plastidic components is hard to explain with the model where IPP is exclusively synthesized in the cytoplasm. If cytoplasmic IPP-synthesis is completely blocked in the presence of high doses of mevinolin[151,183] as indicated by the lack of any incorporation of [2-^{14}C]-acetate into phytosterols -in contrast to ^{3}H-MVA!-, then the observed *de novo* synthesis of carotenoids and chlorophylls can only be explained by the assumption of separated MVA (IPP) synthesizing pathways that exist within the plant cell.

In addition, if cytoplasmic IPP serves as a substrate for the multi-branched and compartmentalized isoprenoid pathway, various prenyltransferases and translocators would have to compete for IPP or its isomer DMAPP. Such sequential enzyme systems should possess substrate affinities that differ by orders of magnitude in order to allow for an unchanged substrate flow into a special end-product in presence of the inhibitor; clear information on this topic, however, is lacking (cf.[27,56]). The enhanced accumulation of plastid prenylquinones in radish seedlings treated with moderate concentrations of mevinolin[177] cannot be explained by the hypothesis that IPP synthesis occurs exclusively in the cytoplasm either. Speculations on a synthetic pathway leading to IPP and not requiring the intermediate HMG-CoA (cf.[67,221]) have not found any clear experimental support. Emmanuel & Robblee[222] have proposed a pathway starting with propionate which was incorporated into cholesterol in rat liver.[223] According to their hypothesis propionyl-CoA could condense with propionaldehyde in a reaction analogous to the formation of malate from acetyl-CoA and glyoxylate, the mutation of a methyl group by a mutase in analogy to the conversion of L-glutamate to threo-β-methyl-L-aspartate. Oxidation of this hypothetical intermediate would yield MVA. Some conversion of propionate *in vivo* into 3-hydroxypropionate (and traces of acetate) in lima beans (*Phaseolus limensis*) with acrylic acid as an intermediate has been described.[224] However, a vitamin B_{12}-dependent mutase activity which would have led to the formation of a methylmalonyl derivative and finally to succinate could not be detected.

The possibility remains of a solution to these apparently differing findings. The dependence of organellar isoprenoid synthesis on cytoplasmic IPP formation might be a function of age. Only during later stages of development do organelles increasingly become dependent on cytoplasmic supply of IPP, whereas in early stages they are fully autonomous (cf.[225]). Then acetate made in the mitochondria and transported into the plastid *via* the cytoplasm and converted to acetyl-CoA within the plastid by a highly active acetyl-CoA synthetase[220] would find its main entry into the *de novo* synthesis of fatty acids with no or little exchange with internally produced acetyl-CoA, which would exclusively serve as the substrate for isoprenoid synthesis (see Fig. 21). Metabolism of pyruvate to acetyl-CoA and the presence of a rudimentary glycolytic pathway in plastids,[226] thereby having a linkage to the Calvin-cycle, would guarantee the supply of prenyllipid synthesis with acetyl-CoA. Independent support that acetyl-CoA serves as the substrate for isoprenoid formation in plastids exists.[69,215] In thiamine-deficient mutants of *Nicotiana sylvestris* the synthesis of carotenes and chlorophylls but not of fatty acids is blocked.[227] Since thiamine is a mandatory cofactor in the conversion of pyruvate to acetyl-CoA, the interruption of pigment synthesis in its absence indicates the tight linkage to the activity of the pyruvate dehydrogenase complex. However, some questions remain as to the formation of acetoacetyl-CoA in

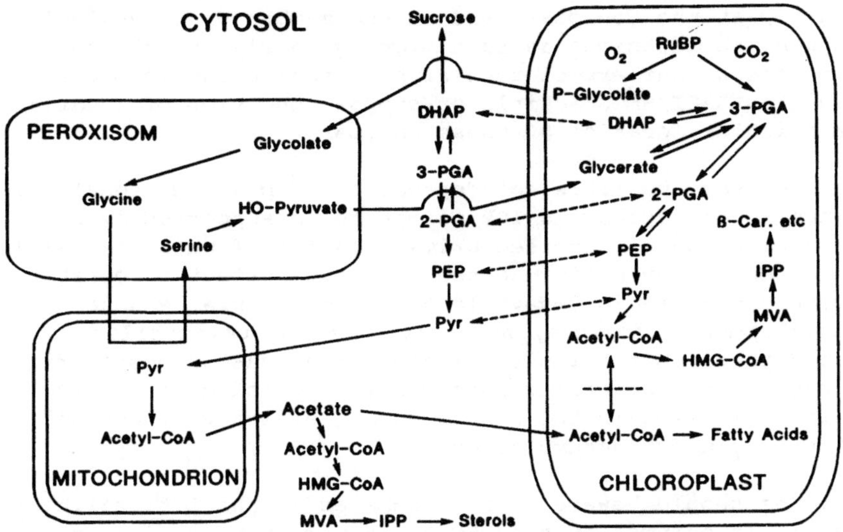

Fig. 21. Modified model concerning the compartmentalization of MVA and isoprenoid biosynthesis in plants. According to Schultz & Schulze-Siebert[185] and literature cited therein.

plastidic preparations.[69,185] The spectrophotometric assay at 303 nm is quite unsuitable for measuring the formation of acetoacetyl-CoA in the presence of membranous material containing high amounts of pigments. In addition, our experiments indicate a channeled synthesis of HMG-CoA from acetyl-CoA without the appearance of appreciable quantitites of acetoacetyl-CoA (see Table 6).

Recently, a cytochemical approach was used to assay HMGS activity in osmophores of the orchid, *Stanhopea anfracta*.[228] Electron-dense precipitates formed through reduction of ferricyanide to ferrocyanide by HS-CoA (released during the condensation of acetoacetyl-CoA and acetyl-CoA) with ferrocyanide finally reacting with uranylacetate to produce insoluble uranylferrocyanide, were detected at the smooth endoplasmic reticulum, between the outer and inner envelope of the mitochondria, at the tonoplast and at the membranes of amyloplasts. Although the technique as such might give rise to the formation of artifacts, our experiences with cytosolic and membrane-associated AACT/HMGS-activities support the view of a variable intracellular location of those enzymes.

The very possible age-dependence of plastid prenyllipid synthesis upon cytosolic IPP formation is supported by the observation that in excised primary leaves of wheat extremely high doses of mevinolin exerted a greater effect on plastid pigment synthesis in 9-day-old seedlings than at younger stages of development.[174] Thus, not only a better permeability of the plastid envelope towards the inhibitor as was suggested first, but rather the increasing auxotrophy of plastids as to IPP synthesis would explain the age-dependent efficacy of mevinolin in affecting pigment accumulation.

One should beware of any dogma as to the intracellular location of enzymes such as HMGR. For many years it was commonly accepted that the mammalian HMGR is exclusively bound to the endoplasmic reticulum of the cytoplasm (cf. Sabine[24] and literature cited therein). However, by the aid of density gradient centrifugation and comparison with marker enzymes as well as with immunocytochemical methods, HMGR activity has been demonstrated in peroxisomes from rat liver.[190,229] When rat liver mRNA was translated in a cell-free system and native HMGR-protein was precipitated with monospecific antibodies in SDS-gels, two bands could be detected with little difference in the molecular mass.[230] It can be thought feasible to assume that primary mRNA transcripts read from the DNA would give rise to the formation of slightly modified forms of the enzyme through differential splicing with the consequence of a variable intracellular location. At least the known complicated structure of the mammalian HMGR gene[80-84,231-233] could allow for such mechanisms. Rat-liver peroxisomes were shown to synthesize cholesterol from MVA in the presence of cytosolic proteins.[234]

In yeast HMGR activity was assayed in mitochondria and microsomes[235] (for older and partially conflicting literature see refs.[35,236]). Two different but related HMGR genes have been identified[237] with an extensive sequence homology in a section comprising the active site -which has been conserved in all eukaryotic HMGR species sequenced so far- and with less conserved spacer regions separating the seven potential membrane-spanning domains.[79] It seems likely that these two genes (*HMG1* and *HMG2*) arose from the duplication of a single gene after the divergence of yeast and mammals, where only one HMGR-gene was detected. Assuming some evolutionary pressure existing even for the linker regions, the 90% divergence observed led to an estimate of about 200 million years since the duplication event.[79]

Based on experiments feeding mevinolin to intact plants and cell cultures and the observation of apparently independent intracellular sites of MVA biosynthesis as well from data of other laboratories (see Table 1), it was postulated that plants may contain as many as three HMGR-genes.[26] Although in plants (*Arabidopsis thaliana*,[238,239] tomato,[46] and radish (Wettstein et al., manuscript in preparation), only one single HMGR-gene has been characterized, some observations indicate the occurrence of a gene family. Differential expression of genes is known to occur following gene duplication, with one gene copy becoming specialized in a function in a particular tissue. This is a common evolutionary strategy. Aspects of molecular biology of MVA biosynthesis in plants will be broadly discussed in another contribution in this volume.

CONCLUSION

Through the concerted efforts of the increasing numbers of research groups all over the world in the field of the enzymology and regulation of early enzymes of the isoprenoid pathway, powerful tools such as monospecific antibodies and heterologous and homologous cDNA probes should facilitate clear answers to some of the questions in the field of MVA biosynthesis in plants within the near future. Although our knowledge of the enzymology of HMG-CoA biosynthesis is limited because the enzymes involved have not yet been purified to homogeneity, the data, even though preliminary to some extent, open new avenues towards a better understanding of the regulation of the isoprenoid pathway in plants. It can be expected that the next few years will provide us with great advances about the mechanisms of enzyme catalysis, the regulation of the enzymes within the complex interplay of various metabolic routes.

ACKNOWLEDGMENTS

Our investigations are supported by grants from the Deutsche Forschungsgemeinschaft (Ba 871/2-2) and from the NATO Scientific Research Division (collaborative research grant No.

538-88). We are grateful to the following scientists for providing us with various enzyme inhibitors: A.W. Alberts, M.D. Greenspan (Merck Sharp & Dohme Research Labs); Professor S. Omura (The Kitasato Institute, Tokyo); Gr. G. Beck (Hoechst A.G). We appreciate the information on partially unpublished observations from several labs as well as the fruitful discussions with experts in this field. Thanks are also due to the various research groups which we have joined in the past, to further some of our research interests.

REFERENCES

1. KLOSTERMAN, H.J., SMITH, F. 1954. The isolation of β-hydroxy-β-methylglutaric acid from the seed of flax. *J. Am. Chem. Soc.* 76: 1229-1230.
2. JOHNSTON, J.A., RACUSEN, D.W., BONNER, J. 1954. The metabolism of isoprenoid precursors in a plant system. *Proc. Natl. Acad. Sci. U.S.A.* 40: 1031-1037.
3. RUDNEY, H. 1955. The synthesis of β,β-dimethylacrylic acid in rat liver homogenates. *J. Am. Chem. Soc.* 77: 1698-1699.
4. RUDNEY, H. 1957. The biosynthesis of β-hydroxy-β-methylglutaric acid. *J. Biol. Chem.* 227: 363-377.
5. RUDNEY, H. 1957. The biosynthesis of β-hydroxy-β-methylglutaryl coenzyme A. *J. Am. Chem. Soc.* 79: 5580-5581.
6. RUDNEY, H., FERGUSON, J.J. 1959. The biosynthesis of β-hydroxy-β-methylglutaryl coenzyme A in yeast. II. The formation of hydroxymethylglutaryl coenzyme A via the condensation of acetyl-coenzyme A and acetoacetyl coenzyme A in yeast. *J. Biol. Chem.* 234: 1076-1080.
7. LYNEN, F., HENNING, U., BUBLITZ, C., SÖRBO, B., KRÖPLIN-RUEFF, L. Der chemische Mechanismus der Acetessigsäurebildung in der Leber. *Biochem. Z.* 330: 269-295.
8. WOLF, D.E., HOFFMAN, C.H., ALDRICH, P.E., SKEGGS, H.R., WRIGHT, L.D., FOLKERS, K. 1956. β-Hydroxy-β-methyl δ-valerolactone (divalonic acid), a new biological factor. *J. Amer. Chem. Soc.* 78: 4499.
9. TAVORMINA, P.A., GIBBS, M.H. 1956. The metabolism of β,δ-dihydroxy-β-methylvaleric acid by liver homogenates. *J. Am. Chem. Soc.* 78: 6210.
10. TAVORMINA, P.A., GIBBS, M.H., HUFF, J.W. 1956. The utilization of β-hydroxy-β-methyl-δ-valerolactone in cholesterol biosynthesis. *J. Am. Chem. Soc.* 78: 4498.
11. LYNEN, F., GRASSL, M. 1958. Darstellung von (-)-

Mevalonsäure durch bakterielle Racematspaltung. *Biochem. Z.* 335: 123-127.
12. GOODWIN, T W. (ed.) 1970. Natural substances formed biologically from mevalonic acid. *Biochem. Soc. Symp. No. 29, Liverpool 1969,* Academic Press, New York.
13. NES, W.R., MCKEAN, M.L. 1977. Biochemistry of steroids and other isopentenoids. University Park Press, Baltimore - London - Tokyo.
14. PORTER, J., SPURGEON, S.L. (eds.) 1981. *Biosynthesis of Isoprenoid Compounds,* Vol. 1, John Wiley and Sons, New York - Chichester - Brisbane - Toronto.
15. FERGUSON, J.J., DURR, I.F., RUDNEY, H. 1959. The biosynthesis of mevalonic acid. *Proc. Natl. Acad. Sci. U.S.A.* 45: 499-504.
16. RUDNEY, H. The biosynthesis of β-hydroxy-β-methyl-glutaryl coenzyme A and its conversion to mevalonic acid. In: *Ciba Foundation Symposium on the Biosynthesis of Terpenes and Sterols.* (G E.W. Wolstenholme, M. O'Connor, eds.), J.A. Churchill, London, pp. 75-94.
17. KIRTLEY, M.E., RUDNEY, H. 1967. Some properties and mechanism of action of the β-hydroxy-β-methylglutaryl coenzyme A reductase of yeast. *Biochemistry U.S.A.* 6: 230-238.
18. LYNEN, F. New aspects of acetate incorporation into isoprenoid precursors. In: *Ciba Foundation Symposium on the Biosynthesis of Terpenes and Sterols.* (G.E.W. Wolstenholme, M. O'Connor, eds.), J.A. Churchill, London, pp. 95-118.
19. KNAPPE, J, RINGELMANN, E., LYNEN, F. 1959. Über die β-Hydroxy-β-methylglutaryl Reduktase der Hefe. Zur Biosynthese der Terpene IX. *Biochem. Z.* 332: 195-213.
20. BRODIE J., PORTER, J.W. 1960. The synthesis of mevalonic acid by non-particulate avian and mammalian enzyme systems. *Biochim. Biophys. Res. Commun.* 3: 173-177.
21. SPURGEON, S.L, PORTER, J.W. 1981. Introduction. In: *Biosynthesis of Isoprenoid Compounds.* (J.W. Porter, S.L. Spurgeon, eds.), Vol. 1, John Wiley and Sons, New York - Chichester - Brisbane - Toronto, pp. 1-46.
22. QURESHI, N., PORTER, J.W. 1981. Conversion of acetyl-coenzyme A to isopentenyl pyrophosphate. In: *Biosynthesis of Isoprenoid Compounds.* (J.W. Porter, S.L. Spurgeon, eds.), Vol. 1, John Wiley and Sons, New York - Chichester - Brisbane - Toronto, pp. 47-94.
23. DUGAN, R.E. 1981. Regulation of HMG-CoA reductase. In:

Biosynthesis of Isoprenoid Compounds. (J.W. Porter, S.L. Spurgeon, eds.), Vol. 1, John Wiley and Sons, New York - Chichester - Brisbane - Toronto, pp. 95-159.
24. SABINE, J.R. 1983. *Monographs on enzyme biology:* HMG-CoA reductase. (J.R. Sabine, ed.), CRC Press, Boca Raton.
25. PREISS, B. 1985. Regulation of HMG-CoA reductase. (B. Preiss, ed.), Academic Press, New York.
26. BACH, T.J. 1987. Synthesis and metabolism of mevalonic acid in plants. *Plant Physiol. Biochem.* 25: 163-178.
27. GRAY, J.C. 1987. Control of isoprenoid biosynthesis in higher plants. *Adv. Bot. Res.* 14: 25-91.
28. BROOKER, J.D., RUSSELL, D.W. 1975. Properties of microsomal 3-hydroxy-3-methylglutaryl-coenzyme A reductase from *Pisum sativum* seedlings. *Arch. Biochem. Biopys.* 167: 723-729.
29. BROOKER, J.D., RUSSELL, D.W. 1979. Regulation of microsomal 3-hydroxy-3-methylglutaryl-coenzyme A reductase from pea seedlings. Rapid post-translational phytochrome-mediated decrease in activity and *in vivo* regulation by isoprenoid products. *Arch. Biophys. Biochem.* 198: 323-334.
30. BACH, T.J., LICHTENTHALER, H.K., RÉTEY, J. 1980. Properties of membrane-bound 3-hydroxy-3-methylglutaryl-coenzyme A reductase from radish seedlings and some aspects of its regulation. In: *Biogenesis and Function of Plant Lipids.* (P. Mazliak et al., eds.), Elsevier, Amsterdam, pp. 355-362.
31. WONG, R.J., MCCORMACK, D.K., RUSSELL, D.W. 1982. Plastid 3-hydroxy-3-methylglutaryl-coenzyme A reductase has distinctive kinetic and regulatory features: properties of the enzyme and positive phytochrome control of activity in pea seedlings. *Arch. Biophys. Biochem.* 216: 613-638.
32. RUSSELL, D.W., KNIGHT, J.S., WILSON, T.M. 1985. Pea seedling HMG-CoA reductase: regulation of activity *in vitro* by phosphorylation and Ca^{2+}, and posttranslational control *in vivo* by phytochrome and isoprenoid hormones. *Curr. Top. Plant Biochem. Physiol.* 4: 191-206.
33. GRUMBACH, K.H., BACH, T.J. 1979. The effect of PSII herbicides, amitrol and SAN 6706 on the activity of 3-hydroxy-3-methylglutaryl-coenzyme A reductase and the incorporation of (2-^{14}C) acetate and (2-^{3}H) mevalonate into chloroplast pigments of radish seedlings. *Z. Naturforsch.* 34c: 941-943.

34. RUSSELL, D.W., DAVIDSON, H. 1982. Regulation of cytosolic HMG-CoA reductase activity in pea seedlings: contrasting responses to different hormones, and hormone-product interaction, suggest hormonal modulation of activity. *Biochem. Biophys. Res. Commun.* 104: 1537-1543.

35. BACH, T.J. 1981. Untersuchungen zur Charakterisierung und Regulation der 3-Hydroxy-3-methylglutaryl-Coenzym A Reduktase (Mevalonat: $NADP^+$ Oxidoreduktase, CoA acylierend, E.C. 1.1.1.34) in Keimlingen von *Raphanus sativus*.- Karlsr. Beitr. Pflanzenphysiol. (*Karlsr. Contrib. Plant Physiol.* ISSN 0173-3133) 10: 1-219.

36. BACH, T.J., LICHTENTHALER, H.K. 1984. Application of modified Lineweaver-Burk plots to studies of kinetics and regulation of radish 3-hydroxy-3-methylglutaryl-CoA reductase from radish seedlings. *Biochim. Biophys. Acta* 794: 152-161.

37. ISA, R.B.M., SIPAT, A.B. 1982. 3-Hydroxy-3-methylglutaryl-coenzyme A reductase of Hevea latex: The occurrence of a heat-stable activator in the C-serum. *Biochem. Biophys. Res. Commun.* 108: 206-212.

38. SUZUKI, H., URITANI, I., OBA, K. 1975. The occurrence of and properties of 3-hydroxy-3-methylglutaryl-coenzyme A reductase in sweet potato roots infected by *Ceratocystis fimbriata*. *Physiol. Plant Pathol.* 7: 265-276.

39 SUZUKI, K., OBA, K., URITANI, I. 1976. Subcellular localization of 3-hydroxy-3-methylglutaryl-coenzyme A reductase and other membrane-bound enzymes in sweet potato roots. *Plant Cell Physiol.* 17: 691-700.

40. ITO, R., OBA, K., URITANI, I. 1979. Mechanisms for the induction of 3-hydroxy-3-methylglutaryl-coenzyme A reductase in $HgCl_2$-treated sweet potato root tissue. *Plant Cell Physiol.* 20: 867-874.

41. OBA, K., KONDO, K., DOKE, N., URITANI, I. 1985. Induction of 3-hydroxy-3-methylglutaryl CoA reductase in potato after slicing, fungal infection or chemical treatment, and some properties of the enzyme. *Plant Cell Physiol.* 26: 873-880.

42. STERMER, B.A., BOSTOCK, R.M. 1987. Involvement of 3-hydroxy-3-methyl-glutaryl coenzyme A reductase in the regulation of sesquiterpenoid phytoalexin synthesis in potato. *Plant Physiol.* 84: 404-408.

43. CHAPPELL, J., NABLE, R. 1987. Induction of sesquiterpenoid biosynthesis in tobacco cell suspension cultures by fungal elicitor. *Plant Physiol.* 85: 469-

473.
44. LEUBE, J., GRISEBACH, H. 1983. Further studies on induction of enzymes of phytoalexin synthesis in soybean and cultured soybean cells. *Z. Naturforsch.* 38c: 730-735.
45. VÖGELI, U., CHAPPELL, J. 1988. Induction of sesquiterpene cyclase and suppression of squalene synthetase activities in plant cell cultures treated with fungal elicitor. *Plant Physiol.* 88: 1291-1296.
46. NARITA, J.O., GRUISSEM, W. 1989 Tomato hydroxymethyl-glutaryl-CoA reductase is required early in fruit development but not during fruit ripening. *Plant Cell* 1: 181-190.
47. LYNEN, F. 1967. Biosynthesis pathways from acetate to natural products. Activity of the enzymes in rubber synthesis. *Pure Appl. Chem.* 14: 137-167.
48. HEPPER, C.M., AUDLEY, B.G. 1969. The biosynthesis of rubber from β-hydroxy-β-methylglutaryl-coenzyme A in *Hevea brasiliensis* latex. *Biochem. J.* 114: 379-386.
49. WITITSUWANNAKUL, R. 1986. Diurnal variation of 3-hydroxy-3-methylglutaryl-coenzyme A reductase activity in latex of *Hevea brasiliensis* and its relation to rubber content. *Experientia* 42: 44-46.
50. SIPAT, A.B. 1982. Hydroxymethylglutaryl CoA reductase (NADPH) in the latex of *Hevea brasiliensis. Phytochemistry* 21: 2613-2618.
51. SIPAT, A. 1985. 3-Hydroxy-3-methylglutaryl-CoA reductase in the latex of *Hevea brasiliensis. Meth. Enzymol.* 110: 40-51.
52. BROOKER, J.D., RUSSELL, D.W. 1975. Subcellular localization of 3-hydroxy-3-methylglutaryl-coenzyme A reductase in *Pisum sativum* seedlings. *Arch. Biochem. Biophys.* 67: 730-737 .
53. SUZUKI, H., URITANI, I. 1977. Effects of bovine serum albumin and phospholipids on activity of microsomal 3-hydroxy-3-methylglutaryl-coenzyme A reductase in sweet potato roots. *Plant Cell Physiol.* 18: 485-495.
54. YU-ITO, R., OBA, K., URITANI, I. 1982. Some problems in the assay method of HMG-CoA reductase activity in sweet potato in the presence of other HMG-CoA utilizing enzymes. *Agric. Biol. Chem.* 46: 2087-2091.
55. DOUGLAS, T.J., PALEG, L.G. 1978. AMO 1618 effects on the incorporation of ^{14}C-MVA and ^{14}C-acetate into sterol in *Nicotiana* and *Digitalis* seedlings and cell-free preparations from *Nicotiana. Phytochemistry* 17: 713-718.

56. BACH, T.J. 1986. Hydroxymethylglutaryl-CoA reductase, a key enzyme in phytosterol synthesis? *Lipids* 21: 82-88.
57. BACH, T.J., ROGERS, D.H., RUDNEY, H. 1986. Detergent-solubilization, purification, and characterization of 3-hydroxy-3-methylglutaryl-CoA reductase from radish seedlings. *Eur. J. Biochem.* 154: 103-111.
58. CAMARA, B., BARDAT, F., DOGBO, O., BRANGEON, J., MONÉGER, R. 1983. Terpenoid metabolism in plastids. Isolation and biochemical characteristics of *Capsicum annuum* chromoplasts. *Plant Physiol.* 73: 94-99.
59. NISHI, A., TSURITANI, I. 1983. Effect of auxin on the metabolism of mevalonic acid in suspension-cultured carrot cells. *Phytochemistry* 22: 399-401.
60. BONHOFF, A., LOYAL, R., FELLER, K., EBEL, J., GRISEBACH, H. 1986. Further investigations of race: cultivar-specific induction of enzymes related to phytoalexin biosynthesis in soybean roots following infection with *Phytophthora megasperma* f.sp. *glycinea*. *Biol. Chem. Hoppe-Seyler* 367: 797-802.
61. AREBALO, R.E., MITCHELL Jr., E.D. 1984. Cellular distribution of 3-hydroxy-3-methylglutaryl coenzyme A reductase and mevalonate kinase in leaves of *Nepeta cataria*. *Phytochemistry* 23: 13-18.
62. KREUZ, K., KLEINIG, H. 1984. Synthesis of prenyl lipids in cells of spinach leaf. Compartmentation of enzymes for formation of isopentenyl diphosphate. *Eur. J. Biochem.* 141: 531-535.
63. CECCARELLI, N., LORENZI, R. 1984. Growth inhibition by competitive inhibitors of 3-hydroxy-3-methylglutaryl coenzyme A reductase in *Helianthus tuberosus* tissue explants. *Plant Sci. Lett.* 34: 269-276.
64. KONDO, K., OBA, K. 1986. Purification and characterization of 3-hydroxy-3-methylglutaryl CoA reductase from potato tubers. *J. Biochem. Tokyo* 100: 967-974.
65. SKRUKRUD, C.L., TAYLOR, S.E., HAWKINS, D.R., CALVIN, M. 1987. Triterpenoid biosynthesis in *Euphorbia lathyris*. In: *The Metabolism, Structure, and Function of Plant Lipids*. (P. K. Stumpf, J. B. Mudd, W. D. Nes, eds.), Plenum Press, New York - London, pp. 115-118.
66. SKRUKRUD, C.L., TAYLOR, S.E., HAWKINS, D.R., NEMETHY, E.K., CALVIN, M. 1988. Subcellular fractionation of triterpenoid biosynthesis in *Euphorbia lathyris* latex. *Physiol. Plant.* 74: 306-316.

67. LÜTKE-BRINKHAUS, F., KLEINIG, H. 1987. Formation of isopentenyl diphosphate via mevalonate does not occur within etioplasts and etiochloroplasts of mustard (*Sinapis alba* L.) seedlings. *Planta* 171: 406-411.
68. RAMACHANDRA REDDY, A., DAS, V. S. R. 1986. Partial purification and characterization of 3-hydroxy-3-methylglutaryl coenzyme A reductase from the leaves of guayule (*Parthenium argentatum*). *Phytochemistry* 25: 2471-2474.
69. RAMACHANDRA REDDY, A., DAS, V.S.R. 1987. Chloroplast autonomy for the biosynthesis of isopentenyl diphosphate in guayule (*Parthenium argentatum* Gray). *New. Phytol.* 106: 457-464.
70. BOLL, M., KARDINAL, A., BERNDT, J. 1987. Properties and regulation of 3-hydroxy-3-methylglutaryl coenzyme A reductase from spruce (*Picea abies*). *Biol. Chem. Hoppe-Seyler* 368: 1024.
71. BOLL, M., MEẞNER, B., KARDINAL, A. 1988. Regulation of phenylalanine ammonia-lyase (PAL) and 3-hydroxy-3-methylglutaryl coenzyme A reductase (HMGR) in spruce (*Picea abies*). *Biol. Chem. Hoppe-Seyler* 369: 799.
72. MAUREY, K.M., GOLBECK, J.H. 1985. Modulation of 3-hydroxy-3-methylglutaryl-coenzyme A reductase (HMGR) activity in *Ochromonas malhamensis*. *Plant Physiol.* 77 (supplement): 45.
73. MAUREY, K., WOLF, F., GOLBECK, J. 1986. 3-Hydroxy-3-methylglutaryl coenzyme A reductase activity in *Ochromonas malhamensis*. A system to study the relationship between enzyme activity and rate of steroid biosynthesis. *Plant Physiol.* 82: 523-527.
74. GOLBECK, J.H., MAUREY, K. M., NEWBERRY, G.A. 1985. Detection of 3-hydroxy-3-methylglutaryl-coenzyme A reductase (HMGR) activity in *Dunaliella salina*. *Plant Physiol.* 77 (supplement): 48.
75. ROGERS, D.H., PANINI, S.R., RUDNEY, H. 1980. Rapid, high-yield purification of rat liver 3-hydroxy-3-methyl-glutaryl-coenzyme A reductase. *Anal. Biochem.* 101: 107-111.
76. BACH, T.J., RUDNEY, H. 1983 Solubilization and partial purification of membrane-bound HMG-CoA reductase from a plant source. *J. Lipid Res.* 24: 1404-1405.
77. BACH, T.J., ROGERS, D.H., RUDNEY, H. 1984. Purification and properties of membrane-bound 3-hydroxy-3-methylglutaryl-CoA reductase from radish seedlings. In: *Structure, Function and Metabolism of Plant Lipids*. (P.A. Siegenthaler and W. Eichenberger,

eds.), Elsevier Science Publishers BV, Amsterdam, pp. 221-224.

78. QURESHI, N., DUGAN, R.E., CLELAND , W.W., PORTER, J.W. 1976. Kinetic analysis of the individual reductive steps catalyzed by β-hydroxy-β-methylglutaryl coenzyme A reductase obtained from yeast. *Biochemistry U.S.A.* 15: 4191-4197.

79. BASSON, M.E, THORSNESS, M, FINER-MOORE, J., STROUD, R.M. RINE, J. 1988. Structural and functional conservation between yeast and human 3-hydroxy-3-methylglutaryl coenzyme A reductases, the rate-limiting enzyme of sterol biosynthesis. *Mol. Cell. Biol.* 8: 3797-3808.

80. CHIN, D.J., GIL, G., RUSSELL, D.W., LISCUM, L , LUSKEY, K.L., BASU, S.K., OKAYAMA, H., BERG, P., GOLDSTEIN, J.L., BROWN, M.S. 1984. Nucleotide sequence of HMG-CoA reductase, a glycoprotein of the endoplasmic reticulum. *Nature (Lond.)* 308: 613-617.

81. REYNOLDS, G.A., BASU, S.K., OSBORNE, T.F., CHIN, D.J., GIL, G., BROWN, M.S., GOLDSTEIN, J.L., LUSKEY, K.L. 1984. HMG-CoA reductase: A negatively regulated gene with unusual promoter and 5' untranslated regions. *Cell* 38: 275-285.

82. BROWN, D.A., SIMONI, R.D. 1984. Biogenesis of 3-hydroxy-3-methylglutaryl-coenzyme A reductase, an integral glycoprotein of the endoplasmic reticulum. *Proc. Natl. Acad. Sci. U.S.A.* 81: 1674-1678.

83. LISCUM, L., FINER-MOORE, J., STROUD, R.M., LUSKEY, K.L., BROWN, M.S., GOLDSTEIN, J.L. 1985. Domain structure of 3-hydroxy-3-methylglutaryl coenzyme A reductase, a glycoprotein of the endoplasmic reticulum. *J. Biol. Chem.* 260: 522-530.

84. LUSKEY, K.L., STEVENS, B. 1985. Human 3-hydroxy-3-methylglutaryl coenzyme A reductase. Conserved domains responsible for catalytic activity and sterol-regulated degradation. *J. Biol Chem.* 260: 10271-12277.

85. WOODWARD, H.D., ALLEN, J.M.C., LENNARZ, W.L. 1988. 3-Hydroxy-3-methylglutaryl-coenzyme A reductase of the sea urchin embryo. Deduced structure and regulatory properties. *J. Biol. Chem.* 263: 18411-18418.

86. GERTLER, F.B., CHIU, C.Y., RICHTER-MANN, L., CHIN, D. J. 1988. Developmental and metabolic regulation of the *Drosophila melanogaster* 3-hydroxy-3-methylglutaryl coenzyme A reductase. *Mol. Cell. Biol.* 8: 2713-2721.

87. ROGERS, D.H., PANINI, S.R., RUDNEY, H. 1983. Properties

of HMG-CoA reductase and its mechanism of action. In: *3-Hydroxy-3-methylglutaryl Coenzyme A Reductase*. (J. R. Sabine, ed.), CRC Press, Inc., Boca Raton, Florida, pp. 58-75.

88. EDWARDS, P.A., KEMPNER, E.S., LAN, S.F., ERICKSON, S.K. 1985. Functional size of hepatic 3-hydroxy-3-methylglutaryl coenzyme A reductase as determined by radiation inactivation. *J. Biol. Chem.* 260: 10278-10282,
89. NESS, G.C., PENDLETON, L.C., MCCREERY, M.J. 1988. In situ determination of the functional size of hepatic 3-hydroxy-3-methylglutaryl-CoA reductase by radiation inactivation analysis. *Biochim. Biophys. Acta* 953: 361-364.
90. EDWARDS, P.A., LAN, S.F., FOGELMAN, A.M. 1983. Alterations in the rates of synthesis and degradation of rat liver 3-hydroxy-3-methylglutaryl coenzyme A reductase produced by cholestyramine and mevinolin. *J. Biol.Chem.* 258: 10219-10222.
91. NESS, G.C., MCCREERY, M.J., SAMPLE, C.E., SMITH, M., PENDLETON, L.C. 1985. Sulhydryl/disulfide forms of rat liver 3-hydroxy-3-methylglutaryl coenzyme A reductase. *J. Biol. Chem.* 260: 16395-16399.
92. KENNELLY, P.J., BRANDT, K.G., RODWELL, V.W. 1983. 3-Hydroxy-3-methylglutaryl-CoA reductase: Solubilization in the presence of proteolytic inhibitors, partial purification, and reversible phosphorylation-dephosphorylation. *Biochemistry USA* 22: 2784-2788.
93. DOTAN, I., SHECHTER, I. 1985. Reduced glutathione in chinese hamster ovary cells protects against inactivation of 3-hydroxy-3-methylglutaryl coenzyme A reductase by 2-mercaptoethanol disulfide. *J. Cell. Physiol.* 122: 14-20.
94. ROITELMAN, J., SHECHTER, I. 1984. Regulation of rat liver 3-hydroxy-3-methylglutaryl coenzyme A reductase. Evidence for thiol-dependent modulation of enzyme activity. *J. Biol. Chem.* 259: 870-877.
95. NESS, G.C., EALES, S.J., PENDLETON, L.C., SMITH, M. 1985. Activation of rat liver microsomal 3-hydroxy-3-methylglutaryl coenzyme A reductase by NADPH. Effects of dietary treatments. *J. Biol. Chem.* 260: 12391-12393.
96. CAPPEL, R.E., GILBERT, H.F. 1988. Thiol/disulfide exchange between 3-hydroxy-3-methylglutaryl-CoA reductase and glutathione. A thermodynamically facile dithiol oxidation. *J. Biol Chem.* 263: 12204-12212.

97. CLELAND, W.W. 1963. The kinetics of enzyme-catalyzed reactions with two or more substrates or products. I. Nomenclature and rate equations. *Biochim. Biophys. Acta* 67: 104-137.

98. BACH, T.J., LICHTENTHALER, H.K. 1983. Mechanisms of inhibition by mevinolin (MK 803) of microsome-bound radish and of partially purified yeast HMG-CoA reductase (E.C. 1.1.1.34). *Z. Naturforsch.* 38c: 212-219.

99. TANZAWA, K., ENDO, A. 1979. Kinetic analysis of the reaction catalyzed by rat-liver 3-hydroxy-3-methylglutaryl-coenzyme A reductase using two specific inhibitors. *Eur. J. Biochem.* 98: 195-201.

100. ROGERS, D.H., RUDNEY, H. 1982. Modification of 3-hydroxy-3-methyl-glutaryl coenzyme A reductase immunoinhibition curves by substrates and inhibitors. Evidence for conformational changes leading to alterations in antigenicity. *J. Biol. Chem.* 257: 10650-10658.

101. CLELAND, W.W. 1963. The kinetics of enzyme-catalyzed reactions with two or more substrates. III. Prediction of initial velocity and inhibition patterns by inspection. *Biochim. Biophys. Acta* 67: 188-196.

102. BACH, T.J. 1989. Zur Biosynthese und physiologischen Funktion der Mevalonsäure in Pflanzen. *Habilitationsschrift, Fakultät für Bio und Geowissenschaften, Universität Karlsruhe*, pp. 1-135.

103. FEYEREISEN, R., FARNSWORTH, D.E. 1987. Characterization and regulation of HMG-CoA reductase during a cycle of juvenile hormone synthesis. *Mol. Cell. Endocrinol.* 53: 227-238.

104. WOODWARD, H.D., ALLEN, J.M.C., LENNARZ, W.J. 1988. 3-Hydroxy-3-methylglutaryl coenzyme A reductase in the sea urchin embryo is developmentally regulated. *J. Biol. Chem.* 263: 2513-2517.

105. GILL Jr., J.F., RODWELL, V.W. 1983. Purification and properties of *Pseudomonas* HMG-CoA reductase. *J. Lipid Res.* 24: 1407.

106. GILL Jr., J.F., BEACH, M.J., RODWELL, V.W. 1985. Mevalonate utilization in *Pseudomonas* sp. M. Purification and characterization of an inducible 3-hydroxy-3-methylglutaryl coenzyme A reductase. *J. Biol. Chem.* 260: 9393-9398.

107. LOWE, D.M., TUBBS, P.K. 1985. 3-Hydroxy-3-methylglutaryl-coenzyme A synthase from ox liver. Purification, molecular and catalytic properties.

Biochem J. 227: 591-599.

108. MIZIORKO, H.M., BEHNKE, C.E. 1985. Active site-directed inhibition of 3-hydroxy-3-methylglutaryl coenzyme A synthase by 3-chloropropionyl coenzyme A. *Biochemistry U.S.A.* 24: 3174-3179.

109. MIZIORKO, H.M., BEHNKE, C.E. 1985. Amino acid seqence of an active site peptide of avian liver mitochondrial 3-hydroxy-3-methylglutaryl-CoA synthase. *J. Biol. Chem.* 260: 13513-13516.

110. GIL, G., GOLDSTEIN, J.L., SLAUGHTER, C.A., BROWN, M.S. 1986. Cytoplasmic 3-hydroxy-3-methylglutaryl coenzyme A synthase from the hamster: I. Isolation and sequencing of a full-length cDNA. *J. Biol. Chem.* 261: 3710-3716.

111. GIL, G., BROWN, M.S., GOLDSTEIN, J.L. 1986. Cytoplasmic 3-hydroxy-3-methylglutaryl coenzyme A synthase from the hamster: II. Isolation of the gene and characterization of the 5' flanking region. *J. Biol. Chem.* 261: 3717-3724.

112. SMITH, J.R., OSBORNE, T.F., BROWN, M.S., GOLDSTEIN, J.L., GIL, G. 1988. Multiple sterol regulatory elements in the promoter for hamster 3-hydroxy-3-methylglutaryl-coenzyme A synthase. *J. Biol. Chem.* 263: 18480-18487.

113. DEQUIN, S., GLOECKLER, R., HERBERT, C.J., BOUTELET, F. 1988. Cloning, sequencing and analysis of the yeast *S. uvarum ERG10* gene encoding acetoacetyl CoA thiolase. *Curr. Genet.* 13: 471-478.

114. OSBORNE, T.F., GIL, G., GOLDSTEIN, J.L., BROWN, M.S. 1988. Operator constitutive mutation of 3-hydroxy-3-methylglutaryl coenzyme A reductase promoter abolishes protein binding to sterol regulatory element. *J. Biol. Chem.* 263: 3380-3387.

115. DAWSON, P.A., HOFMANN, S.L., VAN DER WESTHUYZEN, D.R., SÜDHOFF, T.C., BROWN, M.S., GOLDSTEIN, J.L. 1988. Sterol-dependent repression of low density lipoprotein receptor promoter mediated by 16-base pair sequence adjacent to binding site for transcription factor Sp1. *J. Biol. Chem.* 263: 3372-3379.

116. GIL, G., SMITH, J.R., GOLDSTEIN, J.L., SLAUGHTER, C.A., ORTH, K., BROWN, M.S. 1988. Multiple genes encode nuclear factor 1-like proteins that bind to the promoter for 3-hydroxy-3-methylglutaryl-coentyme A reductase. Proc. Natl. Acad. Sci. 85: 8963-8967.

117. OSHIMA, K., URITANI, I. 1968. Enzymatic synthesis of a β-hydroxy-β-methylglutaric acid-derivative by a cell

free system from sweet potato infected with black rot. *J. Biochem. Tokyo* 63: 617-625.

118. POTTY, V.H. 1969. Occurrence and properties of enzymes associated with mevalonic acid synthesis in the orange. *J. Food Sci.* 34: 231-234.

119. MESSNER, B., EGGERER, H., CORNFORTH, J.W., MALLABY, R. 1975. Substrate stereochemistry of the hxdroxymethylglutaryl-CoA lyase and methylglutaconyl-CoA hydratase reactions. *Eur. J. Biochem.* 53: 255-264.

120. NEMETHY, E.K., SKRUKRUD, C., PIAZZA, G.J., CALVIN, M. 1983. Terpenoid biosynthesis in *Euphorbia* latex. *Biochim. Biophys. Acta* 760: 343-349.

121. BACH, T.J., WEBER, T. 1989. Enzymic synthesis of mevalonic acid in plants. In: *Biological Role of Plant Lipids.* (P.A. Biacs, K. Gruiz, T. Kremmer, eds.), Akademiai Kiado, Budapest and Plenum Publishing Corporation, New York and London, pp. 279-282.

122. CLINKENBEARD, K.D., SUGIYAMA, T., MOSS, J., REED, W.D., LANE, M.D. 1973. Molecular and catalytic properties of cytosolic acetoacetyl coenzyme A thiolase from avian liver. *J. Biol. Chem.* 248: 2275-2284.

123. CLINKENBEARD, K.D., REED, W.D., MOONEY, R.A., LANE, M.D. 1975. Intracellular localisation of the 3-hydroxy-3-methylglutaryl coenzyme A cycle enzymes in liver: Separate cytoplasmic and mitochondrial 3-hydroxy-3-methylglutaryl coenzyme A generating systems for cholesterogenesis and ketogenesis. *J. Biol. Chem.* 250: 2108-3116.

124. CLINKENBEARD, K.D., SUGIYAMA, T., REED, W.D., LANE, M.D. 1975. Cytoplasmic 3-hydroxy-3-methylglutaryl coenzyme A synthase from liver: Purification, properties and role in cholesterol synthesis. *J. Biol. Chem.* 250: 3124-3135.

125. TOMODA, H., KUMAGAI, H., TANAKA, H., OMURA, S. 1987. F-244 specifically inhibits 3-hydroxy-3-methylglutaryl coenzyme A synthase. *Biochim. Biophys. Acta* 922: 351-356.

126 GREENSPAN, M. D., YUDKOVITZ, J.B., LO, C.Y.L., CHEN, J.S., ALBERTS, A.W., HUNT, V.M., CHANG, M.N., YANG, S.S., THOMPSON, K.L., CHIANG, Y.C.P., CHABALA, J.C., MONAGHAN, R.L., SCHWARTZ, R.L. 1987. Inhibition of hydroxymethylglutaryl-coenzyme A synthase by L-659,699. *Proc. Natl. Acad. Sci. U.S.A.* 84: 7488-7492.

127. BUCHER, N.L.R., OVERATH, P., LYNEN, F. 1960. β-Hydroxy-β-methylglutaryl coenzyme A reductase,

cleavage and condensing enzymes in relation to cholesterol formation in rat liver. *Biochim. Biophys. Acta* 40: 491-501.

128. QUANT, P.A., TUBBS, P.K., BRAND, M.D. 1989. Glucagon increases mitochondrial 3-hydroxy-3-methylglutaryl-coenzyme A synthase activity *in vivo* by desuccinylating the enzyme. *Biochem. Soc. Transact.* 17: 147-148.

129. BALASUBRAMANIAM, S., GOLDSTEIN, J.L., BROWN, M.S. 1977. Regulation of cholesterol synthesis in rat adrenal gland through coordinate control of 3-hydroxy-3-methylglutaryl coenzyme A synthase and reductase activities. *Proc. Natl. Acad. Sci. U.S.A.* 74: 1421-1425.

130. CHIN, D.J., LUSKEY, K.L., ANDERSON, R.G.W., FAUST, J.R., GOLDSTEIN, J.L., BROWN, M.S. 1982. Appearance of crystalloid endoplasmic reticulum in compactin-resistant Chinese hamster cells with a 500-fold increase in 3-hydroxy-3-methylglutaryl coenzyme A reductase. *Proc. Natl. Acad. Sci. U.S.A.* 79: 1185-1189.

131. LUSKEY, K.L., FAUST, J.R., CHIN, D.J., BROWN, M.S., GOLDSTEIN, J.L. 1983. Amplification of the gene for 3-hydroxy-3-methylglutaryl coenzyme A reductase, but not for the 53-kDa protein, in UT-1 cells. *J. Biol. Chem.* 258: 8462-8469.

132. LEONARD, S., ARBOGAST, D., GEYER, D., JONES, C., SINENSKY, M. 1986. Localization of the gene encoding 3-hydroxy-3-methylglutaryl-coenzyme A synthase to human chromosome 5. *Proc. Natl. Acad. Sci.* 83: 2187-2189.

133. HUMPHRIES, S.E., TATA, F., HENRY, I., BARICHARD, F., HOLM, M., JUNIEN, C., WILLIAMSON, R. 1985. The isolation, characterisation, and chromosomal assignment of the gene for human 3-hydroxy-3-methylglutaryl coenzyme A reductase, (HMG-CoA reductase). *Hum. Genet.* 71: 254-258.

134. LINDGREN, V., LUSKEY, K.L., RUSSELL, D.W., FRANCKE, U. 1985. Human genes involved in cholesterol metabolism: Chromosomal mapping of the loci for the low density lipoprotein receptor and 3-hydroxy-3-methylglutaryl-coenzyme A reductase with cDNA probes. *Proc. Natl. Acad. Sci.* 82: 8567-8571.

135. KORNBLATT, J., RUDNEY, H. 1971. Two forms of acetoacetyl coenzyme A thiolase in yeast: I. Separation and properties. *J. Biol. Chem.* 246: 4417-

4423.

136. KORNBLATT, J., RUDNEY, H. 1971. Two forms of acetoacetyl coenzyme A thiolase in yeast: II. Intracellular location and relationship to growth. *J. Biol. Chem.* 246: 4424-4430.

137. TROCHA, P.J., SPRINSON, D.B. 1976. Location and regulation of early enzymes of sterol biosynthesis in yeast. *Arch. Biochem. Biophys.* 174: 45-51.

138. SERVOUSE, M., KARST, F. 1986. Regulation of early enzymes of ergosterol biosynthesis in *Saccharomyces cerevisiae*. *Biochem. J.* 240: 541-547.

139. KAWAGUCHI, A. 1970. Control of ergosterol biosynthesis in yeast: Existence of lipid inhibitors. *J. Biochem.* 67: 219-227.

140. BOLL, M., LÖWELL, M., STILL, J., BERNDT, J. 1975. Sterol biosynthesis in yeast. *Eur. J. Biochem.* 54: 435-444.

141. QUAIN, D.E., HASLAM, J.M. 1979. The effects of catabolite derepression on the accumulation of steryl esters and the activity of β-hydroxy-β-methylglutaryl-CoA reductase in *Saccharomyces cerevisiae*. *J. Gen. Microbiol.* 11: 343-351.

142. BERG, D., DRABER, W., VON HUGO, H., HUMMEL, W., MAYER, D. 1981. The effect of clotrimazole and triadimefon on 3-hydroxy-3-methyl-glutaryl-CoA reductase-[EC 1.1.1.34]-activity in *Saccharomyces cerevisiae*. *Z. Naturforsch.* 36c: 798-803.

143. DÉQUIN, S., BOUTELET, F., SERVOUSE, M., KARST, F. 1988. Effect of acetoacetyl CoA thiolase amplification on sterol synthesis in the yeasts *S. cerevisiae* and *S. uvarum*. *Biotechnol. Lett.* 10: 457-462.

144. KRAMER, P.R., MIZIORKO, H.M. 1983. 3-Hydroxy-3-methylglutaryl-CoA lyase: Catalysis of acetyl coenzyme A enolization. *Biochemistry USA* 22: 2353-2357.

145. STEGINK, L.D., COON, M.J. 1968. Stereospecificity and other properties of highly purified β-hydroxy-β-methylglutaryl coenzyme A cleavage enzyme from bovine liver. *J. Biol. Chem.* 243: 5272-5279.

146. BACHHAWAT, B. K., ROBINSON, W.G., COON, M.J. 1955. The enzymatic cleavage of β-hydroxy-β-methylglutaryl-coenzyme A to acetoacetate and acetyl coenzyme A. *J. Biol. Chem.* 216: 727-736.

147. HIGGINS, M.J.P., KORNBLATT, J.A., RUDNEY, H. 1972. Acyl-CoA ligases. The Enzymes, Vol. VII (P.D. Boyer, ed.), Academic Press, New York, pp. 407-434.

148. POPJÀK, G. 1971. Specificity of enzymes of sterol

biosynthesis. *Harvey Lect.* 65: 127-156.

149. LANDAU, B.R., BRUNENGRABER, H. 1985. Shunt pathway of mevalonate metabolism. *Meth. Enzymol.* 110: 100-114.

150. NES, W.D., CAMPBELL, B.C., STAFFORD, A.E., HADDON, W.F., BENSON, M. 1982. Metabolism of mevalonic acid to long chain fatty alcohols in an insect. *Biochem. Biophys. Res. Commun.* 108: 1258-1263.

151. NES, W.D., BACH, T.J. 1985. Evidence for a mevalonate shunt in a tracheophyte. *Proc. Roy. Soc. Lond.* B 225: 425-444.

152. ALLEN, K.G., BANTHORPE, D.V., CHARLWOOD, B.V., EKUNDAYO, O., MANN, J. 1976. Metabolic pools associated with monoterpene biosynthesis in higher plant. *Phytochemistry* 15: 101-107.

153. SINGH, S.S., NEE, T.Y., POLLARD, M.R. 1986. Acetate and mevalonate labeling studies with developing *Cuphea lutea* seeds. *Lipids* 21: 143-149.

154. GROENEVELD, H.W., ELINGS, J.C., JORGE BLANCO, M.S., BRAMWELL, D. 1989. Quantitative aspects of triacylglycerol metabolism and sterol synthesis from ^{14}C-acetate in etiolated seedlings of *Euphorbia lambii*. *Physiol. Plant.* 75: 227-232.

155. SAUVAIRE, Y., TAL, B., HEUPEL, R.C., ENGLAND, R., HANNERS, P.K., NES, W.D., MUDD, J.B. 1987. A comparison of sterol and long chain fatty alcohol biosynthesis in *Sorghum bicolor*. In: The *Metabolism, Structure, and Function of Plant Lipids*. (P.K. Stumpf, J.B. Mudd, W.D. Nes, eds.), Plenum Press, New York - London, pp. 107-110.

156. BAISTED, D.J., NES, W.R. 1963. The relative efficacy of mevalonic and dimethylacrylic acids as isopentenoid precursors in germinating seeds of *Pisum sativum*. *J. Biol. Chem.* 238: 1947-1952.

157. KANNANGARA , C.G., JENSEN, C.G. 1975. Biotin carboxyl carrier protein in barley chloroplast membranes. *Eur. J. Biochem.* 54: 25-30.

158. NIKOLAU, B.J., WURTELE, E.S., STUMPF, P.K. 1984. Subcellular distribution of acetyl-coenzyme A carboxylase in mesophyll cells of barley and sorghum leaves. *Arch. Biochem. Biophys.* 235: 555-561.

159. NIKOLAU, B. J., WURTELE, E.S., STUMPF, P.K. 1985. Use of streptavidin to detect biotin-containing proteins in plants. *Anal. Biochem.* 149: 448-453.

160. HOFFMAN, N.E., PICHERSKY, E., CASHMORE, A.R. 1987. A tomato cDNA encoding a biotin-binding protein. *Nucl. Acids Res.* 15: 3928.

161. ITO, M., FUKUI, T., SAITO, T., TOMITA, K. 1987. Inhibition of acetoacetyl-CoA synthetase from rat liver by fatty acyl-CoAs. *Biochim. Biophys. Acta* 922: 287-293.

162. WONG, G.A., BERGSTROM, J.D., EDMOND, J. 1987. Acetoacetyl-CoA ligase activity in the isolated rat hepatocyte effects of 25-hxdroxycholesterol and high density lipoprotein. *Biosci. Rep.* 7: 217-224.

163. SALAM, W.H., CAGEN, L.M., HEIMBERG, M. 1988. Regulation of hepatic cholesterol biosynthesis by fatty acids: Effects of feeding olive oil on cytoplasmic acetoacetyl-coenzyme A thiolase, β-hydroxy-β-methylglutaryl-CoA synthase, and acetoacetyl-coenzyme A ligase. *Biochem. Biophys. Res. Commun.* 153: 422-427.

164. HARWOOD, J.L. 1987. Lipid metabolism. In: *The Lipid Handbook*. (F.D. Gunstone, J.L. Harwood, F.B. Padley, eds.), Chapman and Hall, London - New York, pp. 485-503.

165. BEHRENDS, W., ENGELAND, K., KINDL, H. 1988. Characterization of two forms of the multifunctional protein acting in fatty acid β-oxidation. *Arch. Biochem. Biophys.* 263: 161-169.

166. ERNST-FONBERG, M.L. 1986. An NADH-dependent acetoacetyl-CoA reductase from *Euglena gracilis*. Purification and characterization, including inhibition by acyl carrier protein. *Plant Physiol.* 82: 978-984.

167. KURIHARA, T., UEDA, M., TANAKA, A. 1988. Occurrence and possible roles of acetoacetyl-CoA thiolase and 3-ketoacyl-CoA thiolase in peroxisomes of an *n*-alkane-grown yeast, *Candida tropicalis*. *FEBS Lett.* 229: 215-218.

168. ENDO, T., HAMAGUCHI, N., HASHIMOTO, T., YAMADA, Y. 1988. Non-enzymatic synthesis of hygrine from acetoacetic acid and from acetonedicarboxylic acid. *FEBS Lett.* 234: 86-90.

169. HAUGHAN, P.A., LENTON, J.R., GOAD, L.J. 1987. Inhibition of growth of celery cells by paclobutrazol and its reversal by added sterols. In: *The Metabolism, Structure, and Function of Plant Lipids*. (P.K. Stumpf, J.B. Mudd, W.D. Nes, eds.), Plenum Press, New York - London, pp. 91-93).

170. GRUNDY, S.M. 1988. HMG-CoA reductase inhibitors for treatment of hypercholesterolemia. *N. Engl. J. Med.* 319: 24-33.

171. BACH, T.J., LICHTENTHALER, H.K. 1987. Plant growth regulation by mevinolin and other sterol biosynthesis inhibitors. In: *Ecology and Metabolism of Plant Lipids*. (G. Fuller and W.D. Nes, eds.), Am. Chem. Soc. Symposium Series 325, American Chemical Society, Washington, pp. 109-139.

172. BACH, T.J., LICHTENTHALER, H.K. 1982 Mevinolin: A highly specific inhibitor of microsomal 3-hydroxy-3-methylglutaryl-coenzyme A reductase of radish plants. *Z. Naturforsch.* 37c: 46-50.

173. BACH, T.J., LICHTENTHALER, H.K. 1982. Inhibition by mevinolin of mevalonate formation and plant root elongation. *Naturwissenschaften* 69: 242.

174. BACH, T.J., LICHTENTHALER, H.K. 1983. Inhibition by mevinolin of plant growth, sterol formation and pigment accumulation. *Physiol. Plant.* 59: 50-60.

175. BACH, T.J. 1985. Selected natural and synthetic enzyme inhibitors of sterol biosynthesis as molecular probes for *in vivo* studies concerning the regulation of plant growth. *Plant Sci.* 39: 183-187.

176. HATA, S., TAKAGISHI, H., EGAWA, Y., OTA, Y. 1986. Effects of compactin, a 3-hydroxy-3-methylglutaryl coenzyme A reductase inhibitor, on the growth of alfalfa (*Medicago sativa*) seedlings and the rhizogenes of pepper (*Capsicum annuum*) explants. *Plant Growth Regul.* 4: 335-346.

177. SCHINDLER, S., BACH, T.J., LICHTENTHALER, H.K. 1985. Differential inhibition by mevinolin of prenyllipid accumulation in radish seedlings. *Z. Naturforsch.* 40c: 208-214.

178. PENNOCK, J.F., THRELFALL, D.R. 1983. Biosynthesis of ubiquinones and related compounds. In: *Biosynthesis of Isoprenoid Compounds*. (J.W. Porter, S.L. Spurgeon, eds.), Vol. 2, J. Wiley, New York, pp. 291-303.

179. CAMARA, B., BARDAT, F., SEYE, A., D'HARLINGUE, A., MONÉGER, R. 1982. Terpenoid metabolism in plastids. Localization of α-tocopherol synthesis in *Capsicum* chromoplasts. *Plant Physiol.* 70: 1562-1563.

180. SCHULTZ, G., SOLL, J., FIEDLER, E., SCHULZE-SIEBERT, D. 1985. Synthesis of prenylquinones in chloroplasts. *Physiol. Plant.* 64: 123-129.

181. SOLL, J., SCHULTZ, G., JOYARD, J., DOUCE, R., BLOCK, M.A. 1985. Localization and synthesis of prenylquinones in isolated outer and inner envelope membranes from spinach chloroplasts. *Arch. Biochem. Biophys.* 238: 290-299.

182. SCHINDLER, S., LICHTENTHALER, H.K. 1982. Distribution and levels of ubiquinone homologues in higher plants. In: *Biochemistry and Metabolism of Plant Lipids.* (J.F.G.M. Wintermans, P.J.C. Kuiper, eds.), Elsevier-North Holland Biomedical Press, Amsterdam, pp. 545-548.

183. BACH, T.J., NES, W.D. 1984. Evidence for a mevalonate shunt pathway in wheat. In: *Structure, Function and Metabolism of Plant Lipids.* (P.A. Siegenthaler, W. Eichenberger, eds.), Elsevier Science Publishers BV, Amsterdam, pp. 217-220.

184. LICHTENTHALER, H.K., BACH, T.J., WELLBURN, A. 1982. Cytoplasmic and plastidic isoprenoid compounds of oat seedlings and their distinct labelling from ^{14}C-mevalonate. In: *Biochemistry and Metabolism of Plant Lipids.* (J.F.G.M. Wintermans, P.J.K. Kuiper, eds.), Elsevier, Amsterdam, pp. 489-500.

185. SCHULZE-SIEBERT, D., SCHULTZ, G. 1987. Full autonomy in isoprenoid synthesis in spinach chloroplasts. *Plant Physiol. Biochem.* 25: 145-153.

186. HATA, S., TAKAGISHI, H., KOUCHI, H. 1986. Variation in the content and composition of sterols in alfalfa seedlings treated with compactin (ML-236B) and mevalonic acid. *Plant Cell Physiol.* 28: 709-714.

187. HATA, S., SHIRATA, K., TAKAGISHI, H., KOUCHI, H. 1987. Accumulation of rare phytosterols in plant cells on treatment with metabolic inhibitors and mevalonic acid. *Plant Cell Physiol.* 28: 715-722.

188. FAUST, J.R., GOLDSTEIN, J.L., BROWN, M.S. 1979. Synthesis of ubiquinone and cholesterol in human fibroblasts: Regulation of a branched pathway. *Arch. Biochem. Biophys.* 192: 86-89.

189. JAMES, M.J., KANDUTSCH, A.A. 1979. Inter-relationships between dolichol and sterol synthesis in mammalian cell cultures. *J. Biol. Chem.* 254: 8442-8446.

190. KELLER, G. A., PAZIRANDEH, M., KRISANS, S. 1986. 3-Hydroxy-3-methyl-glutaryl-coenzyme A reductase localization in rat liver peroxisomes and microsomes of control and cholestyramine-treated animals: quantitative biochemical and immunoelectron microscopical analyses. *J. Cell Biol.* 103: 875-886.

191. FAUST, J.R., BROWN, M.S., GOLDSTEIN, J.L. 1980. Synthesis of Δ^2-isopentenyl tRNA from mevalonate in cultured human fibroblasts. *J. Biol Chem.* 255: 6546-6548.

192. KENDE, H. 1965. Kinetin-like factors in the root

exudate of sunflowers. *Proc. Natl. Acad. Sci. U.S.A.* 53: 1302-1307.

193. WEISS, C., VAADIA, Y. 1965. Kinetin-like activity in root apices of sunflower plants. *Life Sci.* 4: 1323-1326.

194. BUSCHMANN, C., LICHTENTHALER, H.K. 1982: The effects of cytokinins on growth and pigment accumulation of radish seedlings (*Raphanus sativus* L.) grown in the dark and at different light quanta fluence rates. *Photochem. Photobiol.* 35: 217-221.

195. BITTNER, A., BUSCHMANN, C. 1983 Uptake and translocation of K^+, Ca^{2+} and Mg^{2+} by seedlings of *Raphanus sativus* L. treated with kinetin. *Z. Pflanzenphysiol.* 109: 191-189.

196. DÖLL, M., SCHINDLER, S., LICHTENTHALER, H.K., BACH, T.J. 1984. Differential inhibition by mevinolin of prenyllipid accumulation in cell suspension cultures of *Silybum marianum* L. In: *Structure, Function and Metabolism of Plant Lipids.* (P.A. Siegenthaler and W. Eichenberger, eds.), Elsevier Science Publishers BV, Amsterdam, pp. 277-280.

197. SCHINDLER, S. 1984. Verbreitung und Konzentration von Ubichinon-Homologen in Pflanzen. Karls. Beitr. Pflanzenphysiol. (*Karlsr. Contrib. Plant Physiol.*, ISSN 0173 - 3133) 12: 1-240.

198. RYDER, N.S., GOAD, L.J. 1980: The effect of the 3-hydroxy-3-methylglutaryl CoA reductase inhibitor ML-236B on phytosterol synthesis in *Acer pseudoplantanus* tissue culture. *Biochim. Biophys. Acta* 619: 424-427.

199. HASHIZUME, T., MATSUBARA, S., ENDO, A. 1983. Compactin (ML-236B) as a new growth inhibitor of plant callus. *Agric. Biol. Chem.* 47: 1401-1403.

200. MCCHESNEY, J.D. 1970. Stimulation of *in vitro* growth of plant tissues by D,L-mevalonic acid. *Can. J. Bot.* 48: 2357-2359.

201. MURAI, N., ARMSTRONG, D.J., SKOOG, F. 1975: Incorporation of mevalonic acid into ribosylzeatin in tobacco callus ribonucleic acid preparations. *Plant Physiol.* 55: 853-858.

202. CHEN, T.H., SHIAO, M.S., HSIEH, W.L. 1987. Inhibition of somatic embryogenesis in cultured carrot cells by mevinolin. *Bot. Bull. Academia Sinica* 28: 211-217.

203. FAUST, J.R., LUSKEY, K.L., CHIN, D.J., GOLDSTEIN, J.L., BROWN, M.S. 1982. Regulation of synthesis and degradation of 3-hydroxy-3-methylglutaryl-coenzyme A reductase by low density lipoprotein and 25-

hydroxycholesterol in UT-1 cells. *Proc. Natl. Acad. Sci. U.S.A.* 79: 5205-5209.

204. LUSKEY, K.L., CHIN, D.J., MACDONALD, R.J., K,L., GOLDSTEIN, J.L., BROWN, M.S. 1982. Identification of a cholesterol-regulated 53,000-dalton cytosolic protein in UT-1 cells and cloning of its cDNA. *Proc. Natl. Acad. Sci. U.S.A.* 79: 6210-6214.

205. HARDEMAN, E.C., JENKE, H.S., SIMONI, R.D. 1983. Overproduction of a M_r 92,000 protomer of 3-hydroxy-3-methylglutaryl-coenzyme A reductase in compactin-resistant C100 cells. *Proc. Natl. Acad. Sci. U.S.A.* 80: 1516-520.

206. HARDEMAN, E.C., ENDO, A., SIMONI, R.D. 1984. Effects of compactin on the levels of 3-hydroxy-3-methylglutaryl coenzyme A reductase in compactin-resistant C100 and wild-type cells. *Arch. Biochem. Biophys.* 232: 549-561.

207. MASUDA, A., KUWANO, M., SHIMADA, T. 1983. Ultrastructural changes during the enhancement of cellular 3-hydroxy-3-methyl-glutaryl-coenzyme A reductase in a Chinese hamster cell mutant resistant to compactin (ML 236B). *Cell Struct. Function* 8: 309-312.

208. ORCI, L., BROWN, M.S., GOLDSTEIN, J.L., GARCIA-SEGURA, L.M., ANDERSON, R.G.W. 1984. Increase in membrane cholesterol: a possible trigger for degradation of HMG CoA reductase and crystalloid endoplasmic reticulum in UT-1 cells. *Cell* 36: 835-845.

209. HASUMI, K., YAMADA, A., SHIMIZU, Y., ENDO, A. 1987. Overaccumulation of 3-hydroxy-3-methylglutaryl-coenzyme-A reductase in a compactin (ML-236B)-resistant mouse cell line with defects in the regulation of its activity. *Eur. J. Biochem.* 164: 547-552.

210. GOODWIN, T.W. 1958. Incorporation of $^{14}CO_2$, [2-^{14}C]acetate and [2-^{14}C]mevalonic acid into β-carotene in etiolated maize seedlings. *Biochem. J.* 68: 26-27.

211. GOODWIN, T.W. 1958. Studies in carotenogenesis. 25. The incorporation of $^{14}CO_2$, [2-^{14}C]acetate and [2-^{14}C]mevalonic acid into β-carotene by illuminated etiolated maize seedlings. *Biochem. J.* 70: 612-617.

212. GOODWIN, T.W. 1965. The biosynthesis of carotenoids. In: *Biosynthetic Pathways in Plants*. (J.B. Pridham, T. Swain, eds.), Academic Press, London, pp. 37-55.

213. ROGERS, L.J., SHAH, S.P.J., GOODWIN, T.W. 1966. Compartmentation of terpenoid biosynthesis in green

plants. *Biochem. J.* 104: 395-405.

214. SCHULTZ, G., HUCHZERMEYER, Y., REUPKE, B., BICKEL, H. 1976. On the intra-cellular site of biosynthesis of α-tocopherol in *Hordeum vulgare*. *Phytochemistry*, 15: 1253-1255.

215. GRUMBACH, K.H., FORN, B. 1980. Chloroplast autonomy in acetyl-coenzyme A formation and terpenoid biosynthesis. *Z. Naturforsch.* 35c: 645-648.

216. GRAY, J.C., KEKWICK, R.G.O. 1973. Mevalonate kinase in green leaves and etiolated cotyledons of the french bean *Phaseolus vulgaris*. *Biochem. J.* 133: 335-347.

217. HILL, H.W., ROGERS, L.J. 1974. Mevalonate activating enzymes and phosphatases in higher plants. *Phytochemistry* 13: 763-777.

218. KREUZ, K., KLEINIG, H. 1981. On the compartmentation of isopentenyl diphosphate synthesis and utilization in plant cells. *Planta* 153, 578-581.

219. LÜTKE-BRINKHAUS, F., LIEDVOGEL, B., KLEINIG, H. 1984. On the biosynthesis of ubiquinones in plant mitochondria. *Eur. J. Biochem.* 141: 537-541.

220. LIEDVOGEL, B. 1986. Acetyl coenzyme A and isopentenyl pyrophosphate as lipid precursors in plant cells - biosynthesis and compartmentation. *J. Plant Physiol.* 124: 211-222.

221. KLEINIG, H. 1987. Carotenoid biosynthesis and carotenogenic enzymes in plants. In: *The Metabolism, Structure, and Function of Plant Lipids*. (P.K. Stumpf, J.B. Mudd, W.D. Nes, eds.), Plenum Press, New York - London, pp. 37-43.

222. EMMANUEL, B., ROBBLEE, A.R. 1984. Cholesterogenesis from propionate: facts and speculations. *Int. J. Biochem.* 16: 907-911.

223. DAVIS, A.R. 1978. Participation of propionate in cholesterol biosynthesis in rat liver. *Steroids* 31: 593-600.

224. HALARNKAR, P.P., WAKAYAMA, E.J., BLOMQUIST, G.J. 1988. Metabolism of propionate to 3-hydroxypropionate and acetate in the Lima bean *Phaseolus limensis*. *Phytochemistry* 27: 997-999.

225. SCHULTZ, G., SCHULZE-SIEBERT, D. 1989. Chloroplast isoprenoid synthesis in spinach and barley: direct intermediate supply by photosynthetic carbon fixation. In: *Biological Role of Plant Lipids*. (P.A. Biacs, K. Gruiz, T. Kremmer, eds.), Akadémiai Kiado, Budapest, and Plenum Publishing Corporation, New York - London, pp. 313-319.

226. STITT, M., AP REES, T. 1979. Capacities of pea chloroplasts to catalyse the oxidative pentose pathway and glycolysis. *Phytochemistry* 18: 1905-1911.
227. MCHALE, N. A., HANSON, K.R., ZELITCH, I. 1988. A nuclear mutation in *Nicotiana sylvestris* causing a thiamine-reversible defect in synthesis of chlorplast pigments. *Plant Physiol.* 88: 930-935.
228. CURRY, K.J. 1987. Initiation of terpenoid synthesis in osmophores of *Stanhopea anfracta* (Orchidaceae): a cytochemical study. *Amer. J. Bot.* 74: 1332-1338.
229. KELLER, G.A., BARTON, M.C., SHAPIRO, D.J., SINGER, S.J. 1985. 3-Hydroxy-3-methylglutaryl-coenzyme A reductase is present in peroxisomes in normal rat liver cells. *Proc. Natl. Acad. Sci. U.S.A.* 82: 770-774.
230. SHAPIRO, D.J., GOULD, L., MARTIN, G., WILLIAMS, D., WOLSKI, R. 1986. Organization and regulated expression of rat liver HMG-CoA reductase genes. *J. Lipid Res.* 27: 567-568.
231. REYNOLDS, G. A., GOLDSTEIN, J.L., BROWN, M.S. 1985. Multiple mRNAs for 3-hydroxy-3-methylglutaryl coenzyme A reductase determined by multiple transcription initiation sites and intron splicing sites in the 5'-untranslated region. *J. Biol. Chem.* 260: 10369-10377.
232. GIL, G., FAUST, J.R., CHIN, D.J., GOLDSTEIN, J.L., BROWN, M.S. 1985. Membrane-bound domain of HMG CoA reductase is required for sterol-enhanced degradation of the enzyme. *Cell* 41: 249-258.
233. LUSKEY, K.L. 1987. Conservation of promoter sequence but not complex intron splicing pattern in human and hamster genes for 3-hydroxy-3-methylglutaryl coenzyme A reductase. *Mol. Cell. Biol.* 7: 1881-1893.
234. THOMPSON, S.L., BURROWS, R., LAUB, R.J., KRISANS, S.K. 1987. Cholesterol synthesis in rat liver peroxisomes. Conversion of mevalonic acid to cholesterol. *J. Biol. Chem.* 262: 17420-17425.
235. HASLAM, J.M., BIRCH, D.J., CRABBE, T.B. 1986. The control of 3-hydroxy-3-methylglutaryl-CoA reductase activity in *Saccharomyces cerevisiae*. *J. Lipid Res.* 27: 563.
236. BOLL, M. 1983. 3-Hydroxy-3-methylglutaryl coenzyme A reductase in yeast and other microorganismes. In: *3-Hydroxy-3-Methylglutaryl Coenzyme A Reductase*. (J.R. Sabine, ed.), *CRC Series in Enzyme Biology*, CRC Press, Inc., Boca Raton, Florida, pp. 39-53.
237. BASSON, M.E., THORSNESS, M., RINE, J. 1986. *Saccharomyces cerevisiae* contains two functional genes

encoding 3-hydroxy-3-methylglutaryl-coenzyme A reductase. *Proc. Natl. Acad. Sci. U.S.A.* 83: 5563-5567.

238. CAELLES, C., FERRER, A., BALCELLS, L., HEGARDT, F.G., BORONAT, A. 1989. Isolation and structural characterization of a cDNA encoding *Arabidopsis thaliana* 3-hydroxy-3-methylglutaryl coenzyme A reductase. *Plant. Mol. Biol.* (in press).

239. LEARNED, R.M., FINK, G.R. 1989. 3-Hydroxy-3-methylglutaryl-coenzyme A reductase from *Arabidops. thaliana* is structurally distinct from the yeast aı animal enzymes. *Proc. Natl. Acad. Sci. U.S.A.* 86: 2779-2783.

Chapter Two

MOLECULAR CLONING AND CHARACTERIZATION OF PLANT 3-HYDROXY-3-METHYLGLUTARYL COENZYME A REDUCTASE

M. MONFAR, C. CAELLES, L. BALCELLS,
A. FERRER, F.G. HEGARDT AND A. BORONAT

Unitat de Bioquímica
Facultat de Farmacia
Universitat de Barcelona
08028 Barcelona, Spain

INTRODUCTION

Mevalonic acid is the specific precursor of the vast array of isoprenoid compounds present in plants. These compounds play essential roles in plant growth and development and include growth regulators (gibberellins and abscisic acid), side chains of many biologically active molecules (chlorophylls, prenylquinones and ubiquinone), carotenoids, dolichols and sterols. In addition, individual groups of plants have evolved specific isoprenoid compounds for specific functions, such as attractants for insect pollination, or as phytoalexins for defense against fungal or bacterial infection. Mevalonate is synthesized from 3-hydroxy-3-methylglutaryl coenzyme A (HMG-CoA) in a reaction catalyzed by the enzyme HMG-CoA reductase (EC 1.1.1.34). This enzymatic

Biochemistry of the Mevalonic Acid Pathway to Terpenoids
Edited by G.H.N. Towers and H. A. Stafford
Plenum Press, New York

activity has been described in a large number of plants (for recent reviews[1,2]); in all cases the enzyme is membrane bound. The main subcellular location of the enzyme activity appears to be the endoplasmic reticulum, but it has also been reported to be associated with mitochondrial and plastid membranes.[3,4] In fact, the occurrence of HMG-CoA reductase in different cellular compartments is a matter of controversy and represents a major point to be clarified before a better understanding of the control of isoprenoid metabolism in plants is reached.

HMG-CoA reductase has been cloned and extensively characterized in mammals,[5-7] in which it catalyzes the major rate-limiting step in the cholesterol biosynthetic pathway. This control is achieved through complex processes at both transcriptional[8,9] and post-transcriptional levels.[10,11] Recently, cDNA and genomic HMG-CoA reductase clones have been isolated from *Drosophila*,[12] yeast[13] and sea urchin,[14] and from these the primary structure of the proteins has been deduced. At present, it is generally accepted that the reaction catalyzed by the HMG-CoA reductase is also the main rate-limiting step in plant isoprenoid biosynthesis, although more experimental data are needed to support this hypothesis.[1,2] The characterization of plant HMG-CoA reductase, however, has been hampered mainly by difficulties associated with its solubilization and purification.[15-17]

In order to characterize plant HMG-CoA reductase at the molecular level and to define more precisely its role in controlling isoprenoid biosynthesis, we have applied molecular biology techniques to isolate HMG-CoA reductase cDNA and genomic clones from different plant sources. The strategy we devised for cloning plant HMG-CoA reductase was based upon the assumption that enzymes catalyzing key reactions in intermediary metabolism could be conserved during evolution. In the case of HMG-CoA reductase our hypothesis was early reinforced by the observation of Basson et al.,[18] who found that the catalytic site of the enzyme was very conserved between yeast and hamster, two distantly related organisms.

Previous physiological and biochemical studies on plant HMG-CoA reductase have been done in a great variety of systems.[1,2] Nevertheless, in order to clone plant HMG-CoA reductase,we selected *Arabidopsis thaliana* because of its advantages for molecular genetic studies. Its most relevant features are: a very simple genomic organization of only $7x10^4$ kb, being the smallest haploid genomic size reported in plants,[19] a minimal content of repetitive DNA,[20] a short generation time of about 5 weeks[21] and the existence of efficient transformation methods using *Agrobacterium tumefaciens*.[22,23]

Preliminary analysis by Southern blot clearly demonstrated the existence of sequences homologous to hamster HMG-CoA reductase in the genome of *A. thaliana*.[24] This was the starting point for the isolation of an *A. thaliana* genomic clone, which in turn was used as a probe to further isolate cDNA and genomic clones from *A. thaliana*,[25] pea (Monfar and Boronat, manuscript in preparation) and radish (Wettstein et al., manuscript in preparation). Recently, the isolation and partial characterization of a cDNA clone has also been reported from tomato.[26]

PRIMARY STRUCTURE OF PLANT HMG-CoA REDUCTASE

From the nucleotide sequence of the cDNA clones, it is possible to know the general structure of plant HMG-CoA reductase. For the moment, however, the complete amino acid sequence is only available for the *A. thaliana* enzyme.[25] *A.thaliana* HMG-CoA reductase is a protein of 592 amino acids with a molecular weight of 63,605 daltons (Fig.1). The partial amino acid sequences available from pea, radish and tomato suggest that the general structure of HMG-CoA reductase is very similar among plants.

In order to discuss the main features of plant HMG-CoA reductase we will distinguish three different structural regions in the protein: the N-terminal or membrane domain, the linker region and the C-terminal or soluble domain.

Characteristics of the N-terminal Domain

The hydropathy profile of the protein (Fig. 2), shows the presence of two hydrophobic sequences located within the first 117 residues (residues 47 to 69 and 83 to 117), each of which is long enough to span a membrane bilayer (regions 1 and 2 in Fig. 2). These hydrophobic sequences are separated and flanked by hydrophilic regions rich in charged residues. Since it is well known that plant HMG-CoA reductase is a membrane-bound protein,[1,2] it is reasonable to postulate that at least one of these N-terminal hydrophobic regions could correspond to membrane-spanning regions. Nevertheless, more

1 M D L R R R P P K P P V T N N N N S N G S F R S Y Q P R T S D D D H R R R A T T

41 I A P P P K A S D A L P L P L Y L T N A V F F T L F F S V A Y Y L L H R W R D K

81 I R Y N T P L H V V T I T E L G A I I A L I A S F I Y L L G F F G I D F V Q S F

121 I S R A S G D A W D L A D T I D D D D H R L V T C S P P T P I V S V A K L P N P

161 E P I V T E S L P E E D E E I V K S V I D G V I P S Y S L E S R L G D C K R A A

201 S I R R E A L Q R V T G R S I E G L P L D G F D Y E S I L G Q C C E M P V G Y I

241 Q I P V G I A G P L L L D G Y E Y S V P M A T T E G C L V A S T N R G C K A M F

281 I S G G A T S T V L K D G M T R A P V V R F A S A R R A S E L K F F L E N P E N

321 F D T L A V V F N R S S R F A R L Q S V K C T I A G K N A Y V R F C C S T G D A

361 M G M N M V S K G V Q N V L E Y L T D D F P D M D V I G I S G N F C S D K K P A

401 A V N W I E G R G K S V V C E A V I R G E I V N K V L K T S V A A L V E L N M L

441 K N L A G S A V A G S L G G F N A H A S N I V S A V F I A T G Q D P A Q N V E S

481 S Q C I T M M E A I N D G K D I H I S V T M P S I E V G T V G G G T Q L A S Q S

521 A C L N L L G V K G A S T E S P G M N A R R L A T I V A G A V L A G E L S L M S

561 A I A A G Q L V R S H M K Y N R S S R D I S G A T T T T T T T T

Fig. 1. Amino acid sequence of *A. thaliana* HMG-CoA reductase. The two putative trans-membrane regions are underlined (continuous line). The PEST sequence is indicated by a dashed line. The asterisk shows the start point of sequence homology among HMG-CoA reductases (see text). Arrows indicate the relative position of introns in HMG1 gene.

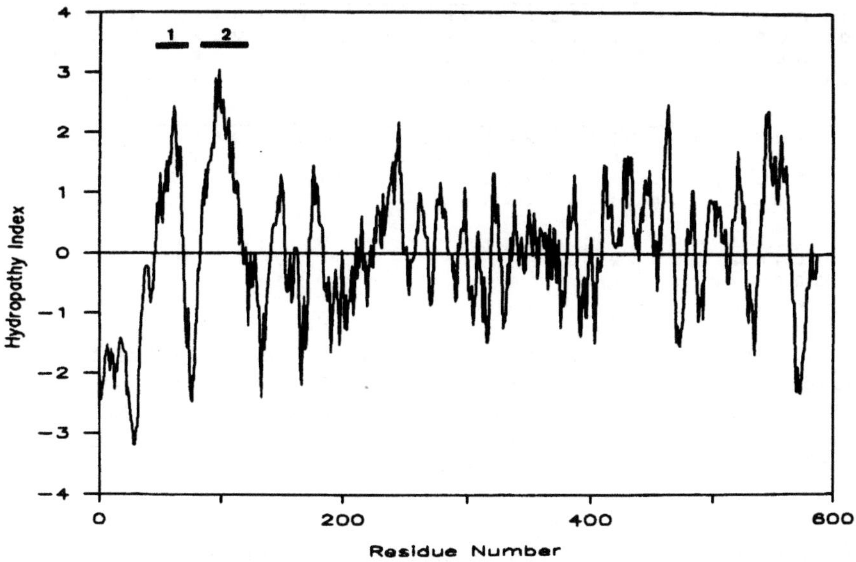

Fig. 2. Hydropathy index plot of the *A. thaliana* HMG-CoA reductase. The algorithm of Kite and Doolittle[27] was used with a window size of 9 residues. Positive values indicate hydrophobic regions. Bars 1 and 2 indicate putative trans-membrane regions.

experimental data are needed to precisely define the structure of the trans-membrane domain.

It is interesting to point out the different structural organization of the N-terminal region of the plant enzyme, when compared to the HMG-CoA reductases characterized from animals and yeast.[25] The most striking difference is its much simpler structure with only two hydrophobic regions, in contrast with the general occurrence of seven trans-membrane domains in the enzymes from other organisms (mammals,[5-7] insects[12] and yeast[13]). Furthermore, the sequence of the N-terminal region is also completely different. The N-terminal amino acid sequence of HMG-CoA reductase has recently been obtained from pea (Monfar and Boronat, manuscript in preparation) and radish (Wettstein et al., manuscript in preparation). From these data, it is clear that these sequences have diverged among the different plant HMG-CoA reductases, except for the hydrophobic regions, which are highly conserved (Fig. 3).

```
AT1: I A P P P K A S D A L P L P L Y L T N A V F F T L F F S V A Y Y L L H R W R D K
AT2: D P . L D . . . . . . . . . . . . . . W F . L S . . . A T V . F . . S . . . E .
PSa: - K N E K H D . Q S Q . - S . . . . . . . . . G . . . . . . . F . . . . . . E .
RS : D Q P R T . . . . . . . . . . . . . . . . . . . . . . . . . . . . R . . . D .

     I R Y N T P L H V V T I T E L G A I I A L I A S F I Y L L G F F G I D F V - Q S
     . . N S . . . . . . D L S . I C . L . G F V . . . . . . . . . C . . . L I F R .
     . . T S . . . . . L . . S . T V . L L S . . . . I F . . M A . . S . S . . L H P
     . . H S . . . . . M . . . . . . . . L . . V . . . . . . . . . . . . . . . - . .

     F - I S R A - S G D A W D
     S S D D D V W V N . G M I
     . P V . . . V Q Y . E E .
     . - . . . . D T N N E F E
```

Fig. 3. Amino acid sequence alignment of the putative transmembrane regions of HMG-CoA reductases from *A. thaliana* (AT1 and AT2, corresponding respectively to the HMG1 and HMG2 gene products), pea (PSa) and radish (RS). Dots indicate residues identical to the A. thaliana enzyme AT1.

From this observation it is tempting to speculate that this conservation may reflect the importance of this region either in anchoring the enzyme to the membrane or in mediating some regulatory mechanism in response to physiological or environmental stimuli.

Characteristics of the Linker Region

The linker region, joining the N- and the conserved C-terminal domains (see below), is also highly divergent among all HMG-CoA reductases, including the plant enzymes. Still, a common feature shared by them all is the presence of different versions of a "PEST" sequence, described for proteins having rapid turnover rates.[28] The general occurrence of this particular sequence in all HMG-CoA reductases suggests that it could be involved in a common regulatory mechanism, possibly through modification of the half-life of the enzyme, as described for the hamster enzyme.[10,11]

Characteristics of the C-terminal Domain

From the alignment of all the available HMG-CoA reductase amino acid sequences it has become evident that they share a high level of similarity in the C-terminal region,

```
b1

AT: R E A L Q - R V - - - T G R S I E G L P L D G F D Y E S I L G Q C C E M P V G Y
HA: . Q L . S T K L - - P E P S . L Q Y . . Y R D Y N . S L V M . A . . . N V I . .
DM: . Q I I G S . A K M P V . . - L D V . . Y E H . . . R K V . N A . . . N V L . .
Y1: . K . . S I L A E A P V L A . - D R . . Y K N Y . . D R V F . A . . . N V I . .

    I Q I P V G I A G P L L L D G - - Y E Y S V P M A T T E G C L V A S T N R G C K
    M P . . . . V . . . . C . . . K - - . . Q . . . . . . . . . . . . . . . . . . R
    V P . . . . Y . . . . . . . . . E T . - . - . . . . . . . . A . . . . . . . . .
    M P L . . . V I . . . V I . . - T S - . H I . . . . . . . . . . . . . A M . . . .

    A M F I S G G A T S T V L - K D G M T R A P V V R F
    . I G L G . . . S . R . . - A . . . . . G . . . . L
    . L S V R . - V R . V . E - D V . . . . . . C . . .
    . I N A G . . . . - . . . T . . . . . . G . . . . .

b2

AT: N L A G S A V A G S L G G F N A H A S N I V S A V F I A T G Q D P A Q N V E S S
HA: . . V . . . M . . . I . . Y . . . . A . . . T . I Y . . C . . . A . . . . G . .
DM: . M G . . . M . . . I . . N . . . . A . M . T . . . L . . . . . . . . . . T . .
Y1: . . V . . . M . . . V . . . . . . . A . L . T . . . L . L . . . . . . . . . . .

    Q C I T M M E A - I N D G K D I H I S V T M P S I E V G T V G G G T Q L A S Q S
    N . . . L . . . S G P T N E . L Y . . C . . . . . . I . . . . . . . . N . L P . Q
    N . S . A . . C W A E N S E . L Y M T C . . . . L . . . . . . . . . . G . P G . .
    N . . . L . K E - V - . . - . L R . . . S . . . . . . . . . I . . . . . V . E P . G

    A C L N L L G V K G A
    . . . Q M . . . Q . .
    . . . E M . . . R . .
    . M . D . . . . R . P
```

Fig. 4. Amino acid sequence alignment of the b1 and b2 regions (see text) of HMG-CoA reductases from *A. thaliana* (AT, HMG1 gene product, b1: residues 204 to 302 and b2: residues 442 to 531), hamster (HA, b1: residues 495 to 595 and b2: residues 735 to 825), *D. melanogaster* (DM, b1: residues 518 to 618 and b2: residues 758 to 848) and yeast (Y1, HMG1 gene product, b1: residues 650 to 751 and b2: residues 893 to 980). Dots indicate residues identical to the *A. thaliana* enzyme.

which contains the catalytic site. The C-terminal part of the *A. thaliana* enzyme (residues 172 to 579) shows a high level of similarity to the HMG-CoA reductase from animals and yeast, with values ranging from 56 to 58% identical and 12 to 14% conserved residues.[25] In additon, 40% of the amino acids are identical and 16% of the changes are conservative in all proteins. In particular the most conserved region among the different HMG-CoA reductases lies in the ß-domains b1 and b2 (Fig. 4), that have been postulated to be part of the active site of the hamster enzyme.[29]

When the C-terminal domains of the plant enzymes are compared (Fig. 5) the level of similarity is much higher, with values of 71% identical and 16% conservative residues. The

```
AT1:  D E E I V K S V I D G V I P S Y S L E S R L G D C K R A A S I R R E A L Q R V T
PSa:  . . . V . Q A . V S . S . . . . . . . T K . . . . R . . . A . . N Q . V E . . .
PSb:            N . . V E . T T . . . A . . . . . . . . F . . . A . . . . . . . . T S
PSc:            N . . V S . S . . . . . . . . . . A . . R K . . A . . . S . V . T I S

      G R S I E G L P L D G F D Y E S I L G Q C C E M P V G Y I Q I P V G I A G P L L
      . . . L . D . . M E N . . . D . . . . . . . . . . I . F V . . . . . V . . . . .
      . K . L K . . . . E . . N . D A . . . . . . . . . . . F . . . . . . V . . . . .
      . K . L . . . . . E . . . . D . . . . . . . . . . I . F V . . . . . V . . . . .

      L D G Y E Y S V P M A T T E G C L V A S T N R G C K A M F I S G G A T S T V L K
      . . . K . F T . . . . . . . . . . . . . . . . . . . . I Y V . . . . S A V . . R
      . . . F . . T . . . . . . . . . . . . . . . . . . . . I Y A . . . . E . V . . R
      . . . I . . T . . . . . . . . . . . . . . . . . . . . I H A C . . . S . V L . R

      D G M T R A P V V R F A S A R R A S E L K F F L E N P E N F D T L A V V F N R S
      . . . . . . . . . . . N . . K . . . Q . . . . . . D . Q . . . S . . H T . . K .
      . A . S . . . . . . . S . . K . . . Q . . . . . . D . L . S E . . . . . . . . .
      . . . . . . . . . . . S . . K . . A Q . . . Y . . D . L . . . E . S L S . . K .
Tom:                                                                        . . K .

      S R F A R L Q S V K C T I A G K N A Y V R F C C S T G D A M G M N M V S K G V Q
      . . . . . . . N I . A . . . . . . L . I . . T . . . . . . . . . . . . . . . . .
      . . . . . . L G I Q . A M . . . . L . M . . T . . . . . . . . . . . . . . . . .
      . . . . . . . . I Q P A . . . . . L . I . . R . . . . . . . . . . . . . . . . .
      . . . . . . . G I Q . A . . . . . L . M . . S . . . . . . . . . . . . . . . . .

      N V L E Y L T D D F P D M D V I G I S G N F C S D K K P A A V N W I E G R G K S
      . . . D F . Q S . . . . . . . . . . . . . . . . . . . A . . . . . . . . . . . .
      . . . D F . R N . . . . . . . . . . . . . Y . A . . . . . . . . . . . . . . . .
      . . . D F . Q N . . F . . E . . . . . . . . . . . . . A . . . . . . . . . . . .
      . . . D . . Q S E Y . . . . . . . . . . . . . . . . . . . . . . . . . . . . . .

      V V C E A V I R G E I V N K V L K T S V A A L V E L N M L K N L A G S A V A G S
      . . . . . . . K E . V . K . . . . . N . E S . . . . . . . . . . T . . . M . . A
      . . . . . . . K E . . . K . . . . . . . . . . . . . . . . . . . . . . . I . . A
      . . . . . . . K E . V . Q . . . . . N . E . . . . . . . . . . . T . . . I . . A
      . . . . . I . K E D V . K . . . . . E . . . . . . . . . . . . . T . . . M . . A

      L G G F N A H A S N I V S A V F I A T G Q D P A Q N V E S S Q C I T M M E A I N
      . . . . . . . . . . . . . . I Y L . . . . . . . . . . . . . H . M . . . . . V .
      . . . . . . . . . . . . . . I . . . . . . . . . . I . . . . H . . . . . . . V .
      . . . . . . . . . . . . . . I . L . . . . . . . . . . . . . H . . . . . . . V .
      . . . . . . . . . . . . . . . Y L . . . . . . . . . . . . . H . . . . . . . V .

      D G K D I H I S V T M P S I E V G T V G G G T Q L A S Q S A C L N L L G V K G A
      . . . . L . V . . . . . . . . . . . . . . . . E . . . . . . . . . . . . . . . .
      . . . . L . . . . . . . . . . . . . . . . . . . . . . . . . . . . . . . . . . S
      . . . . L . . . . . . . . . . . . . . . . . . . . . . . . . . . . . . . . . . .
      . . . . L . . . . . . . . . . . . . . . . . . . . . . . . . . . . . . . . . . .

      S T E S P G M N A R R L A T I V A G A V L A G E L S L M S A I A A G Q L V R S H
      . K . . . . A . S . Q . . I . . . A S . . . . . . . . . . . . . . . . . . K . .
      . K . . . . S . S . L . . . . . . . S . . . . . . . . . . . . . . . . . . N . .
      . K D . . . A . S . L . . . . . . . S . . . . . . . . . . . . . . . . . . K . .
      N R D A . . S . . . L . . . . . . . S . . . . . . . . . . . . . S . . . . K . .

      M K Y N R S S R D I S G A T T T T T T T T
      . . . . . . . K . V T K I A S
      . . . . . . C K . V T K I S S
      . . . . . . . . . . . K V V S
      . . . . .
```

Fig. 5. Amino acid sequence alignment of the C-terminal region of the HMG-CoA reductases from *A. thaliana* (AT1, HMG1 gene product), pea (PSa, PSb and PSc) and tomato (Tom). Dots indicate residues identical to the *A. thaliana* enzyme AT1.

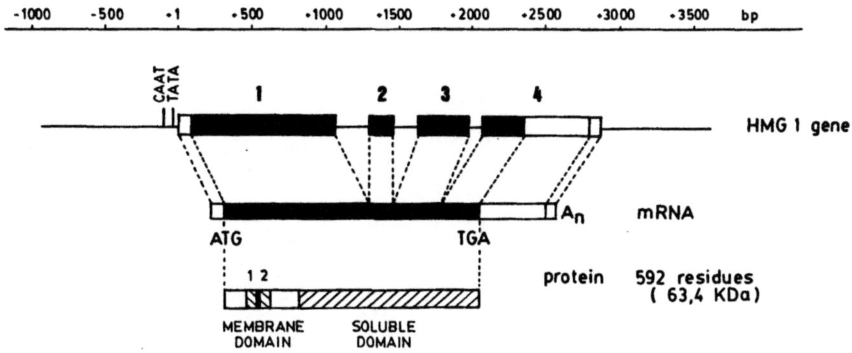

Fig. 6. Schematic representation of the *A. thaliana* HMG1 gene (exons are numbered from 1 to 4), transcript, and protein (1 and 2 indicate the putative trans-membrane regions). Black boxes indicate the coding regions.

comparison of the three different enzymatic forms found in pea leaves shows the existence of 79% identical and 12% conservative residues. At present, the significance of these forms of pea HMG-CoA reductase is not known, although they could correspond to different isoenzymes serving specific functions and possibly having different intracellular locations, as previously reported.[3,4]

The comparison data presented above reveal the existence of a high degree of conservation in the C-terminal region of HMG-CoA reductases. In particular, there is a high number of identical amino acids maintained at specific positions in the different enzymes. This observation suggests the existence of a strong evolutionary pressure to preserve the structure and function of the catalytic domain.

GENOMIC ORGANIZATION AND EXPRESSION OF PLANT HMG-CoA REDUCTASE GENES

A. thaliana is the only plant for which the organization of HMG-CoA reductase genes has been studied in detail. We have shown the existence of two HMG-CoA reductase genes in *A. thaliana*, designated HMG1 and HMG2[25] The HMG1 gene, coding for the protein shown in Figure 1, has been cloned and sequenced and its general structure is shown in Figure 6. The gene is organized in four exons, separated by three small

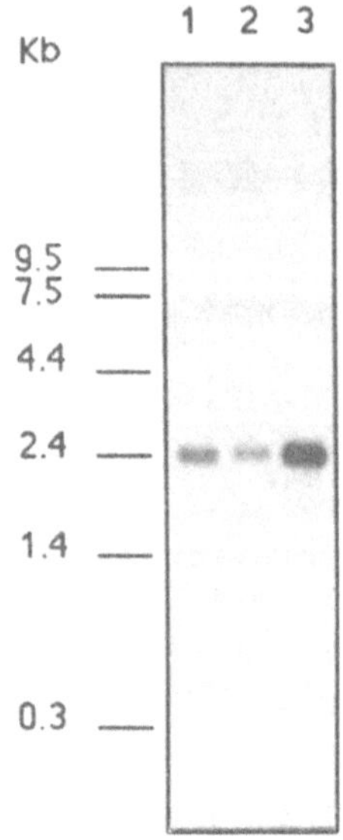

Fig. 7. Northern blot-hybridization of *A. thaliana* poly A^+ RNA (0.5 μg/lane) from 4-5 week old rosette leaves (lane 1), 7-8 day old seedlings grown in the presence of light (lane 2) or in the dark (lane 3). The probe used was a cDNA corresponding to HMG1 gene. The mobility of the molecular size markers is indicated.

introns of 225, 159 and 91 bp, and contains typical TATA and CAAT boxes preceding the transcription start point. As for the HMG2 gene, it has also been cloned, although its nucleotide sequence has not yet been completed. Nevertheless, preliminary results indicate that it codes for a different form of HMG-CoA reductase, although its structure is similar to that of the HMG1 gene.

Northern blot analysis, using either HMG1 or HMG2 genes as a probe, under moderate hybridization conditions, reveals a single transcript of 2.4 kb (Fig. 7). However, under high-stringency conditions the transcript is only detected when HMG1 gene is used as a probe. These results show that in *A. thaliana* the HMG1 gene is actively expressed in seedlings and leaves under normal growth conditions, whereas the expression product of the HMG2 gene has not yet been identified. Since the expression studies have been restricted to seedlings and product of the HMG2 gene has not yet been identified. Since the expression studies have been restricted to seedlings and leaves, it cannot be excluded that the HMG2 gene would be expressed in other tissues, in other stages of plant development or in response to specific stimuli. In this respect it is interesting to note that the promoter sequences of both genes are quite different, thus reinforcing the idea of a differential expression of the two HMG-CoA reductase genes.

Several authors have reported that etiolated plants contain higher levels of HMG-CoA reductase activity than light-grown plants.[3,30,31] These results correlate with those presented in Figure 7, in which it was shown that the level of expression of the HMG1 gene in etiolated seedlings was much higher than the level found in seedlings grown in the presence of light.

Although *A. thaliana* presents a simple genomic organization for the HMG-CoA reductase genes, the situation appears to be much more complex in other plants. In particular, three different cDNA clones have been isolated from a pea leaf cDNA library (Monfar and Boronat, maunscript in preparation), each coding for a different form of HMG-CoA reductase. The C-terminal regions of these proteins are shown in Figure 5. Northern blot experiments revealed the presence of a single transcript of 2.5 kb. In addition, when the same cDNA probes were used in Southern blot experiments, a complex pattern of bands appeared (not shown). These results, in combination with those obtained using specific probes derived from the 3'-untranslated region of the cDNAs, suggest that pea HMG-CoA reductase is encoded by a multigene family of at least 5 members.

CONCLUSION

The control of mevalonate synthesis is pivotal in the biosynthesis of isoprenoids in plants. It is widely assumed that the enzyme HMG-CoA reductase, which catalyzes mevalonate synthesis, plays a key role in controlling this process. The plant enzyme has been a subject of great interest for many years, although the first data about its structure are just starting to emerge. Through the cloning and analysis of HMG-CoA reductase cDNA clones from different plants (*A. thaliana*, pea, radish and tomato), a general pattern of the structure of the enzyme is being defined. It seems clear that the structure of the membrane domain of the plant enzyme is distinct from both the animal and the yeast enzyme. The former is shorter and has only two putative membrane-spanning regions. Nonetheless, there is a great similarity in the C-terminal part of the enzyme in all the reductases characterized so far, thus suggesting a strong evolutionary pressure in order to preserve the catalytic function. At present, the precise cellular location of the cloned plant enzymes is not known; the enzyme activity has been reported to be associated with different subcellular membrane fractions (endoplasmic reticulum, mitochondria and plastids). Of course, further experiments are needed to clarify the intracellular location of the enzyme, an aspect that will be speeded up with the use of specific antibodies. Up to now, a single enzymatic form of the enzyme has been found in *A. thaliana*, although in other plants like pea the presence of different isoenzymatic forms has been detected. While *A. thaliana* contains two genes encoding HMG-CoA reductase, only one is actively expressed in seedlings and leaves under normal growth conditions. Presently, it is not known whether the other HMG-CoA reductase gene is really an active gene or whether it represents a pseudogene. In contrast with the simple organization of HMG-CoA reductase genes in *A. thaliana*, the enzyme is encoded by a complex multigene family in pea.

It is expected that in a very near future the powerful techniques of molecular biology will lead to new insights on the role of HMG-CoA reductase in controlling isoprenoid biosynthesis in plants.

ACKNOWLEDGMENTS

We thank Dr. M. Arró for her assistance in preparing the figures. This research was supported by Grants PB87-5703 from the Dirección General de Investigación Científica y Técnica, Spain (to A.B.) and AR/88 from the CIRIT, Generalitat de Catalunya (to C.C.). M.M. and C.C. were supported by predoctoral fellowships (FPI) from the Ministerio de Educación y Ciencia, Spain. L.B. holds a postdoctoral fellowship from the Ministerio de Educación y Ciencia, Spain.

REFERENCES

1. GRAY, J.C. 1987. Control of isoprenoid biosynthesis in

higher plants. Adv. Botan. Res. 7:25-91.
2. BACH, T.J. 1987. Synthesis and metabolism of mevalonic acid in plants. Plant Physiol. Biochem. 25: 163-178.
3. BROOKER, J.D., D.W. RUSSELL. 1975. Subcellular localization of 3-hydroxy-3methylglutaryl coenzyme A reductase in *Pisum sativum* seedlings. Arch. Biochem. Biophys. 167: 730-737.
4. RUSSELL, D.W. 1985. 3-Hydroxy-3-methylglutaryl-CoA reductases from pea seedlings. Methods Enzymol. 110: 26-40.
5. CHIN, D.J., G. GIL, D.W. RUSSELL, L. LISCUM, K.L. LUSKEY, S.K. BASU, H. OKAYAMA, P. BERG, J.L. GOLDSTEIN, M.S. BROWN. 1984. Nucleotide sequence of 3-hydroxy-3-methylglutaryl coenzyme A reductase, a glycoprotein of endoplasmic reticulum. Nature 308: 613-617.
6. SKALINK, D.G., R.D. SIMONI. 1985. The nucleotide sequence of syrian hamster HMG-CoA reductase cDNA. DNA 4: 439-444.
7. LUSKEY, K.L., B. STEVENS. 1985. Human 3-Hydroxy-3-methylglutaryl coenzyme A reductase. Conserved domains responsible for catalytic activity and sterol-regulated degradation. J. Biol. Chem. 260: 10271-10277.
8. REYNOLDS, G.A., J.L. GOLDSTEIN, M.S. BROWN. 1985. Multiple mRNAs for 3-hydroxy-3-methylglutaryl coenzyme A reductase determined by multiple transcription initiation sites and intron splicing in the 5'-untranslated region. J. Biol. Chem. 260: 10369-10377.
9. OSBORNE, T.F., J.L. GOLDSTEIN, M.S. BROWN. 1985. 5' End of HMG-CoA reductase gene contains sequences responsible for cholesterol-mediated inhibition of transcription. Cell 42: 203-212.
10. GIL, G., J.R. FAUST, D.J. CHIN, J.L. GOLDSTEIN, M.S. BROWN. 1985. Membrane-bound domain of HMG-CoA reductase is required for sterol-enhanced degradation of the enzyme. Cell 41: 249-258.
11. CHIN, D.J., G. GIL, J.R. FAUST, J.L. GOLDSTEIN, M.S. BROWN, K. LUSKEY. 1985. Sterols accelerate degradation of hamster 3-hydroxy-3-methylglutaryl coenzyme A reductase encoded by a constitutively expressed cDNA. Mol. Cell. Biol. 5: 634-641.
12. GERTLER, F.B., C.-Y. CHIU, L. RICHTER-MANN, D.J. CHIN. 1988. Developmental and metabolic regulation of the *Drosophila melanogaster* 3-hydroxy-3-methylglutaryl coenzyme A reductase. Mol. Cell. Biol. 8: 2713-2721.
13. BASSON, M.E., M. THORSNESS, J. FINER-MOORE, R.M. STROUD, J. RINE. 1988. Structural and functional conservation

between yeast and human 3-hydroxy-3-methylglutaryl coenzyme A reductases, the rate-limiting enzyme of sterol biosynthesis. Mol. Cell. Biol. 8: 3797-3808.

14. WOODWARD, H.D., J.M.C. ALLEN, W.J. LENNARZ. 1988. 3-Hydroxy-3-methylglutaryl coenzyme A reductase in the sea urchin embryo is developmentally regulated. J. Biol. Chem. 263: 2513-2517.

15. BACH, T.J., D.H. ROGERS, H. RUDNEY. 1986. Detergent solubilization, purification and characterization of 3-hydroxy-3-methylglutaryl-CoA reductase from radish seedlings. Eur. J. Biochem. 154: 103-111.

16. KONDO, K., K. OBA. 1986. Purification and characterization of 3-hydroxy-3-methylglutaryl CoA reductase from potato tubers. J. Biochem. 100: 967-974.

17. REDDY, A.R., W.S.R. DAS. 1986. Purification and partial characterization of 3-hydroxy-3-methylglutaryl coenzyme A reductase from the leaves of guayule (*Partenium argentatum*). Phytochemistry 25: 2471-2474.

18. BASSON, M., M. THORSNESS, J. RINE. 1986. *Saccaromyces cerevisiae* contains two functional genes encoding 3-hydroxy-3-methylglutaryl-coenzyme A reductase. Proc. Natl. Acad. Sci. USA 83: 5563-5567.

19. LEUTWILER, L.S., B.R. HOUGH-EVANS, E.M. MEYEROWITZ. 1984. The DNA of *Arabidopsis thaliana*. Mol. Gen. Genet. 194: 15-23.

20. PRUITT, R.E., E.M. MEYEROWITZ. 1986. Characterization of the genome of *Arabidopsis thaliana*. J. Mol. Biol. 187: 169-183.

21. REDEY, G.P. 1975 Arabidopsis as a genetic tool. Ann. Rev. Genet. 9: 111-127.

22. LLOYD, A.M., A.R. BARNASON, S.G. ROGERS, M.C. BYRNE, R.T. FRALEY, R.B. HORSCH. 1986. Transformation of *Arabidopsis thaliana* with *Agrobacterium tumefaciens*. Science 234: 464-466.

23. VALVEKENS, D., M. VAN MONTAGU, M. VAN LIJSEBETTENS. 1988. *Agrobacterium tumefaciens*-mediated transformation of *Arabidopsis thaliana* root explants by using kanamycin selection. Proc. Natl. Acad. Sci. USA 85: 5536-5540.

24. CAELLES, C., A. FERRER, F.G. HEGARDT, A. BORONAT. 1987. Characterization of a genomic fragment of *Arabidopsis thaliana* with homology to a HMG-CoA reductase clone from hamster. Third International Meeting on Arabidopsis. Michigan State University. Abstract no. 120.

25. CAELLES, C., A. FERRER, L. BALCELLS, F.G. HEGARDT, A.

BORONAT. 1989. Isolation and characterization of a cDNA encoding *Arabidopsis thaliana* 3-hydroxy-3-methylglutaryl coenzyme A reductase. Plant. Mol. Biol. 13: 627-638.

26. NARITA, J., W. GRUISSEM. 1989. Tomato hydroxymethylglutaryl-CoA reductase is requited early in fruit development but not during ripening. The Plant Cell 1: 181-190.
27. KYTE, J., R.F. DOOLITTLE. 1982. A simple method for displaying the hydropathic character of a protein. J. Mol. Biol. 157: 105-132.
28. ROGERS, S., R. WELLS, M. RECHSTEINER. 1986. Amino acid sequences common to rapidly degraded proteins: the PEST hypothesis. Science 234: 364-368.
29. LISCUM, L., J. FINER-MOORE, R.M. STROUD, K.L. LUSKEY, M.S. BROWN, J.L. GOLDSTEIN. 1985. Domain structure of 3-hydroxy-3-methylglutaryl coenzyme A reductase, a glycoprotein of the endoplasmic reticulum. J. Biol. Chem. 260: 522-530.
30. BROOKER, J.D., D.W. RUSSELL. 1975. Properties of microsomal 3-hydroxy-3-methylglutaryl Coenzyme A reductase from *Pisum sativum* seedlings. Arch. Biochem. Biophys. 167: 723-729.
31. BACH, T.J., H.K. LICHTENTHALER, Y. RETEI. 1980. Properties of membrane bound 3-hydroxy-3-methylglutaryl-coenzyme A reductase from radish seedlings and some aspects of its regulation. In: Biogenesis and function of plant lipids, P. Mazliak, P. Benveniste, C. Costes, R. Douce, eds., Elsevier, Amsterdam, pp. 355-362.

Chapter Three

REGULATION OF MONOTERPENE BIOSYNTHESIS IN HIGHER PLANTS

JONATHAN GERSHENZON AND RODNEY CROTEAU

Institute of Biological Chemistry
Washington State University
Pullman, WA 99164-6340, USA.

INTRODUCTION

Among the low molecular weight products of the mevalonate pathway in plants are the monoterpenes, the C_{10} representatives of the terpenoid family of natural products. Monoterpenes are colorless, lipophilic, volatile substances responsible for the characteristic odors of many plants. They have been reported from nearly 50 families of flowering plants,[1] being best known as constituents of the essential oils of pines, mints and citrus fruits. Monoterpenes are classified as secondary metabolites because they do not appear

Biochemistry of the Mevalonic Acid Pathway to Terpenoids
Edited by G.H.N. Towers and H. A. Stafford
Plenum Press, New York

to have any direct role in the basic processes of growth and development. Their functions in plants are still obscure, although, like other secondary metabolites, they may have ecological roles, serving as attractants to pollinators, allelopathic agents or defenses against predators and pathogens.

A striking feature of plant monoterpenes is that they are usually synthesized and accumulated in complex secretory structures, such as resin ducts, resin cavities or glandular trichomes.[2] These specialized structures appear only in particular organs at particular stages of plant development, and their formation is greatly influenced by environmental conditions. Monoterpene biosynthesis is thus restricted to certain times and places in the life of the plant and, consequently, seems to be a carefully regulated process.

Unfortunately, little is known about how plants regulate the synthesis of monoterpenes, since this topic has rarely been investigated directly. In the past, monoterpenes and other plant secondary metabolites were widely believed to be "waste products" and, therefore, their production was not thought to be under tight controls. Until recently, there has also been a lack of basic information on the later stages of monoterpene metabolism. However, in the last ten years, the use of cell-free extracts has promoted significant advances in understanding the steps following the formation of geranyl pyrophosphate (GPP), the branch point intermediate at which monoterpene biosynthesis diverges from the pathways to other terpenoids. The complete sequence of steps from GPP to several different types of monoterpenoid end products has been determined, a number of enzymes have been partially purified and characterized, and the mechanisms of some key cyclization reactions have been studied in detail.[3] These investigations provide a foundation which now permits the problem of monoterpene regulation to be approached directly.

This chapter provides a survey of the available evidence on how monoterpene biosynthesis is regulated in plants, with some discussion of relevant work on related classes of terpenoids. The overall accumulation of monoterpenes in plants depends upon the balance between synthetic and catabolic processes. The catabolism of monoterpenes has recently been reviewed[4] and will not be discussed here. In the remainder of the introduction, we present a brief overview of the biosynthetic routes to monoterpenes and discuss the need for regulatory controls.

The Pathway of Monoterpene Biosynthesis

The biosynthesis of monoterpenes proceeds from acetyl coenzyme A (acetyl-CoA) to geranyl pyrophosphate (GPP) via the well-documented steps of the mevalonate pathway (Fig. 1). This sequence of metabolic transformations was first discovered through investigation of steroid biosynthesis in mammals and yeast, but almost all of the steps have now been demonstrated in plants.[5] Three units of acetyl-CoA are involved in the early stages of the pathway. Two of these condense to form acetoacetyl-CoA by reversal of the thiolase reaction, while the third is joined via an aldol-type condensation, with loss of CoA, to yield (3*S*)-3-hydroxyl-3-methylglutaryl-CoA (HMG-CoA).

Next, reduction of HMG-CoA by a two-step, NADPH-dependent process generates (3*R*)-mevalonic acid, under catalysis of an enzyme called HMG-CoA reductase. The formation of mevalonate is the principal rate-controlling step of terpenoid biosynthesis in animals and microorganisms[6] and, as will be discussed later, may have similar importance in plants. In the following steps, mevalonic acid is phosphorylated twice in sequence to give mevalonate 5-pyrophosphate, which is then subjected to an ATP-driven, decarboxylative elimination forming isopentenyl pyrophosphate (IPP), the C_5 building block of terpenoid biosynthesis. IPP can be reversibly isomerized to 3,3-dimethylallyl pyrophosphate (DMAPP), an allylic pyrophosphate, which then alkylates IPP to give GPP, an acyclic, allylic C_{10} compound. The sequential addition of further IPP moieties gives C_{15} and C_{20} products which serve as precursors of the other major classes of terpenoids (Fig. 2). These condensations of IPP with allylic pyrophosphates of increasing chain length are catalyzed by a family of enzymes known as prenyltransferases.[7]

Monoterpene biosynthesis diverges from the pathways leading to other isoprenoids at the acyclic, C_{10} intermediate GPP. Although the later steps of monoterpene synthesis are not as well studied as those in the pathway leading to GPP, considerable progress has been made in recent years in understanding how GPP is converted to the enormous variety of monoterpenes present in higher plants. The vast majority of natural plant monoterpenes are cyclic compounds based on the cyclohexanoid ring skeleton. Cyclohexanoid monoterpenes are derived from GPP via cyclization reactions catalyzed by enzymes known as cyclases (Fig. 3).[8] GPP itself cannot

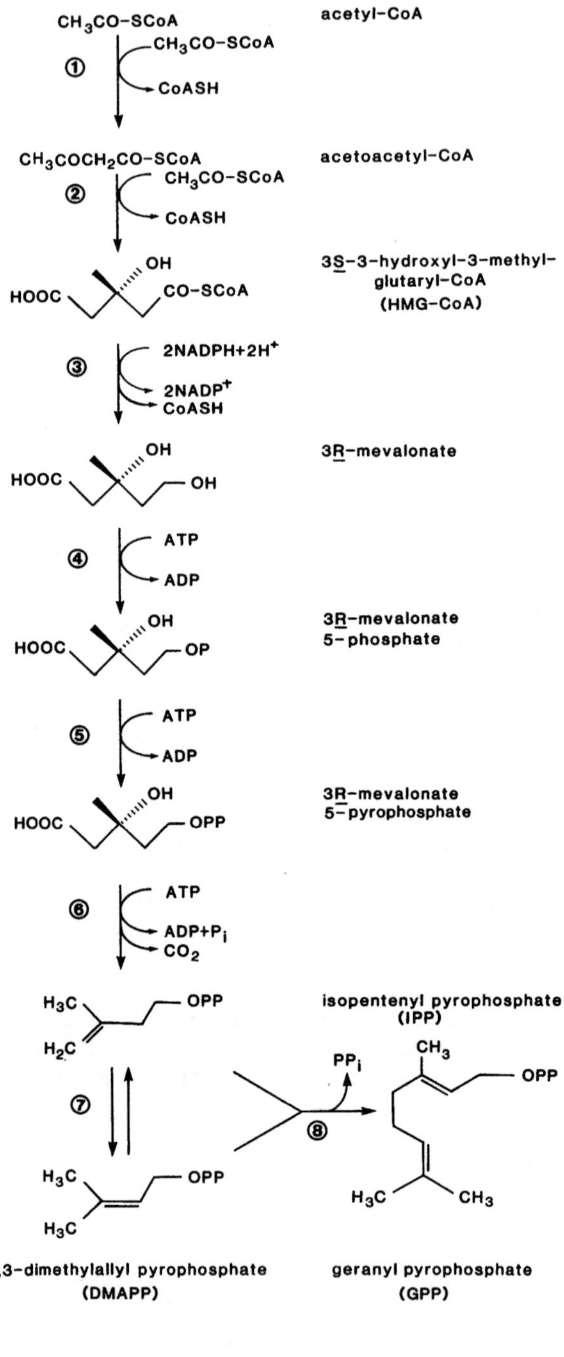
acetyl-CoA
acetoacetyl-CoA
3S-3-hydroxyl-3-methyl-glutaryl-CoA (HMG-CoA)
3R-mevalonate
3R-mevalonate 5-phosphate
3R-mevalonate 5-pyrophosphate
isopentenyl pyrophosphate (IPP)
3,3-dimethylallyl pyrophosphate (DMAPP)
geranyl pyrophosphate (GPP)

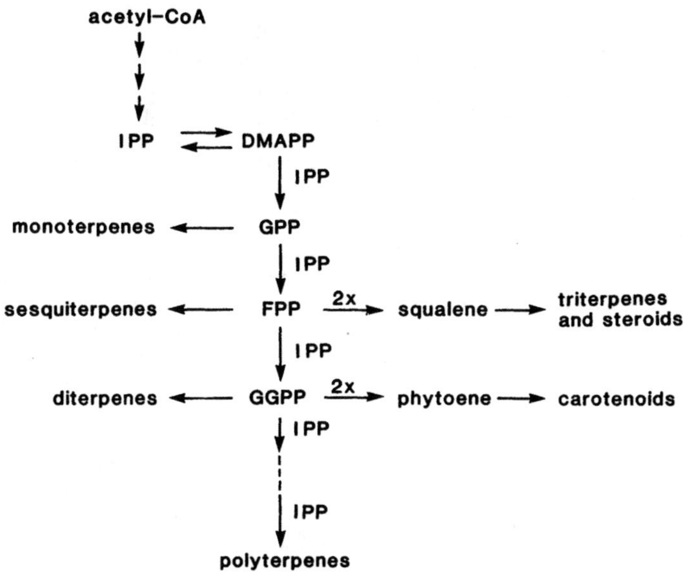

Figure 2. Outline of the major branches of terpenoid biosynthesis. Abbreviations: IPP, DMAPP, GPP, FPP and GGPP are the pyrophosphate esters of isopentenol, dimethylallyl alcohol, geraniol, farnesol and geranylgeraniol, respectively.

cyclize directly to form a six-membered ring because of the *trans*-geometry of the double bond at C-2. However, all cyclases that have been carefully studied to date first isomerize GPP to a bound intermediate, thought to resemble the tertiary isomer linalyl pyrophosphate (Fig. 4), which is then cyclized.[3]

The products of cyclization reactions are in some cases end products of monoterpene synthesis, such as α- and -pinene

---

Figure 1. Pathway of isoprenoid biosynthesis from acetyl-CoA to GPP. Key to enzymes: (1) = acetoacetyl-CoA thiolase; (2) = 3-hydroxy-3-methylglutaryl-CoA synthase; (3) = 3-hydroxy-3-methylglutaryl-CoA reductase; (4) = mevalonate kinase; (5) = mevalonate 5-phosphate kinase; (6) = mevalonate 5-pyrophosphate decarboxylase; (7) = isopentenyl pyrophosphate isomerase; (8) = geranyl pyrophosphate synthase (a prenyltransferase).

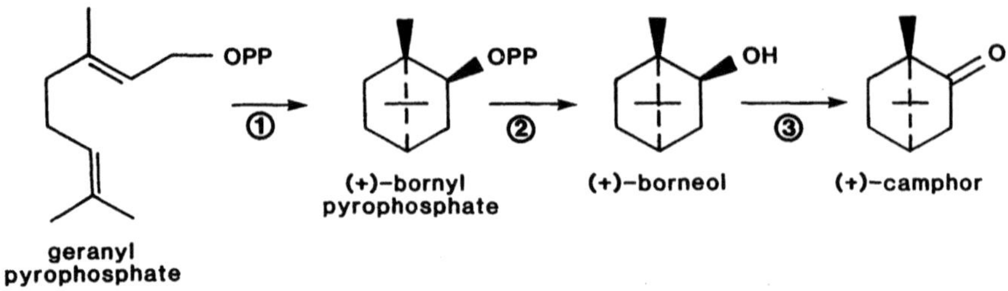

Figure 3. Pathway of (+)-camphor biosynthesis from geranyl pyrophosphate in *Salvia officinalis*. Key to enzymes: (1) = (+)-bornyl pyrophosphate synthase (a cyclase); (2) = (+)-bornyl pyrophosphate phosphohydrolase; (3) = (+)-borneol dehydrogenase.

in *Salvia officinalis* (garden sage) (Fig. 4).[9] In other cases, the initial cyclization products undergo a variety of subsequent transformations, including oxidations, reductions, isomerizations and hydrations leading to a diverse array of final products. For example, in *S. officinalis*, the initial cyclization product (+)-bornyl pyrophosphate[10] is hydrolyzed to (+)-borneol[11] and then oxidized to (+)-camphor (Fig. 3).[12] The oxygen function can be introduced in the cyclization step, as in the formation of (+)-bornyl pyrophosphate or 1,8-cineole (Fig. 4), or can result from a cytochrome P-450-dependent mixed-function oxygenase reaction, as in the conversion of (+)-sabinene to (+)-*cis*-sabinol (Fig. 4).[13]

The Need for Regulation

The basic steps of monoterpene biosynthesis prior to GPP are part of the general pathway of terpenoid biosynthesis. Since plants produce an enormous variety of terpenoid compounds in different amounts, some of which are restricted to specialized cell types and some of which are found in all cells (sterols, ubiquinone) or in all green tissues (carotenoids, phytol), regulatory mechanisms must exist to insure the selective operation of different branches of terpenoid synthesis in different parts of the plant at different times. Monoterpene synthesis itself requires the operation of the mevalonate pathway up to GPP and then diversion of GPP to monoterpene formation, rather than to the synthesis of higher isoprenoids. Thus, promotion of monoterpene biosynthesis requires, at minimum, stimulation of the basic pathway and branch point control at GPP.

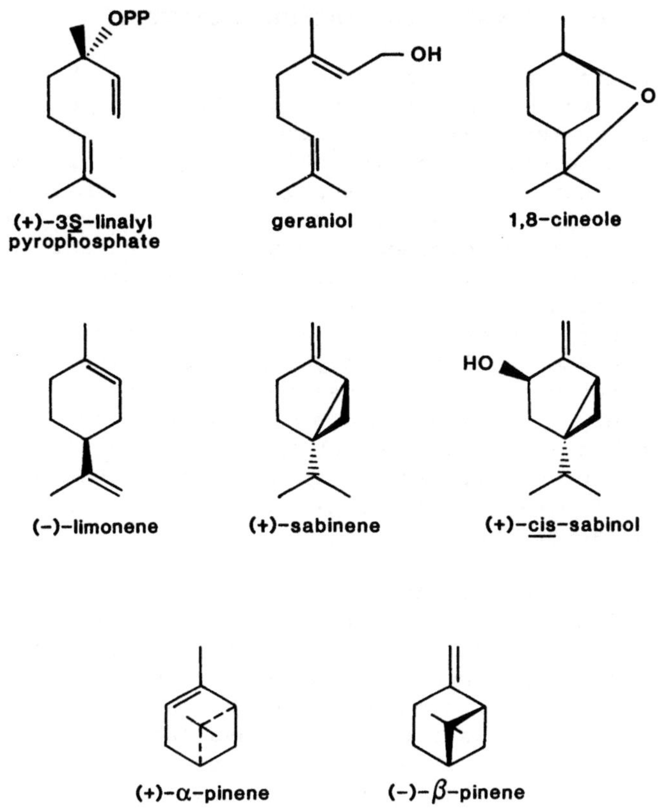

Fig. 4. Structures of various monoterpenes mentioned in the text.

Regulatory controls could operate in a hierarchy from the biosynthetic enzyme up to the level of the whole organism. The first part of this review will be focused at the enzyme level, surveying how changes in enzyme activity could control the production of monoterpenes. This will be followed by an examination of compartmentation phenomena of potential regulatory significance at the cellular level. Next, the role of assimilate partitioning in regulating monoterpene biosynthesis at the tissue and organ level will be considered. The final section will discuss the relationship between morphological differentiation and monoterpene production in the whole plant.

REGULATION BY CHANGES IN ENZYME ACTIVITY

Metabolic pathways are often regulated by changes in the activities of key enzymes. This generalization seems to especially apply to secondary metabolism,[14] though much of the evidence to date comes from investigations on microorganisms rather than plants. The regulatory role of enzymes in biosynthesis is usually inferred from studies that correlate the levels of activity determined *in vitro* with the rate of biosynthesis measured *in vivo* under environmental or developmental conditions where the rate changes. The validity of such correlations depends on the absence of significant changes in enzyme stability, enzyme extractability and concentrations of inhibitory substances throughout the time period of measurement. In these studies, it is also assumed that the rate of synthesis, as measured by incorporation of isotopically-labeled precursor, is not influenced by changes in uptake, sizes of metabolic pools or rates of competing pathways over the course of study.

For an enzyme that regulates monoterpene biosynthesis, a strong correlation should exist between the level of its activity and the overall rate of monoterpene synthesis. This section surveys experimental reports of such correlations covering the enzymes of the monoterpene pathway in biosynthetic order. In addition, we also examine available information on how enzyme activity is itself controlled. Alterations in the levels of activity may be caused by changes in the absolute amounts of enzymes, as a result of variations in the rates of enzyme synthesis or degradation, or by changes in catalytic capability, as determined by allosterism or covalent modification.

In considering the regulatory potential of an enzyme, it would be useful to know the total amount of enzyme protein present in addition to the quantity that is catalytically active. However, until very recently, no enzyme of plant terpene biosynthesis had been sufficiently purified to permit quantitative assessment by immunochemical methods. Thus, there is no information on the relationships between the dynamics of monoterpene synthesis and changes in total enzyme levels, and all enzyme measurements reported here are based solely on catalytic activity.

Few attempts have been made to investigate the enzymology of the earliest steps of terpenoid biosynthesis in plants.

Uritani and collaborators have examined the activities of two "pre-mevalonate pathway" enzymes, acetyl-CoA synthase (EC 6.2.1.1) and ATP citrate lyase (EC 4.1.3.8).[15,16] When sweet potato (*Ipomoea batatas*) roots are infected by the fungus *Ceratocystis fimbriata*, furanoid sesquiterpenes, such as ipomeamarone, are synthesized in adjacent non-infected regions, presumably as phytoalexins. Increases in both acetyl-CoA synthase and ATP citrate lyase activities were shown to be closely correlated with the rise in sesquiterpene content following infection, suggesting that both enzymes may be important in controlling the availability of acetyl-CoA as a substrate for terpenoid biosynthesis.

HMG-CoA Reductase

The rate-limiting step of isoprenoid synthesis in mammals is the formation of mevalonic acid from 3-hydroxy-3-methylglutaryl-CoA (HMG-CoA), catalyzed by HMG-CoA reductase (EC 1.1.1.34).[6,17] This enzyme has attracted much interest because of its regulatory role in cholesterol biosynthesis. The importance of HMG-CoA reductase in mammalian metabolism, plus the availability of a convenient *in vitro* assay,[18] have stimulated efforts to determine whether this enzyme also constitutes a controlling step in plant terpene production.

Although there has been no direct work on monoterpene biosynthesis, several research groups have examined the relationship between the level of HMG-CoA reductase activity and the rate of sesquiterpene biosynthesis. High amounts of HMG-CoA reductase activity were observed just prior to the accumulation of sesquiterpenoid phytoalexins in sweet potato (*Ipomoea batatas*) roots[19] and cell cultures,[20] in potato (*Solanum tuberosum*) tubers,[21,22] and in tobacco cell suspension cultures.[23,24] In addition, treatments that reduced HMG-CoA reductase activity, such as irradiation with white light or application of mevinolin (a fungal metabolite which is a very specific inhibitor of HMG-CoA reductase[25]), also reduced the accumulation of sesquiterpenes in potato tubers and tobacco cultures.[22,24] However, further work on the tobacco system indicated that, while fungal elicitors which induced sesquiterpene synthesis did cause a general rise in HMG-CoA reductase activity, elevation of such activity was transient, returning to control levels by the time the sesquiterpene biosynthetic rate was maximum.[23] In addition, the pattern of HMG-CoA reductase activity did not correspond well to changes in the rate of sesquiterpene biosynthesis over a ten-day cell culture cycle.[26]

HMG-CoA reductase activity levels have also been examined in relation to the synthesis of rubber, an isoprenoid polymer. The role of HMG-CoA reductase as the rate-limiting step in rubber synthesis was first suggested by Lynen[27] in a frequently cited report which showed that the *in vitro* activity of HMG-CoA reductase in *Hevea brasiliensis* (rubber tree) latex was lower than that of any of the other enzymes of the pathway. However, Lynen only measured the HMG-CoA reductase activity present in a 20,000 g supernatant, whereas, the majority of this activity in *H. brasiliensis* latex has since been shown to be membranous, sedimenting at only 600 g.[28] HMG-CoA reductase may, nevertheless, be a controlling step in rubber formation in *H. brasiliensis* latex, since diurnal variations in its activity coincided closely with changes in the rubber content of the latex.[29] Studies on the biosynthesis of primary plant isoprenoids, including sterols[30] and carotenoids,[31] have also indicated the regulatory importance of this enzyme (see Bach *et al.*, Chapter 1.).

On balance, the information available is consistent with the notion that HMG-CoA reductase catalyzes a key regulatory step in the biosynthesis of all plant isoprenoids, including monoterpenes. Proof of this assertion will require further study comparing the activity of HMG-CoA reductase with that of all the other pathway enzymes in a single tissue, while measuring the overall flux through the pathway and the actual concentrations of intermediates *in vivo*. The regulatory significance of HMG-CoA reductase in plants is unlikely to parallel that in mammals where this activity appears to represent the rate-limiting step in isoprenoid formation. The multiple branching of the terpenoid biosynthetic pathway in plants would seem to require additional control points for efficient regulation (in the absence of compartmentation effects). Therefore, it should not be surprising if changes in the rate of synthesis of a particular class of terpenoids occur without any corresponding change in HMG-CoA reductase activity levels, due to regulation at later stages of the biosynthetic pathway. Similarly, changes in HMG-CoA reductase activity could take place that do not affect the rate of formation of every class of terpenoid.

The regulatory picture in plants is further complicated by the presence of more than one species of HMG-CoA reductase. Various studies indicate that plant tissues appear to possess multiple forms of this enzyme, each having a separate subcellular location (the chloroplast, the mitochondrion or the cytosol) and distinct properties.[32-39] Each form may be associated with the biosynthesis of different classes of terpenoids.[22,31,36] Thus, it may be unreasonable to expect that total HMG-CoA reductase activity measured in crude extracts would show a close correspondence with the production of one particular type of terpenoid.

The possibility that HMG-CoA reductase may control plant terpenoid synthesis should stimulate interest in learning how the level of HMG-CoA reductase activity is itself controlled. In mammals, copious information is available on this topic. HMG-CoA reductase activity is subject to end product feedback inhibition by sterols, that is believed to be mediated by both changes in the rates of enzyme synthesis and, more rapidly, by covalent modification via reversible phosphorylation.[6,17] There are preliminary indications that these control mechanisms may also operate in plants.

Substantial increases in the rate of HMG-CoA reductase synthesis are apparently responsible for the rise in enzyme activity that precedes sesquiterpene phytoalexin production in both potato and sweet potato tubers. Work by Uritani's group[21,40,41] showed that treatment with cycloheximide or blasticidin S could almost completely block both the rise in HMG-CoA reductase activity and the appearance of sesquiterpenes, indicating that enzyme activity is contingent upon *de novo* protein synthesis. During the production of carotenoid pigments in ripening tomato fruit, HMG-CoA reductase activity also appears to be regulated by the rate of enzyme synthesis, since activity is closely correlated with the level of HMG-CoA reductase mRNA.[42]

Several reports implicate reversible phosphorylation as a mechanism for controlling the level of catalytically active HMG-CoA reductase in plants. For example, reductase activity in the chloroplast membranes of peas is inhibited by a combination of ATP, Mg^{++} and a crude mixture of soluble stromal proteins, but restored by treatment with phosphatase.[18] These findings, and similar results from *Hevea brasiliensis* latex,[43] could be explained by assuming that plant HMG-CoA reductase is inactivated by a kinase (present in the soluble stromal extract) and reactivated by a phosphatase. Alternatively, Sipat[43,44] has proposed that added Mg^{++}-ATP may facilitate the removal of newly-formed mevalonate from the reaction mixture by promoting mevalonate kinase activity, the next enzyme in the biosynthetic pathway. This would reduce the apparent level of HMG-CoA reductase activity by removing the measured product of the reductase assay. However, in a more recent experiment, Bach[39] controlled for mevalonate loss due to mevalonate kinase and still showed that Mg^{++}-ATP inhibited HMG-CoA reductase activity in tobacco microsomes. Nevertheless, the physiological relevance of these observations is still in doubt, particularly since the Mg^{++}-ATP concentration required for significant inhibition (4-5 mM) appears to exceed typical cellular values.[39] The role of phosphorylation as a means of regulating HMG-CoA reductase activity in plants clearly requires further investigation.

Control of HMG-CoA reductase in plants could also be effected by fluctuations in the concentrations of various products of the reaction. Both NADP and free coenzyme A were demonstrated to be inhibitors of HMG-CoA reductase activity *in vitro*,[35,43,45,46] but mevalonate had no significant effect except at relatively high concentrations (above 2 mM).[28,35,43,45-47] Unfortunately, it is difficult to assess the biological significance of these results, because estimates of the typical cellular concentrations of these substances are not available. However, the K_i for NADP (65.5 μM) is considered "rather small,"[45] raising the possibility that NADP:NADPH ratios could, in fact, modulate the action of HMG-CoA reductase *in vivo*. Both NADP and coenzyme A were competitive inhibitors (for the substrates NADPH and HMG-CoA, respectively), so there is no indication that HMG-CoA reductase is allosterically-regulated by these metabolites.

In mammals, it has long been known that HMG-CoA reductase activity is regulated by the feedback inhibition of later metabolites of the isoprenoid pathway, such as sterols[6,17] or, interestingly, certain plant monoterpenes, including limonene, borneol, geraniol and 1,8-cineole (Fig. 4).[48-50] No comparable regulatory system has been discovered in plants. Such control can be envisioned only at the cellular or subcellular level in plants, since the production of most terpenes is highly localized, and these compounds do not circulate in the vascular system. Conceivably, cellular concentrations of terpene intermediates and end products could modulate the activity of HMG-CoA reductase, and this possibility deserves study.

Other mechanisms for the regulation of HMG-CoA reductase activity in plants have been proposed (e.g., stimulation by a non-enzyme protein[51]), but these have little experimental support. Much additional work is needed before the control of HMG-CoA reductase activity in plants is understood to the same depth as that in mammals. A plant HMG-CoA reductase gene has recently been cloned from *Arabidopsis thaliana*, and its nucleotide sequence shows certain similarities to the HMG-CoA reductase genes of several mammalian species, particularly in the region coding for the catalytic site (Monfar *et al.*, Chapter 2).[52] However, significant differences are apparent between the *A. thaliana* and mammalian genes in the amino terminal domain, which codes for the membrane-binding region implicated in regulatory phenomena in mammals. Thus, the plant enzyme might possess regulatory features very different from those in mammals.

Mevalonic Acid to IPP and DMAPP

The product of the reaction catalyzed by HMG-CoA reductase, mevalonic acid, is converted to the C_5 intermediate isopentenyl pyrophosphate (IPP) by a three-step biosynthetic sequence (Fig. 1). Mevalonate is first pyrophosphorylated in two steps by the successive action of mevalonate kinase (EC 2.7.1.36) and mevalonate 5-phosphate kinase (EC 2.7.4.2). The resulting mevalonate 5-pyrophosphate is then transformed to IPP by the elimination of a carboxyl and a hydroxyl group mediated by mevalonate 5-pyrophosphate decarboxylase (EC 4.1.1.33). Compared to HMG-CoA reductase, there are only a few reports concerning the potential regulatory significance of these enzymes.

None of the three activities involved in converting mevalonate to IPP showed any change in rate during the induction of the biosynthesis of casbene, a diterpene phytoalexin, in castor bean (*Ricinus communis*) seedlings,[53] or during the increased production of carotenoids and steroids in auxin-treated cell suspensions of carrot.[31] In castor bean extracts, the small pool sizes of the intermediates mevalonate 5-phosphate and mevalonate 5-pyrophosphate, measured after mevalonate feeding, suggested that neither of the first two enzymes of the sequence, the kinases, limit the rate of casbene production.[53] However, several studies have attributed regulatory properties to the third enzyme, the decarboxylase. During the induction of furanosesquiterpene phytoalexin biosynthesis in sweet potato roots, a significant rise (greater than 5-fold) in mevalonate 5-pyrophosphate decarboxylase activity was observed which began just before the accumulation of sesquiterpenes.[41] Based on experiments with cycloheximide, the level of this activity appeared to be controlled by the rate of enzyme synthesis.[41] Mevalonate 5-pyrophosphate decarboxylase is an ATP-requiring enzyme, and so its catalytic activity might also be influenced by the energy status of the cell, as indicated by the adenylate energy charge, the molar ratio (ATP + 1/2 (ADP)):(AMP + ADP + ATP).[54] Biosynthesis of the diterpene kaurene from mevalonate in extracts of *Marah macrocarpus* endosperm was stimulated by increases in the value of adenylate energy charge above 0.8, and the site of control was demonstrated to be the decarboxylation of mevalonate 5-pyrophosphate.[55] Response to energy charge was principally due to the inhibitory effect of ADP, as was described for mevalonate 5-pyrophosphate decarboxylase from *Hevea brasiliensis* latex.[56] Regulatory enzymes in many other energy-utilizing biosynthetic sequences show a sharp increase in rate as adenylate energy charge is increased above 0.8.[54,55]

The isomerization of IPP to DMAPP is catalyzed by isopentenyl pyrophosphate Δ^3-Δ^2-isomerase (EC 5.3.3.2). DMAPP, the first allylic pyrophosphate in the terpene pathway, serves as a C_5 "primer" to which varying numbers of IPP moieties can be added in the successive chain elongation steps of the terpenoid pathway (Fig. 2). By changing the proportion of IPP to DMAPP available for subsequent biosynthetic steps, IPP isomerase could exert some general control over the class of terpenes produced. For example, in monoterpene biosynthesis, only a single IPP moiety is needed to condense with DMAPP to give GPP. In polyisoprene biosynthesis, on the other hand, many IPP units are required for each DMAPP primer, so considerably lower relative rates of IPP isomerase activity would be needed. IPP isomerase is inhibited by inorganic phosphate and various prenyl pyrophosphates *in vitro*,[57] so these might serve as modulators of enzyme activity.

Studies in mammals have shown that IPP is not irreversibly committed to terpene biogenesis, but can be converted back to HMG-CoA and then to acetate by a series of transformations known as the mevalonate shunt, first described by Popjak and co-workers.[58] Recently, the operation of the mevalonate shunt has also been documented in wheat seedlings.[59] In addition, the activity of HMG-CoA lyase, one of the steps of the shunt, has been demonstrated in the latices of *Hevea brasiliensis*[28] and *Euphorbia lathyris*.[60] Although the physiological importance of the mevalonate shunt in plants has not been assessed, this pathway could divert a significant quantity of IPP away from terpenoid biosynthesis, and thus modulate regulation by HMG-CoA reductase or other enzymes of the pathway prior to mevalonate pyrophosphate decarboxylase. Additional research to determine under what conditions and in what tissues the mevalonate shunt operates would be most welcome.

Prenyltransferases

The C_5 intermediate IPP participates in a series of chain elongation steps forming allylic pyrophosphates of 10, 15, 20 or more carbon atoms which are precursors of the various major classes of terpenes (Fig. 2). These conversions are catalyzed by prenyltransferases (EC 2.5.1.1), a group of enzymes that mediate the condensation of IPP with an allylic pyrophosphate, generating the next higher C_5-homologue of the allylic substrate. Since prenyltransferases are branch point enzymes,[61] the reactions they catalyze could be important regulatory steps in the biosynthesis of monoterpenes and other isoprenoids. Metabolic pathways are often regulated at branch points so that flux can be selectively directed among several possible routes.

The enzymology of the prenyltransferase reaction has been investigated with cell-free preparations from a variety of plant species.[7] Although our knowledge is still somewhat fragmentary, and caution must be exercised in extrapolating from the results of *in vitro* experiments to the actual behavior of enzymes in the intact plant, the prenyltransferases involved in the basic elongation steps of terpenoid biosynthesis appear to show certain similarities in substrate and product specificity. Initially, one might have supposed that individual prenyltransferases would catalyze only a single elongation step, such as DMAPP + IPP → GPP or GPP + IPP → FPP (Fig. 5A). However, most of the prenyltransferases characterized to date apparently catalyze a multi-step sequence beginning with the substrates DMAPP and IPP and terminating with a specific end product (but see reference 62) (Fig. 5B). For example, specific FPP synthases have been reported that convert DMAPP and IPP to GPP and then condense the newly-formed GPP with another IPP moiety to give FPP, without appreciable accumulation of the intermediate GPP. These enzymes have been isolated from pumpkin fruit,[63] castor bean seedlings,[62] orange rind[64,65] and various animal and fungal sources.[7] Specific plant GGPP synthases are also known that convert DMAPP and IPP directly to GGPP without accumula-

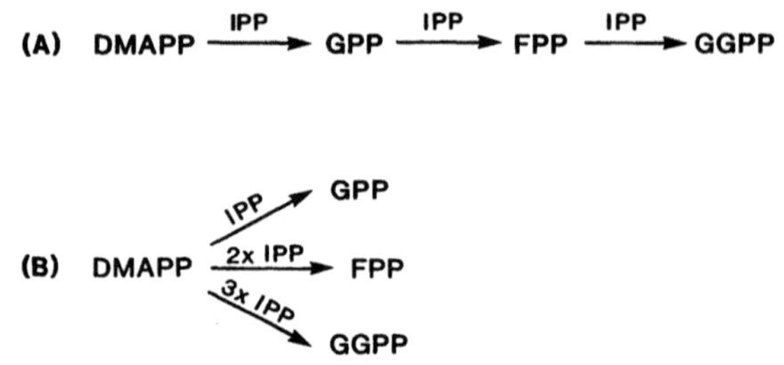

Figure 5. Two alternate schemes for the substrate and product specificities of the prenyltransferases in terpenoid biosynthesis. In (A), the individual prenyltransferases each catalyze only a single step in the elongation sequence, beginning with a unique substrate and forming a unique end product. In (B), the individual prenyltransferases each catalyze an entire sequence of 1-3 steps without the appearance of free intermediates. All of the sequences begin with DMAPP and IPP and end with a unique final product. Abbreviations: IPP, DMAPP, GPP, FPP and GGPP are the pyrophosphate esters of isopentenol, dimethylallyl alcohol, geraniol, farnesol and geranylgeraniol, respectively.

tion of GPP or FPP.[63,66]

Monoterpene biosynthesis requires an ample supply of the intermediate GPP. Therefore, one might expect to find specific GPP synthases associated with plant tissues involved in monoterpene synthesis that produce GPP from IPP and DMAPP without further elongation. Such an activity has recently been described from leaf epidermal extracts of garden sage (*Salvia officinalis*).[67] Preliminary experiments with this species demonstrated the synthesis of GPP from DMAPP and IPP in whole leaf extracts, but the major product of this preparation was FPP (Table 1, line A). Chromatographic separation of the whole leaf extract suggested the presence of a GPP synthase activity that was distinguishable from the much more abundant FPP synthase activity. Since the monoterpenes in sage and other herbaceous species are synthesized and stored in glandular trichomes found on the leaf surface,[68,69] a procedure was employed to selectively extract the contents of these leaf surface structures by gentle abrasion of the epidermis.[70] This technique removed 70-100% of the glandular trichomes and extracted about 70% of the GPP synthase activity present in the whole leaf (although only about 20% of the total prenyltransferase activity), suggesting that GPP synthase is localized primarily in the glandular trichomes (Table 1, compare line B vs. line C). Partial purification of this activity from epidermal extracts clearly showed the selective synthesis of GPP (Table 1, line D). The properties of this enzyme, including molecular weight, pH optimum, kinetic constants and metal ion requirement, resembled those of previously characterized prenyltransferases.[7,62-66,71-74] Thus, a plant tissue specialized for monoterpene production appears to possess a particular prenyltransferase that synthesizes predominantly GPP from IPP and DMAPP and does not catalyze further condensations with IPP.

Other plant tissues that produce copious amounts of monoterpenes, such as the rind of citrus fruits, might also be expected to contain GPP synthase. Extracts from the rind (= exocarp or flavedo) of young *Citrus unshiu* and *Citrofortunella mitis* fruits were separated on sucrose gradients to obtain fractions enriched in leucoplasts that biosynthesized monoterpenes from mixtures of IPP and DMAPP.[75] These preparations synthesized primarily GPP, rather than FPP, from exogenous IPP and DMAPP, implying the operation of a specific GPP synthase. In contrast, prenyltransferases isolated from the rind of ripe *Citrus paradisii*[65] and *C. sinensis* fruit[64] made mostly FPP. However, the absence of GPP-synthesizing activity in these extracts may be a simple consequence of the fact that ripe fruit does not make monoterpenes.[76] Neither the *C. paradisii* nor the *C. sinensis* extract was able to synthesize the monoterpene hydrocarbons characteristic of the intact fruit. Specific GPP synthases have also been found in *Lithospermum erythrorhizon* cell cultures where the enzyme supplies the geranyl moiety to produce geranyl *p*-hydroxybenzoate, an intermediate in shikonin biosynthesis,[74,77] and in the bacterium *Micrococcus lysodeikticus* (now called *M. luteus*), where the enzyme functions to generate the primer for solanesyl pyrophosphate en route to menaquinone.[72]

The synthesis of monoterpenes and other isoprenoids in plants may thus involve several types of prenyltransferases, all of which catalyze the synthesis of specific chain-length prenyl pyrophosphates starting from the substrates IPP and DMAPP. In the absence of compartmentation effects, such prenyltransferases would all be branch point enzymes at the same node of the pathway competing for the same pools of IPP and DMAPP (Fig. 5B). If different prenyltransferases have

Table 1. Product specificities of prenyltransferase preparations from *Salvia officinalis* incubated with [1-^{3}H]IPP and DMAPP. Results show the existence of an activity localized on the leaf surface that synthesizes predominantly GPP. Data derived from reference 67.

Preparation	% of each product formed		
	GPP	FPP	GGPP
A. Whole leaf extract	6	84	4
B. Leaf surface extract	22	70	5
C. Extract of leaf tissue remaining after surface extraction	2	97	2
D. Partially-purified GPP synthase from leaf surface extract	90	2	8

widely divergent substrate affinities, flux into different branches of isoprenoid synthesis may be regulated by changing concentrations of IPP and DMAPP. Therefore, accurate measurements of local IPP and DMAPP concentrations, using a methodology such as that of Bruenger and Rilling,[78] could be quite useful in understanding how plant terpenoid biosynthesis is regulated.

The regulatory significance of prenyltransferases can be assessed by examining the correlation between levels of prenyltransferase activity and changes in the rate of terpenoid biosynthesis. Unfortunately, there is only one investigation of direct relevance to monoterpene biosynthesis. Banthorpe and co-workers[79] reported that GPP synthesizing activity in cell-free extracts of *Pelargonium graveolens* (rose geranium) showed pronounced seasonal variation, being about 400 times higher in summer than in winter. These fluctuations paralleled *in vivo* changes in the rate of synthesis of geraniol (the principal monoterpene in the plant) (Fig. 4) from IPP or mevalonate. Interestingly, the actual amount of geraniol in *P. graveolens* was practically constant over much of the year, suggesting that significant seasonal changes in the rate of geraniol catabolism might accompany changes in the rate of synthesis.

Studies with other classes of terpenes also indicate that prenyltransferase action could serve as a rate-controlling step in biosynthesis. The macrocyclic diterpene casbene is produced in castor bean seedlings upon fungal elicitation. During the induction of casbene synthesis, there is a 25-fold increase in GGPP synthase activity (from FPP and IPP).[53] By contrast, there is no change in HMG-CoA reductase activity after elicitation, and only a small increase (2-fold) in FPP synthase activity (from IPP and either DMAPP or GPP) (See West *et al.*, Chapter 6).

The biosynthesis of the C_{30} compound squalene, a key intermediate on the pathway to triterpenes and steroids, was investigated in germinating peas.[80,81] Of all the enzymes catalyzing steps between mevalonate and squalene, FPP synthase exhibited the most substantial increase in activity during germination, along with the lowest relative activity[81] and the best correlation with squalene synthesis from exogenous mevalonic acid.[80] All these observations are consistent with a regulatory role for FPP synthase in the synthesis of triterpenes and steroids in germinating pea seeds (See also Chapters 7 and 8).

The biosynthesis of rubber, an isoprenoid polymer, requires the action of a "polyprenyltransferase" that catalyzes the addition of thousands of IPP units to a DMAPP primer.[82] When rubber biosynthesis in guayule (*Parthenium argentatum*) is stimulated by water stress,[83] low temperature, or application of the growth regulator DCPTA [2-(3,4-dichlorophenoxy)tri-ethylamine],[84] there are corresponding increases in the activity of rubber polyprenyltransferase. Therefore, this enzyme may regulate the rate of elongation of existing rubber oligopolymers, although IPP isomerase, which produces the C_5 primer DMAPP, may be the rate-limiting step in the initiation of new rubber molecules.[84,85]

A different class of prenyltransferases catalyzes reactions referred to as prenylations in which a dimethylallyl, geranyl or larger terpenyl moiety is transferred to an aromatic nucleus of non-isoprenoid origin, giving rise to products such as ubiquinone, furanocoumarins or prenylated flavonoids.[7] These enzymes also appear to have regulatory importance. For example, the formation of glyceollin, a prenylated isoflavonoid phytoalexin produced in soybean cotyledons and cell cultures, in response to treatment with fungal elicitor, is associated with a significant rise in dimethylallyl transferase activity, whereas there is little change in HMG-CoA reductase activity.[86] Dimethylallyl transferase activity also increases substantially (more than 6-

fold) during the elicitation of furanocoumarin synthesis in parsley cell cultures.[87] A geranyl transferase from *Lithospermum erythrorhizon* cell cultures which produces geranyl *p*-hydroxybenzoate, an intermediate in shikonin biosynthesis, has 35 times greater activity in cultures that make shikonin than in cultures that do not produce this natural product.[88]

Further support for the regulatory role of prenyltransferases comes from studies on the incorporation of isotopically-labeled precursors into monoterpenes *in vivo*. Investigators have frequently noted that the two C_5 units of monoterpenes are labeled asymmetrically; that is the portion of the molecule derived from IPP is usually more heavily labeled than that derived from DMAPP.[68,89,90] Such asymmetric labeling is usually attributed to the existence of a metabolic pool of DMAPP that dilutes the labeled DMAPP formed from an exogenous precursor. The accumulation of an intermediate such as DMAPP suggests that the next enzyme of the pathway, in this case the prenyltransferase, is rate-limiting. If catalysis by prenyltransferase is indeed a rate-limiting step, one would expect a much larger pool of DMAPP than IPP, since equilibrium of the IPP-DMAPP isomerization favors DMAPP.[91] In all probability, the phenomenon of asymmetric labeling is the result of a non-steady state process, observable only when incorporation times are brief or when incorporation rates of exogenous precursors are very low, as is frequently the case with monoterpenes.[68,90]

In summary, several lines of evidence implicate prenyltransferases as important rate-limiting catalysts in the biosynthesis of monoterpenes and other types of plant terpenoids. First, prenyltransferases are situated at the primary branch points of the isoprenoid pathway. Second, the rate of terpenoid formation *in vivo* shows a close relationship with prenyltransferase activity in many experimental systems, often a much closer relationship than that with HMG-CoA reductase activity. Third, the transformations mediated by prenyltransferases seem to be the slow steps in the pathway based on the existence of a metabolic pool of DMAPP, as revealed by isotopic tracer studies, and the results of Green and Baisted,[80] which showed that the prenyltransferase reaction in germinating pea seeds had the lowest velocity of any of the steps between mevalonate and squalene. Finally, prenyltransferases may catalyze the first truly committed step of the isoprenoid pathway, since the operation of the mevalonate shunt allows IPP to be diverted away from terpene production.

Plants may employ a variety of mechanisms to control the levels of prenyltransferase activity, such as covalent modification, changing concentrations of inhibitory substances, or altered rates of enzyme synthesis and degradation. Unfortunately, almost no information is available on this topic. Various prenyl pyrophosphates and their synthetic analogs have been demonstrated to act as prenyltransferase inhibitors in several *in vitro* systems.[57,64,92] Conceivably, substances of this type could function to modulate enzyme activity in the intact plant. Given the presumptive regulatory importance of prenyltransferases in terpenoid biosynthesis, these enzymes clearly deserve more intensive scrutiny.

Monoterpene Cyclases

Following the formation of GPP, the biosynthesis of most monoterpenes can be divided into two parts: cyclization of GPP to one of a number of skeletal types, and secondary transformations of the initial cyclic compound. The cyclization reactions that produce the parent monoterpene skeletons from GPP are catalyzed by enzymes called monoterpene cyclases. These enzymes have received considerable attention in recent years[3,8] because the nature of the cyclization processes determines the basic character of the monoterpene end products.

From a regulatory standpoint, monoterpene cyclases are good candidates for being rate-limiting enzymes in monoterpene biosynthesis because they catalyze the first step leading specifically to monoterpenes. However, whether or not cyclases represent actual branch point enzymes is still unresolved. GPP has traditionally been viewed as a branch point metabolite (Fig. 2); it can either be cyclized and thus irreversibly committed to monoterpene biosynthesis, or it can be converted to FPP and from there to all other types of higher terpenoids. As discussed in the previous section, recent investigations of several prenyltransferases indicate that GPP may not be a free intermediate in the construction of higher isoprenoids since many plants seem to possess specific FPP and GGPP synthases which use only DMAPP as the allylic co-substrate, instead of GPP or FPP (Fig. 5B). Thus, monoterpene cyclases may not function at an actual metabolic branch point, although they could still catalyze the rate-limiting step in monoterpene biosynthesis.

Indeed, several studies suggest that monoterpene cyclases are key regulatory enzymes in the biosynthesis of monoterpenes. For instance, the formation of the monoterpene camphor was examined in relation to leaf development in *Salvia officinalis* (garden sage).[93] Camphor, a bicyclic ketone, is synthesized from GPP in three steps (Fig. 3). First, GPP is cyclized to bornyl pyrophosphate, which is then hydrolyzed to the secondary alcohol borneol and finally oxidized to camphor. The activity of the cyclase, bornyl pyrophosphate synthase, correlated very closely with changes in the rate of camphor synthesis in developing leaves, as measured by $^{14}CO_2$ incorporation (Fig. 6). Bornyl pyrophosphate synthase activity was high in immature, rapidly-expanding leaves (3-4 weeks old), but declined to nearly undetectable levels by 6 weeks of age. Likewise, camphor was synthesized most rapidly in leaves 3 weeks old, but the rate declined precipitously by 7 weeks of age. The activities of the other enzymes of camphor biosynthesis, bornyl pyrophosphate phosphohydrolase and borneol dehydrogenase, had developmental profiles somewhat similar to that of bornyl pyrophosphate synthase, but peaked and declined later in leaf maturation and, therefore, did not correlate as well with the time course of camphor formation (Fig. 6).

Two other findings further substantiate the regulatory role of the cyclization step in the biosynthesis of camphor in *S. officinalis*. First, bornyl pyrophosphate synthase had significantly lower activity than either the prior (prenyltransferase) or the two subsequent enzymes of the camphor pathway, as determined by *in vitro* assay. At peak level, bornyl pyrophosphate synthase activity was only about 10% of GPP synthase activity,[67] 20% of bornyl pyrophosphate phosphohydrolase activity and 40% of borneol dehydrogenase activity.[93] In addition, the product of the cyclization step, bornyl pyrophosphate, was found at lower concentrations in the plant than all other camphor pathway intermediates, including GPP and borneol,[93] indicating that cyclization is likely to be the slowest step in camphor biosynthesis.

The production of large quantities of monoterpenoid and diterpenoid resin components in pine trees attacked by bark beetles provides an excellent system for studying the regulation of terpene synthesis in plants. Increased amounts of monoterpenes and diterpene resin acids are formed in *Pinus contorta* (lodgepole pine) in response to infestation by *Dendroctonus ponderosae* (mountain pine beetle) and its associated blue-stain fungus, *Ceratocystis clavigera*.[94]

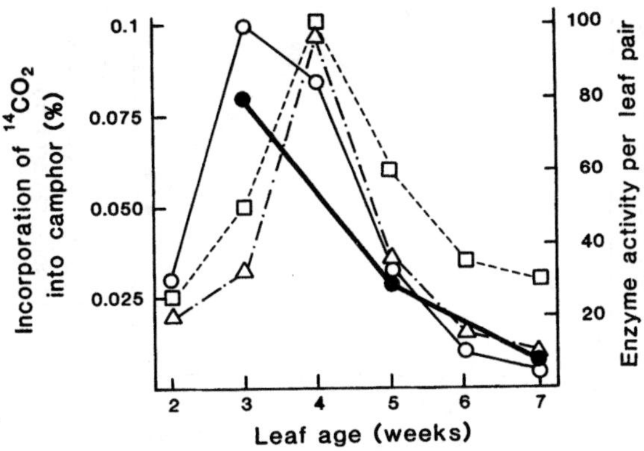

Figure 6. Correlation of the rate of camphor biosynthesis with the activities of several biosynthetic enzymes during leaf development in *Salvia officinalis*. Camphor biosynthesis was measured by $^{14}CO_2$ incorporation. Data are from reference 93. Legend (●——●) = rate of camphor synthesis; (○——○) = bornyl pyrophosphate synthase activity; (□-------□) = bornyl pyrophosphate phosphohydrolase activity; (△-·-·-△) = borneol dehydrogenase activity.

Elevated levels of both of these classes of terpenes can be induced experimentally by inoculation with *C. clavigera* hyphae or by treatment with various oligosaccharide elicitors.[95]

Elicitor application to stem wounds of *P. contorta* saplings stimulated the rate of monoterpene and diterpene synthesis, as measured by [U-^{14}C]-sucrose incorporation.[96] This stimulation was closely paralleled by increased activity of monoterpene and diterpene cyclases (Fig. 7),[96] suggesting that cyclases may exert significant control over monoterpene and diterpene biosynthesis in pines.

Application of cytokinins to the foliage of several mints (species of *Mentha*, *Lavandula* and *Salvia*) caused a two-fold rise in monoterpene content compared to untreated controls.[97] This increase was accompanied by a rise in monoterpene cyclase levels of approximately the same magnitude, providing further evidence of the regulatory significance of cyclases in monoterpene biosynthesis.

Cyclases may also serve as rate-controlling enzymes in sesquiterpene and diterpene biosynthesis. An excellent

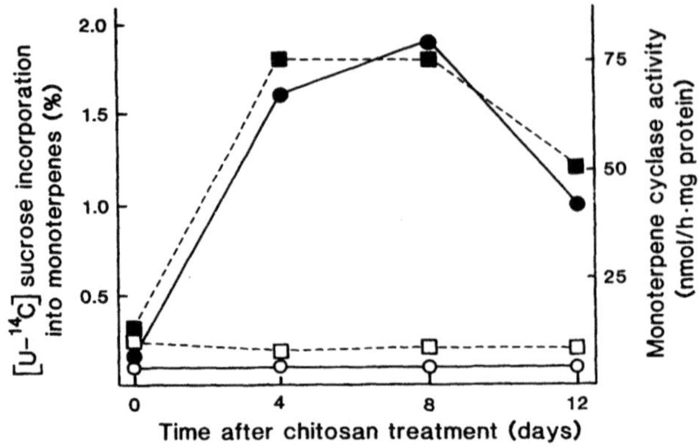

Figure 7. Correlation of the rate of monoterpene synthesis with monoterpene cyclase activity in *Pinus contorta* saplings following wounding and application of chitosan elicitor. Monoterpene synthesis was measured by the incorporation of [U-^{14}C]sucrose. Data are from reference 96. Legend: (●——●) = rate of monoterpene synthesis and (■-------■) = cyclase activity after wounding and chitosan treatment; (○——○) = rate of monoterpene synthesis and (□-------□) = cyclase activity in unwounded controls.

example comes from work with the bacterium *Streptomyces arenae*, which produces the sesquiterpene antibiotic pentalenolactone.[98] Several enzymes of pentalenolactone biosynthesis were measured during culture of *S. arenae*, including HMG-CoA reductase, IPP isomerase, GPP and FPP synthases and pentalenolactone synthase (an FPP cyclase). Of these enzymes, only the activity of the cyclase corresponded well with the time course of pentalenolactone production in *S. arenae*.

In higher plants, cyclase levels have been examined in relation to the production of both sesquiterpene phytoalexins in tobacco cell cultures[23,26,99] and diterpene phytoalexins in castor bean seedlings.[53] Formation of these compounds is triggered by treatment with cellulase, fungal spores or fungal elicitor preparations. While treated tissue had substantial levels of cyclase activity, no cyclase activity was detected in untreated controls.[23,53,99] However, in tobacco cultures, sesquiterpene accumulation is clearly not under the simple control of cyclase activity alone. Vogeli and Chappell[23] observed that the rate of sesquiterpene synthesis from radiolabeled acetate or mevalonate decreased significantly 24 hours after elicitor addition, with a resulting drop in sesquiterpene concentration, but sesquiterpene cyclase activity remained high for an additional 12 hours. Furthermore, Chappell *et al.*[26] measured levels of sesquiterpene cyclase and the rate of sesquiterpene synthesis after elicitor treatment at intervals over a 10-day cell culture cycle and found that, whereas 0- to 2-day-old cultures showed good correlations between sesquiterpene synthesis and cyclase level, 3- to 5-day-old cultures exhibited high cyclase levels but low apparent rates of sesquiterpene synthesis. Changes in the rate of sesquiterpene catabolism may be responsible for these observations.

One can envision the possibility that the biosynthesis of a particular class of terpenes could be controlled, at least in part, by a regulatory enzyme of a competing pathway, and there is evidence for this type of control in the formation of sesquiterpene phytoalexins. FPP is a branch point intermediate that can be utilized by cyclases for sesquiterpene production or channeled to sterol and triterpene biosynthesis via conversion to squalene (Fig. 2). Elegant investigations with tobacco and potato cell cultures, and potato tubers, have demonstrated that the addition of fungus or fungal elicitor preparations, in addition to stimulating the rate of sesquiterpene phytoalexin synthesis, also inhibited sterol and triterpene production.[23,99,100] In tobacco, elicitor treatment caused a decline in squalene synthase activity that paralleled the rise in sesquiterpene cyclase activity.[23,99] Thus, promotion of sesquiterpene phytoalexin synthesis in these species seems to be facilitated by an increase in sesquiterpene cyclase activity coordinated with a suppression of squalene synthase activity.

The potential regulatory role of terpene cyclases makes it important to determine how levels of cyclase activity might be controlled. When terpene phytoalexin production is induced by fungal infection, there is a time lag of 3-4 hours between fungal application and the first appearance of cyclase activity,[100,101] suggesting that cyclase must be newly synthesized. The site of control might be at the level of transcription, translation, or the result of post-translational events. In fungus-infected castor bean seedlings, which produce the diterpene phytoalexin casbene, the appearance of mRNA for casbene cyclase just precedes the rise in activity of this enzyme, indicating that the quantity of active casbene cyclase may be controlled by the amount of translatable mRNA.[101]

More rapid mechanisms for controlling terpene cyclase levels might operate in plants, but there is as yet no evidence for such processes. Commercial preparations of protein kinases and protein phosphatases did not influence the catalytic activity of various monoterpene cyclases (R. Croteau, unpublished results) leading to the suggestion that these enzymes are not modulated by phosphorylation/dephosphorylation events. End-product feedback inhibition of sesquiterpene cyclases has been demonstrated in *Streptomyces*,[98,102] but the activities of various monoterpene cyclases in higher plants were not specifically affected by their end products.[103-105] Finally, there is no evidence for monoterpene cyclase regulation by adenylate energy charge (R. Croteau, unpublished results).

Secondary Transformations

The cyclic monoterpenes initially formed from GPP are subject to an assortment of further enzymatic transformations, including oxidations, reductions and isomerizations. These modifications produce the large number of derivatives of each monoterpene skeletal type which naturally occur in plants. For example, in *S. officinalis*, the initial cyclic product, bornyl pyrophosphate, is converted by a phosphohydrolase to borneol, which is then oxidized by a dehydrogenase to camphor (Fig. 3). As a group, the enzymes catalyzing these secondary transformations are not well studied, so it is difficult to determine their overall regulatory significance.

As discussed in the previous section, the enzyme activities responsible for the three steps of camphor biosynthesis from GPP were compared to the rate of camphor formation during leaf development in *S. officinalis*.[93] The levels of all three activities showed a general correspondence with the rate of camphor biosynthesis, being high in young, expanding leaves and declining as leaves aged (Fig. 6). However, neither the phosphohydrolase nor the dehydrogenase showed as close a correlation with the time course of camphor formation as did the cyclase, implying that neither of these secondary transformations is as important as the cyclization step in controlling the rate of camphor biosynthesis.

The eremophilane sesquiterpene capsidiol is one of several sesquiterpene phytoalexins found in tobacco cell cultures. Capsidiol biosynthesis appears to proceed from FPP by cyclization to a bicyclic olefin followed by two sequential hydroxylation steps.[106] The activity of one of these hydroxylases seems to be induced by cellulase treatment[106] along with cyclase activity,[99] and this may help to promote the biosynthesis of capsidiol at the expense of debneyol, another eremophilane sesquiterpene formed from the same bicyclic olefin intermediate.[106]

Much additional study is needed to evaluate the regulatory status of the enzymes catalyzing secondary transformations. These enzymes are important in determining the specific types of terpenoid derivatives produced, although it seems unlikely that they mediate rate-limiting steps in biosynthesis, since most of the fixed carbon, ATP and NADPH required for the construction of these compounds has already been committed in previous steps.

The earlier sections of this review have presented ample evidence for control of plant monoterpene biosynthesis by changes in enzyme activity. Research to date implicates three specific steps of the monoterpene pathway as having particular regulatory significance: the reduction of HMG-CoA, the condensation of IPP and DMAPP to form GPP, and the cyclization of GPP. HMG-CoA reductase is the crucial rate-limiting step in sterol production in mammals and may well have a similar role in monoterpene formation in plants. However, the much more prolific branching of terpenoid biosynthetic pathways in plants suggests the need for additional points of control. Prenyltransferases, branch point enzymes that synthesize allylic pyrophosphates of various chain lengths, could serve as additional regulatory steps. Cyclases catalyze the first steps leading specifically to monoterpenes, and might also influence the rate of monoterpene biosynthesis. Clearly, more than one enzyme may have regulatory significance, each controlling a different segment of the pathway.

A thorough understanding of the regulation of monoterpene biosynthesis will require experiments that accurately measure the rate of product synthesis and the activities of all pathway enzymes in a single tissue. Care must also be exercised to distinguish among different forms of enzymes, such as HMG-CoA reductase, which have multiple subcellular locations. Once the chief regulatory enzymes have been identified, studies can proceed to elucidate the mechanisms by which the levels of catalytic activity are controlled.

REGULATION BY CELLULAR COMPARTMENTATION

The regulation of monoterpene biosynthesis by control of enzymatic activity can be superseded by the effects of cellular compartmentation. Plant metabolism is extensively compartmented at the cellular level. Certain reaction sequences are physically separated from each other, occurring only in specific organelles or other isolated intracellular sites. Within a compartment, the direction and flux of metabolic conversion are governed by the nature of the enzymes present and the permeability of intracellular membranes to precursors, intermediates and products.

Several types of compartmentation phenomena can be envisioned which could affect the way monoterpene formation is regulated. For instance, a given cellular compartment might not contain the biosynthetic machinery for all branches of the terpenoid pathway, and, therefore, would produce only a limited range of terpenoid products compared to the plant as a whole. Such a compartment might be provisioned with only one type of prenyltransferase, eliminating the need for branch point control at IPP/DMAPP or GPP and permitting the pathway to be regulated at an earlier step, such as HMG-CoA reductase. Multienzyme complexes could also obviate the requirement for branch point control by channeling substrates through a particular sequence of reactions without allowing intermediates to freely diffuse towards enzymes of a competing branch of the pathway.

As another type of compartmentation, the allocation of precursors (acetyl-CoA, ATP, NADPH, etc.) to sites involved in monoterpene biosynthesis could be tightly controlled, enabling the rate of monoterpene formation to be closely regulated. This mode of regulation is discussed in the next section of the chapter.

Clearly, cellular compartmentation could play a critical role in controlling terpenoid production in plants. Unfortunately, there has not yet been any direct investigation of this kind of regulation in connection with monoterpene formation. In this section, after a brief review of current theories on the general subcellular organization of terpenoid metabolism, we critically examine evidence concerning the subcellular location of monoterpene metabolism and consider the possible involvement of multienzyme complexes in monoterpene formation.

Subcellular Compartmentation of Terpenoid Metabolism

Cellular compartmentation is a well-established fact in plant terpenoid biosynthesis. Of the common plant isoprenoids, carotenoids and the side chains of chlorophyll and plastoquinone are made in chloroplasts, ubiquinone is formed in mitochondria, and sterols are produced in the cytoplasm. However, the location of the early steps in the biosynthesis of these compounds, prior to the formation of IPP, is still a matter of considerable controversy.

One view holds that each subcellular site of synthesis (the cytoplasm, the chloroplasts and possibly the mitochondria) is capable of the complete isoprenoid pathway starting from acetyl-CoA. This hypothesis was first advanced in the 1960s,[107] based on *in vivo* tracer studies in which radiolabeled CO_2 was readily incorporated in the light into carotenoids and other chloroplast terpenoids, but not into sterols, which are made in the cytoplasm. In contrast, labeled acetate or mevalonate was extensively incorporated into sterols, but not into carotenoids. The notion that chloroplasts possess a complete terpenoid biosynthetic pathway was not widely accepted at first, but recent studies showing that 1) mevinolin (an HMG-CoA reductase inhibitor) blocks the synthesis of cytoplasmic terpenoids but has almost no effect on the synthesis of plastidial terpenoids,[30] 2) HMG-CoA reductase is present in isolated intact chloroplasts,[32,33,35] and 3) chloroplasts possess all the enzymes necessary to synthesize acetyl-CoA from Calvin cycle intermediates,[108,109] support the autonomy of the chloroplast in terpenoid formation.

An opposing theory on the intracellular organization of terpenoid biosynthesis asserts that all the steps in the pathway prior to IPP are located solely in the cytoplasm. The newly-produced IPP is then transported to various subcellular compartments for the formation of specific terpenoid metabolites. This model was first put forth by Kreuz and Kleinig,[110] based on experiments with spinach extracts which showed that chloroplasts do not incorporate exogenous mevalonate, mevalonate phosphate or mevalonate pyrophosphate into terpenoids, but will incorporate IPP. A cytoplasmic supernatant, on the other hand, incorporated all of these intermediates into newly synthesized terpenes. In further experiments,[111] these investigators found no evidence, in spinach chloroplasts, for any of the enzymes involved in the conversion of HMG-CoA to IPP (HMG-CoA reductase, mevalonate kinase, mevalonate phosphate kinase and mevalonate pyrophosphate decarboxylase). Reports of HMG-CoA reductase localization in chloroplasts[32,33,35] are ascribed simply to cytoplasmic contamination.

Subcellular Sites of Monoterpene Metabolism

Evidence for the subcellular location of monoterpene synthesis has been obtained from several different types of investigations. Numerous ultrastructural studies have examined the secretory structures in which monoterpenes are produced and accumulated. Cells thought to be involved in monoterpene synthesis typically possess an abundance of non-pigmented plastids known as leucoplasts, which have no organized system of thylakoids and few apparent other internal membranes.[112] Leucoplasts have been frequently observed to contain droplets of lipid-like (osmiophilic) material or to be associated with such material at their margins,[2,113] leading to the belief that these organelles are involved in monoterpene biosynthesis (e.g., see reference 114). An extensive survey of the secretory structures of nearly 50 species of higher plants showed a strong correlation between the production of monoterpenes and the presence of leucoplasts in secretory cells.[115] Monoterpene-synthesizing secretory cells also commonly contain a well-developed network of smooth endoplasmic reticulum,[113] so this subcellular compartment might also participate in monoterpene production.

Direct experimental support for the involvement of plastids in monoterpene synthesis comes from work with the rind of young *Citrofortunella mitis* fruits.[75,76] Several membranous fractions from a sucrose density gradient separation of rind extract were able to incorporate IPP into a mixture of monoterpene olefins characteristic of the intact fruit rind. The fraction displaying the highest rate of IPP incorporation was heavily enriched in leucoplasts. Other subcellular fractions, both soluble and particulate, showed little or no incorporation. The ability of leucoplast preparations to synthesize monoterpenes from IPP indicates that, of the enzymes of the monoterpene pathway, these organelles contain at least prenyltransferase and cyclase activities. Since IPP incorporation into monoterpenes required the addition of DMAPP, IPP isomerase activity is apparently not present in these preparations.

Another type of plastid has also been demonstrated to take part in monoterpene biosynthesis. Stromal extracts from the chromoplasts of *Narcissus pseudonarcissus* (daffodil) flowers were able to incorporate IPP into monoterpenes,[116] although the pattern of monoterpenes produced was not the same as that produced by the intact flowers.

The validity of the localization of monoterpene biosynthesis in these studies depends on the purity of the plastid preparations used. The chromoplast fraction from *N. pseudonarcissus* was examined by electron microscopy,[117] and more than 90% of the structures present were found to be intact chromoplasts.[118] The leucoplast fraction from *C. mitis* was analyzed by both ultrastructural and marker enzyme methods. By electron microscopy, this preparation seemed to be made up of double-membraned vesicles which originated from the fragmentation of native leucoplasts during the isolation procedure.[76] Because these organelles were structurally-altered during extraction, assays of marker enzymes are necessary to adequately assess the possibility of contamination. The absence of mitochondria in these preparations was inferred from the lack of fumarase and succinate dehydrogenase activities. However, marker enzymes for the presence of endoplasmic reticulum (e.g., NADH-cytochrome C reductase) or adhering cytoplasmic components (e.g., hexokinase) were not measured. Until more information is available on the purity of the leucoplast fraction from *C. mitis*, the localization of monoterpene biosynthesis in these organelles should be viewed with caution.

In contrast to the work with *C. mitis* and *N. pseudonarcissus*, other investigations have found that monoterpene cyclase activities in cell-free systems are operationally-soluble, and thus presumably found in the cytoplasm or only loosely associated with membranous structures *in situ*. Numerous studies with species of mints,[3] pines[96] and citrus[119] have all shown that, after high speed centrifugation, cyclase activities are present almost exclusively in the supernatant. Prenyltransferase activities associated with monoterpene synthesis also seem to be soluble.[64,65,67,89] Indeed, both leucoplasts and a well-developed endoplasmic reticulum are present in the monoterpene-synthesizing cells of mints, pines and citrus.[115,120-122] However, in extracts made from these cells, cyclase and prenyltransferase activities are not associated with either of these organelles. It might be argued that prenyltransferases and cyclases are in fact located in plastids or the endoplasmic reticulum in intact cells, but are solubilized by conventional extraction methods that rupture these organelles, though not by the techniques used for *C. mitis* and *N. pseudonarcissus*.

Of the enzymes catalyzing secondary transformations, most have also been found to be operationally-soluble, including double bond reductases,[123] isomerases,[123] acetyltransferases[124] and dehydrogenases.[125-130] Only the cytochrome P-450-dependent hydroxylases are particulate, being localized in microsomes (presumed to originate from the endoplasmic reticulum) like other mixed-function oxidases of the cytochrome P-450 type.[13]

Information on the subcellular location of sesquiterpene biosynthesis in plants is also available. A survey of many species showed that sesquiterpene production, unlike monoterpene production, was not correlated with the presence of leucoplasts in secretory cells.[115] Several studies implicate another organelle, the endoplasmic reticulum, as the site of FPP synthase (prenyltransferase) and sesquiterpene cyclase activities. In *Pinus pinaster*, a microsomal pellet incorporated radiolabeled IPP into sesquiterpene olefins[131] and, in *C. mitis*, sesquiterpene biosynthesis from IPP was highest in a subcellular fraction enriched in endoplasmic reticulum membranes.[132] Another investigation purported to demonstrate the involvement of endoplasmic reticulum in sesquiterpene biosynthesis[133] was methodologically-flawed. ^{14}C-Acetate was used to radiolabel sesquiterpenes in the intact primary leaves of *P. pinaster*, and after homogenization of the tissue, subcellular fractions showing the presence of labeled sesquiterpene hydrocarbons were isolated and characterized. Although labeled sesquiterpenes were observed to be associated with endoplasmic reticulum membranes, it seems extremely unwise to use the site of sesquiterpene accumulation after cell disruption as an indication of the site of sesquiterpene synthesis. Indeed, it would not be surprising to find that highly hydrophobic molecules like sesquiterpene hydrocarbons readily bind to certain classes of membranes.

Other studies of sesquiterpene formation have shown that key biosynthetic enzymes are operationally-soluble. In cell-free preparations from two species of the Lamiaceae, *Salvia officinalis*[134,135] and *Pogostemon cablin* (patchouli),[136] sesquiterpene cyclases were located in high-speed supernatants. In addition, soluble extracts of *Artemisia annua* incorporated IPP into the anti-malarial sesquiterpene lactone artemisinin,[137] indicating the presence of IPP isomerase, prenyltransferase and cyclase activities in these preparations.

Several steps of diterpene biosynthesis seem to be localized in plastids. Kaurene synthase, a cyclase participating in gibberellin formation that converts GGPP to kaurene, appears to reside in the chloroplasts of a variety of species.[138,139] In *Ricinus communis* (castor bean) seedlings, four enzymes of the casbene pathway (IPP isomerase, FPP synthase, GGPP synthase and the cyclase, casbene synthase) were found to be restricted to proplastids.[53]

To date, evidence regarding the site of plant monoterpene biosynthesis (and that of related classes of terpenes) does not clearly point to a single subcellular location. It is difficult to reconcile the demonstrated ability of plastids to incorporate IPP into monoterpenes (or the demonstrated ability of the endoplasmic reticulum to incorporate IPP into sesquiterpenes) with the many reports of prenyltransferases and cyclases in soluble extracts. The competing theories on the subcellular organization of general terpenoid metabolism do not help to resolve this problem, since both theories postulate that the steps of terpenoid biosynthesis after the formation of IPP can occur in the plastids or in the cytosol. The compartmentation of monoterpene biosynthesis may well vary from species to species. Alternatively, monoterpenes might be produced in invaginations of the plastid surface filled with cytoplasm (the site of flavonoid biosynthesis in the gland cells of *Populus*[140]), which become bound to the plastid under certain extraction conditions, but are solubilized under others.

The subcellular compartmentation of monoterpene biosynthesis could have profound effects on the ways by which the pathway is regulated. One possibility is that different segments of the pathway occur in separate locations, such as the plastids and the cytosol. In this case, the overall rate of synthesis might be controlled by the transport of intermediates from one site to another, rather than solely by the rate of any individual metabolic conversion. The transport of intermediates could in turn be influenced by changes in membrane permeability.[140] For example, the permeability of the plastid membrane to acetate and mevalonate is substantially altered during the transformation of chloroplasts to chromoplasts in *Capsicum annuum*.[141]

Another scenario is that all of the steps of the monoterpene pathway occur in a compartment specialized for monoterpene biosynthesis. If such a compartment does not produce the large variety of terpenoids made in the plant as a whole, then branch point regulation at the prenyltransferase and/or cyclase level might not be required, and the overall rate of synthesis could be controlled by any of a number of enzymes in the sequence or by the availability of precursors (see next section of this chapter). Based on our current knowledge, the only possible candidates for such specialized monoterpene-synthesizing subcellular compartments are the leucoplasts of *C. mitis*. The biosynthetic capabilities of these organelles have been explored by Gleizes and coworkers.[75,76,142] Preparations of *C. mitis* leucoplasts incorporated IPP into GPP, FPP and GGPP, but apparently did not carry out any further conversions in terpenoid biosynthesis besides the production of monoterpenes. These organelles were unable to produce carotenoids, to esterify chlorophyllides with phytol, or to carry out relevant steps in the biosynthesis of other plastid isoprenoids, such as α-tocopherol and phylloquinone, indicating that the leucoplasts of *C. mitis* may indeed represent a specialized compartment for the formation of monoterpenes. Thus, the production of FPP and GGPP by these structures appears to be anomalous.

Multienzyme Complexes

Enzymes catalyzing sequential steps in a metabolic pathway are sometimes observed to form non-covalent aggregates, called multienzyme complexes, which are capable of channeling the flow of substrates through a linked series of metabolic transformations.[143,144] Channeling of metabolites acts to increase catalytic efficiency by facilitating the transfer of intermediates from one enzyme in the sequence to another. With respect to regulation, multienzyme complexes can help segregate competing pathways, thereby eliminating the need for branch point controls, and they may also permit the activities of a single aggregate to be modulated in unison, enabling metabolic sequences to be controlled more closely.

Several examples of possible multienzyme complexes are known in isoprenoid biosynthesis. Mevalonate kinase, mevalonate 5-phosphate kinase and mevalonate 5-pyrophosphate decarboxylase from the leaves of *Nepeta cataria* co-purify through several chromatographic steps.[32] These activities could form a multienzyme aggregate or, fortuitously, have similar chromatographic properties. Non-covalent associations of several enzymes of carotenoid biosynthesis, including IPP isomerase, GGPP synthase and phytoene synthase, have been described from the chromoplasts of tomato,[145] *Capsicum annuum*[66] and *N. pseudonarcissus*.[146]

At present, there is no evidence for the involvement of multienzyme complexes in monoterpene biosynthesis, although little investigation has been performed in this area. In the camphor pathway of *S. officinalis*, no associations were observed between the cyclase, bornyl pyrophosphate synthase, and either the preceding enzyme, GPP synthase, or the succeeding enzyme, bornyl pyrophosphate hydrolase (R. Croteau, unpublished results). Multienzyme complexes are difficult to detect in higher plants, possibly because they are easily destroyed during conventional extraction procedures.[144] However, the potential significance of these entities in metabolic regulation justifies efforts to look for them.

REGULATION BY ASSIMILATE PARTITIONING

Plant monoterpene biosynthesis can be thought of as starting prior to the mevalonate pathway itself with the formation of the basic precursor acetyl-CoA and the co-factors, ATP and NADPH. These metabolites must be generated at the site of monoterpene synthesis since they are not intercellularly transported. For making these metabolites, monoterpene-producing cells, being non-photosynthetic, require substrate for glycolysis, the TCA cycle and the pentose phosphate pathway. Therefore, the rate of monoterpene biosynthesis could be limited by the availability of assimilated carbon. The partitioning of assimilate among cells thus provides another mechanism for regulating monoterpene production, superimposable on controls at the subcellular and enzyme levels.

Several different lines of evidence suggest that the availability of assimilated carbon may limit the biosynthesis of monoterpenes. Feeding studies with isotopically-labeled precursors have demonstrated that the site of monoterpene biosynthesis is relatively isolated from the mainstream of plant metabolism and, therefore, may not be readily accessible to assimilate. The incorporation of label from exogenous precursors into monoterpenes is notoriously low.[90,147] For most precursors tested, including acetate and mevalonate, incorporation is less than 0.1%.[147,148] Glucose and CO_2 (administered in the light) were the most efficient precursors of monoterpenes in peppermint cuttings,[89,147] indicating that sugars are preferentially transported to the site of synthesis, and that acetate, mevalonate and other intermediates are probably derived from sugars *in situ*, as expected.

Although the sites of biosynthesis of other classes of terpenoids are also relatively inaccessible to exogenous precursors,[68,149] monoterpenes often display lower incorporation rates than do other types of terpenoids when compared in the same tissue. For example, in peppermint cuttings, glucose and CO_2 are incorporated into monoterpenes and sesquiterpenes in proportion to their natural abundances,[89] but mevalonate is a much less efficient precursor of monoterpenes than of sesquiterpenes. Monoterpenes incorporated only one-tenth as much total label from [2-^{14}C]mevalonate as did sesquiterpenes, despite the fact that, in peppermint, monoterpenes are fifty times as abundant as sesquiterpenes.[148] The contrast between monoterpenes and other classes of terpenoids is even greater. Mevalonate incorporation into triterpenes in peppermint leaves was over one-hundred times greater than incorporation into monoterpenes, although the amount of triterpenes and sterols in this tissue is only 2% of the amount of monoterpenes.[150] Differential labeling of monoterpenes and sesquiterpenes has also been observed in *Pinus pinaster*,[151] *P. palustris*[152] and *Poncirus trifoliata*.[153] These findings lend support to the concept that monoterpene biosynthesis occurs in a physiologically-isolated site, relatively inaccessible to most exogenous precursors. They also indicate that each class of terpenoid could be formed in a distinct location, as previously discussed under the heading of cellular compartmentation.

The physiological isolation of monoterpene biosynthesis has a tangible physical basis. Monoterpene formation in plants is restricted to certain cell types located in specialized secretory structures, such as glandular trichomes, resin ducts and secretory cavities. The morphology of these secretory structures suggests that the sites of synthesis are, to some degree, physically separated from the rest of the plant. For example, the glandular trichomes of peppermint and other Lamiaceae project above the leaf surface and are attached to

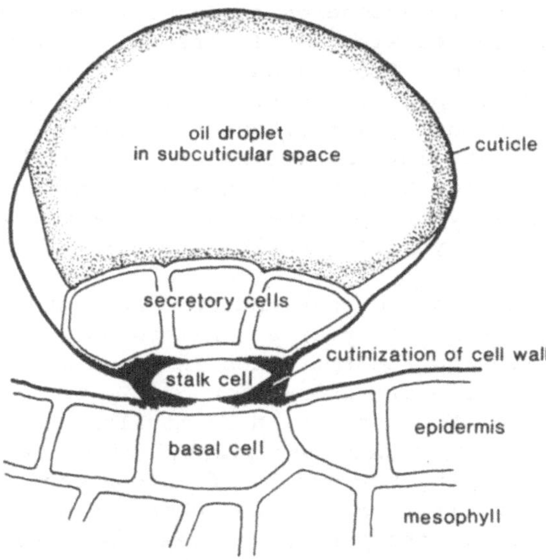

Figure 8. Cross-sectional sketch of a peltate glandular trichome from the leaf surface of *Mentha piperita*. Monoterpenes are presumed to be synthesized in the secretory cells and transported into the subcuticular space for storage. The secretory cells appear isolated, since they are farther away from the rest of the leaf than any other cells of the trichome, and cutinization of the stalk cell walls restricts apoplastic transport to and from these cells.

the epidermal layer by only a single cell (Fig. 8).[120] All exchange of material between the secretory cells, which synthesize the monoterpenes, and the remainder of the leaf must take place through the basal cell and the stalk cell. The walls of the stalk cell are heavily cutinized (Fig. 8), severely restricting apoplastic transport. In fact, the stalk cell is sometimes referred to as a "barrier cell".[113] Thus, the site of monoterpene synthesis in glandular trichomes appears to be physically isolated from the rest of the leaf. Although this arrangement probably arose as an adaptation for sequestering plant defense substances in a location conspicuous to potential herbivores, it may also facilitate the control of metabolic processes in the gland by restricting the transport of assimilate.

Other types of terpenoids produced in glandular trichomes, such as the diterpenes of tobacco,[149] also show poor incorporation of label from exogenous precursors, and may likewise be formed in a remote site. Interestingly, in rose petals, an exceptional plant tissue that readily incorporates exogenous mevalonate into monoterpenes,[154] the site of monoterpene synthesis is located, not in glandular trichomes or resin ducts, but in papillate epidermal cells[155] which do not appear to be morphologically isolated from the rest of the petal.[156]

Additional evidence that monoterpene synthesis is limited by the availability of assimilate comes from investigations of mevalonate incorporation in peppermint cuttings. Co-feeding of sucrose or exposure to 5% CO_2 in the light markedly increased the conversion of mevalonate into monoterpenes.[156] These treatments most likely stimulated monoterpene formation by enhancing the supply of carbohydrate reserves at the site of synthesis. Mevalonate, when fed alone, may be a poor precursor of monoterpenes, because insufficient ATP or NADPH is available to serve as co-factors in monoterpene biosynthesis. Added assimilate (sucrose or the products of CO_2 fixation) provides substrate for the production of these co-factors, permitting monoterpene production to proceed at a faster rate.

In principle, the supply of assimilate used in monoterpene formation could come from either current photosynthate or stored reserves. The rapid incorporation of label into monoterpenes after exposure of plants to $^{14}CO_2$[89,151,157,158] indicates that at least some of the carbon used in monoterpene biosynthesis comes from current photosynthate. Thus, one might expect a close relationship between photosynthesis and monoterpene biosynthesis, even though monoterpene-producing tissue itself is non-photosynthetic.

Experimental manipulations that affect the rate of photosynthesis, such as varying the light regime, do seem to affect monoterpene production. For instance, increasing light intensity enhanced the rate of monoterpene biosynthesis in the primary needles of *Pinus pinaster* seedlings.[159] Sodium carbonate and sodium acetate were the isotopically-labeled precursors used in this investigation. Since, in the dark, no incorporation of these substances occurred, light seems to be essential for monoterpene formation in *P. pinaster*, even from acetate. Increasing light intensity promoted monoterpene accumulation in several members of the Lamiaceae, including *Thymus vulgaris,*[160] *Hedeoma drummondii*[161] and peppermint (*Mentha piperita*).[162,163] However, in *Satureja douglasii,*[164] varying irradiance had no effect on the amount of monoterpenes accumulated per leaf. Under high light intensities, the amount of monoterpenes per leaf dry weight actually declined in this species, since high light caused an increase in leaf dry weight.

In some cases, light clearly seems to be acting through effects on photosynthesis. For example, the action spectrum of the light stimulation of monoterpene biosynthesis in *P. pinaster* seedlings was very similar to that of photosynthesis,[159] and inhibitors of photosynthesis, such as paraquat, reduced monoterpene accumulation in *T. vulgaris.*[165] However, evidence for phytochrome involvement in monoterpene production has also been reported. When etiolated seedlings of *T. vulgaris* were exposed to 30 minutes of red light, there was a 150% increase in monoterpene accumulation, which could be completely nullified by 30 minutes of irradiation with far-red light.[166]

Carbon dioxide concentration, another variable which significantly influences the rate of photosynthesis, also affects monoterpene formation. Peppermint grown under high concentrations of CO_2 was observed to have larger leaves and to accumulate greater amounts of monoterpenes than that grown at low CO_2 concentrations,[167] implying a direct correlation between the rate of photosynthesis and the rate of monoterpene biosynthesis. However, the increase in monoterpenes seen under high CO_2 levels was proportional to the increase in leaf weight, resulting in the lack of any significant differences in monoterpene *concentration* (the amount of monoterpenes per leaf dry weight) among plants grown under different CO_2 conditions. Similar results were obtained from a study on the influence of CO_2 concentration on the level of iridoid-type glycosides in *Plantago lanceolata.*[168] These findings suggest that, under varying CO_2 regimes, the synthesis of monoterpenes *relative to other constituents* is not regulated by the supply of assimilate, at least at the level of the whole plant. Allocation to monoterpenes simply represents a fixed proportion of the total carbohydrate reserves available. However, even under these conditions, assimilate partitioning might still serve as an important control of monoterpene production at the level of the individual organ or tissue.

REGULATION BY MORPHOLOGICAL DIFFERENTIATION

As discussed earlier, monoterpenes in plants usually accumulate in specialized secretory structures, such as the glandular trichomes of the Lamiaceae (the mint family), the resin ducts of conifers and the resin cavities of the Rutaceae (the citrus family).[2] Recent evidence indicates that these secretory structures are also likely to be the primary, if not the exclusive, sites of monoterpene biosynthesis.[69,166,169] Thus, the differentiation of such structures could provide another layer of control over monoterpene formation. This section reviews the evidence for the association of monoterpene biosynthesis with secretory structures, and explores the link between morphological differentiation and monoterpene production, by surveying reports on developmental changes in monoterpene formation and studies of monoterpene production in cell cultures.

For many years, it has been assumed that monoterpene biosynthesis occurs in secretory tissue, since the cells of such tissues display many ultrastructural features indicative of active lipid metabolism.[2,113] However, the difficulties of identifying terpene secretion products microscopically preclude rigorous proof of this assumption using ultrastructural methods.

In mints, biochemical evidence for the capacity of glandular trichomes to synthesize monoterpenes has been obtained from both *in vivo* and *in vitro* investigations. First, it was demonstrated that the isolated epidermis of *Majorana hortensis* (sweet marjoram) leaves, which contains the intact glandular trichomes, possessed the same ability to convert [U-^{14}C] sucrose to monoterpenes as did the whole leaf, whereas the isolated mesophyll was inactive in synthesizing monoterpenes from the same precursor.[169] Thus, monoterpene production could at least be localized in the epidermis. Experiments with *Thymus vulgaris* (garden thyme) seedlings specifically implicated the glandular trichomes as the site of monoterpene biosynthesis, since the rate of monoterpene formation from [U-^{14}C]sucrose in excised cotyledons and hypocotyls was roughly proportional to the abundance of peltate glandular trichomes on the surface of these organs.[166] In addition, cotyledons from which a known percentage of gland heads had been artificially removed with the aid of adhesive tape displayed a commensurate reduction in [U-^{14}C]sucrose incorporation into monoterpenes.

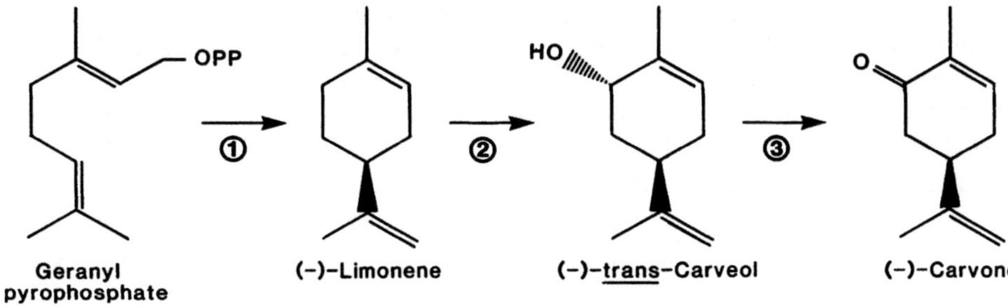

Figure 9. Pathway of (-)-carvone biosynthesis from geranyl pyrophosphate in *Mentha spicata*. Key to enzymes: (1) = geranyl pyrophosphate:(-)-limonene cyclase; (2) = (-)-limonene:*trans*-carveol hydroxylase; (3) = (-)-*trans*-carveol dehydrogenase. From reference 69. Reproduced by permission.

The role of glandular trichomes in the biosynthesis of monoterpenes in *Mentha spicata* (spearmint) leaves was established by *in vitro* studies.[69] Carvone, the principal monoterpene of this species, is biosynthesized from GPP in three steps (Fig. 9). First, GPP is cyclized to the olefin limonene,[170] which is then hydroxylated by a cytochrome P-450-dependent monooxygenase to *trans*-carveol.[171] *trans*-Carveol is subsequently dehydrogenated to carvone. The enzymes catalyzing all three of these transformations were shown to be present in cell-free extracts containing, specifically, the contents of the glandular trichomes (Table 2).[70] Essentially all of the GPP:limonene cyclase and the limonene:*trans*-carveol hydroxylase activities present in the leaf were associated with the glandular trichomes, while only 30% of the *trans*-carveol dehydrogenase activity resided in the glandular trichomes, with the remainder located in the rest of the leaf. Electrophoretic analyses indicated the presence of two *trans*-carveol dehydrogenase species in *M. spicata* leaves, one unique to the glandular trichomes, and the other found in both the glandular trichomes and the rest of the leaf. Since the first two enzymes of the pathway, the cyclase and the hydroxylase, are found exclusively in the glandular trichomes, the biosynthesis of carvone from GPP seems also to be restricted to the glandular trichomes. It can, therefore, be inferred that the *trans*-carveol dehydrogenase activity found throughout the leaf probably utilizes *trans*-carveol only as an adventitious substrate.

Table 2. Localization of enzymes of carvone biosynthesis in *Mentha spicata* leaf extracts. Glandular trichomes were first removed by abrasion of the leaf with glass beads, and the remaining leaves were then homogenized. Enzyme activities in the extracts were assayed by standard methods. Reprinted with permission, from reference 69.

Enzyme	Activity: Glandular trichomes	Activity: Remaining leaf	Activity: Glandular trichomes
	nkat/kg tissue		%
GPP:limonene cyclase	5220	25.0	99.5
Limonene: *trans*-carveol hydroxylase	150	5.6	96.4
trans-Carveol dehydrogenase	8080	18,100	30.9

Monoterpenes in pines and other conifers accumulate in resin ducts found in the leaves and stem. A comparison of ultrastructural and biochemical studies of *Pinus pinaster* needles indicated that monoterpene biosynthesis is almost completely confined to the layer of epithelial cells lining the resin ducts.[151] Electron microscopic analyses showed that the epithelial cells are very active in the synthesis and secretion of lipid material. This activity was found to be transient, and active cells were located primarily near the meristematic region at the base of the needle, whereas epithelial cells found in the upper and middle parts of the needle had degenerated. Subsequent experiments with $^{14}CO_2$ revealed that monoterpene formation occurred predominantly at the base of the needle.

In the Rutaceae, the principal sites of monoterpene accumulation are the resin cavities. To date, there has been no direct proof that resin cavity cells are capable of monoterpene biosynthesis. However, ultrastructural parallels between the resin cavity epithelium and the resin duct epithelium[2,153] strongly suggest that monoterpene biosynthesis occurs in resin cavities of this plant family.

Morphological differentiation appears to dictate not only the location but also the timing of monoterpene biosynthesis. In mints, the rate of monoterpene formation from basic precursors, such as $^{14}CO_2$ and [U-^{14}C]sucrose, is very high in young leaves, but declines precipitously as leaves mature.[93,169] The period of rapid synthesis is correlated with the presence of glandular trichomes with metabolically active secretory cells. Ultrastructural studies have shown that most glandular trichomes are initiated very early in leaf development.[120,122] In peppermint, secretory cells are most active in the synthesis and secretion of lipid material before the leaves are 5 mm long[120] (J. Gershenzon, M. Maffei and R. Croteau, unpublished results). By the time the leaves are nearing full expansion, the secretory cells of most glandular trichomes have begun to degenerate and no longer appear capable of normal metabolic function. Hence, it is not surprising that the rate of monoterpene biosynthesis in mature leaves is very low. Similarly, in resin cavities and resin ducts, monoterpene production seems to occur only when a metabolically active epithelial layer is present.[131,151]

Further evidence for the importance of morphological differentiation in regulating monoterpene biosynthesis comes from studies with plant cell cultures. Undifferentiated callus or suspension cultures derived from monoterpene-producing species usually exhibit no measurable accumulation of monoterpenes.[172-174] However, when differentiated cells are present in the culture, substantial monoterpene accumulation is frequently observed. For example, undifferentiated callus of *Mentha piperita* showed no trace of monoterpene accumulation,[175] while *M. piperita* callus with adventitious shoots, containing leaflets with glandular trichomes, produced significant quantities of monoterpenes.[176,177] The occurrence of complex structures, like glandular trichomes, is not always necessary for monoterpene accumulation,[173] but some degree of differentiation or organization is usually required.

The general absence of monoterpene accumulation in undifferentiated cultures could be due to the lack of significant biosynthetic activity or to the presence of efficient catabolic processes. Monoterpenes formed in undifferentiated cells and secreted into the medium would appear to be much more susceptible to enzymatic degradation than those sequestered in glandular trichomes or resin ducts, since, in culture, plant cells excrete large amounts of hydrolytic and oxidative enzymes into the medium.[178] If cells are unable to store monoterpenes in organized structures, both extra- and intracellular degradation may in fact be critically important in order to avoid the toxic effects of monoterpenes on cell growth and viability.[179] Recent studies have shown that plant cell cultures actively catabolize added monoterpenes[172,180-182] (K. Falk, J. Gershenzon and R. Croteau, unpublished results).

Despite this demonstrated capacity for catabolism, the very low accumulation of monoterpenes in cell culture could be due, at least in part, to a greatly reduced rate of biosynthesis. In experiments with undifferentiated suspension cultures of *Salvia officinalis*, very little evidence for biosynthetic capability was found (K. Falk, J. Gershenzon and R. Croteau, unpublished results). No measurable accumulation of monoterpenes was observed (limits of detection were ~ 1 nmol/g), and the biosynthesis of monoterpenes from [U-^{14}C]sucrose was virtually undetectable. In addition, the activities of specific enzymes of monoterpene biosynthesis, measured *in vitro*, were many orders of magnitude lower than those in the intact plant. Although the ability of these cells to catabolize added monoterpenes was pronounced, the lack of monoterpene accumulation in these cultures seemed principally due to the lack of monoterpene formation.

In summary, the results discussed in this section suggest that specialized secretory structures are a prerequisite for monoterpene synthesis under most conditions. Therefore, morphological differentiation appears to be of crucial importance in monoterpene production. Further attempts to enhance the yield of monoterpenes in cell cultures should provide additional information on the relationship between cell differentiation and monoterpene biosynthesis.

CONCLUSION

The restriction of monoterpene biosynthesis to particular locations and times in the life of the plant implies that monoterpene formation is a tightly regulated process. In this review, we have touched on a wide range of mechanisms which could control monoterpene biosynthesis. Unfortunately, direct evidence bearing on these mechanisms is scant. Researchers to date have generally been more concerned with establishing the basic pathways of monoterpene biosynthesis than with how these pathways are regulated. Thus, we have drawn heavily from the results of biosynthetic studies on other classes of terpenoids, particularly sesquiterpenes and diterpenes, to help understand how monoterpene biosynthesis might be regulated. Our major conclusions can be summarized as follows:

1) Morphological differentiation is of paramount importance in controlling the location and timing of monoterpene biosynthesis. In most plants investigated, monoterpene formation occurs only in the cells of specialized secretory structures, and only during the relatively brief periods when these cells are metabolically active. Studies with callus and suspension cultures have generally confirmed this trend; some degree of differentiation is usually required before significant monoterpene accumulation can be observed.

2) Feeding studies with isotopically-labeled precursors have demonstrated that the site of monoterpene formation is physiologically isolated from the mainstream of plant metabolism, suggesting that monoterpene biosynthesis could be subject to control by the availability of assimilate. In fact, increases in monoterpene production are observed in response to feeding of supplemental sucrose or manipulations that enhance the rate of photosynthesis, supporting the notion that the supply of assimilate can potentially limit monoterpene biosynthesis.

3) Cellular compartmentation could have a profound impact on the way in which monoterpene biosynthesis is regulated. Unfortunately, not enough is yet known about the subcellular location of monoterpene formation to permit much speculation about the regulatory significance of compartmentation phenomena. Several reports implicate the involvement of plastids in monoterpene biosynthesis, at least in the later stages of the pathway.

4) Regulation of monoterpene biosynthesis at the enzyme level is virtually unstudied. However, based on biosynthetic investigations of related classes of terpenes, several enzymes of the pathway seem to be good candidates for catalyzing the rate-limiting step. HMG-CoA reductase could represent the rate-limiting step in monoterpene biosynthesis, as it does in mammalian isoprenoid synthesis. However, the multiple branching of the terpenoid pathway in plants suggests that branch point enzymes, such as the prenyltransferases, may also be important in the regulation of monoterpene production. In addition, monoterpene cyclases, which catalyze the first dedicated step of monoterpene biosynthesis, could also influence the overall rate of monoterpene formation.

Clearly, the rate of monoterpene biosynthesis in plants could be modulated by a complex interplay of different regulatory controls operating at the organ, cellular, subcellular and enzyme levels. However, it must be emphasized again that there is only limited experimental support for most of these mechanisms. In addition, since work to date has focused on only a few of the thousands of plant species that make monoterpenes, it is difficult to predict if the species studied so far are at all representative of monoterpene producing species in general. Increasing awareness of the role of monoterpenes in plant defense and other ecological interactions, and the growing commercial importance of monoterpenes in the flavor, fragrance and pharmaceutical industries, should encourage further investigation of the factors that regulate the biosynthesis of this class of natural products.

Note Added in Proof

A very interesting series of reports by D.R. Light and coworkers has recently appeared (1989, J. Biol. Chem. 264: 18589-18597, 18598-18607, 18608-18617, 18618-18626) describing the purification and properties of a prenyltransferase from the latex of *Hevea brasiliensis*. When assayed with DMAPP and IPP, this enzyme acts as an FPP synthesis catalyzing the *trans* addition of IPP to DMAPP abnd GPP successively. However, in the presence of a protein called rubber elongation factor, which is normally bound to rubber particules, this enzyme alters its stereospecificity and chain-length specificity catalyzing the *cis* addition of IPP to elongating *cis*-polyisoprene rubber molecules. Curiously, purified FPP synthases from yeast and avain liver are also capable of ating as "*cis*-adding" polyprenyltransferase in the presence of rubber elongation factor.

ACKNOWLEDGEMENTS

We thank Nancy Madsen for typing the manuscript, and the National Science Foundation, the U.S. Department of Energy, and the U.S. Department of Agriculture for support.

REFERENCES

1. BANTHORPE, D.V., B.V. CHARLWOOD. 1980. The terpenoids. In: Encyclopedia of Plant Physiology, New Series, Vol. 8, Secondary Plant Products. (E.A. Bell, B.V.

Charlwood, eds.), Springer-Verlag, Berlin, pp. 185-220.
2. FAHN, A. 1979. Secretory Tissues in Plants. Academic Press, London, 302 pp.
3. CROTEAU, R. 1987. Biosynthesis and catabolism of monoterpenoids. Chem. Rev. 87: 929-954.
4. CROTEAU, R. 1988. Catabolism of monoterpenes in essential oil plants. In: Flavors and Fragrances: A World Perspective. (B.M. Lawrence, B.D. Mookherjee and B.J. Willis, eds.), Elsevier, Amsterdam, pp. 65-84.
5. PORTER, J.W., S.L. SPURGEON, eds. 1981. Biosynthesis of Isoprenoid Compounds, Vol. 1. John Wiley and Sons, New York, 558 pp.
6. DUGAN, R.E. 1981. Regulation of HMG-CoA reductase. In: Biosynthesis of Isoprenoid Compounds, Vol. 1. (J.W. Porter, S.L. Spurgeon, eds.), John Wiley and Sons, New York, pp. 95-159.
7. POULTER, C.D., H.C. RILLING. 1981. Prenyltransferases and isomerase. In: Biosynthesis of Isoprenoid Compounds, Vol. 1. (J.W. Porter, S.L. Spurgeon, eds.), John Wiley and Sons, New York, pp. 161-224.
8. CROTEAU, R. 1986. Biosynthesis of cyclic monoterpenes. In: Biogeneration of Aromas, ACS Symposium Series No. 317. (T.H. Parliment, R. Croteau, eds.), American Chemical Society, Washington, D.C., pp. 134-156.
9. GAMBLIEL, H., R. CROTEAU. 1984. Pinene cyclases I and II: Two enzymes from sage (*Salvia officinalis*) which catalyze stereospecific cyclizations of geranyl pyrophosphate to monoterpene olefins of opposite configuration. J. Biol. Chem. 259: 740-748.
10. CROTEAU, R., F. KARP. 1979. Biosynthesis of monoterpenes: Preliminary characterization of bornyl pyrophosphate synthetase from sage (*Salvia officinalis*) and demonstration that geranyl pyrophosphate is the preferred substrate for cyclization. Arch. Biochem. Biophys. 198: 512-522.
11. CROTEAU, R., F. KARP. 1979. Biosynthesis of monoterpenes: Hydrolysis of bornyl pyrophosphate, an essential step in camphor biosynthesis, and hydrolysis of geranyl pyrophosphate, the acyclic precursor of camphor, by enzymes from sage (*Salvia officinalis*). Arch. Biochem. Biophys. 198: 523-532.
12. CROTEAU, R., C.L. HOOPER, M. FELTON. 1978. Biosynthesis of monoterpenes: Partial purification and characterization of a bicyclic monoterpenol dehydrogenase from sage (*Salvia officinalis*). Arch. Biochem. Biophys. 188: 182-193.

13. KARP, F., J.L. HARRIS, R. CROTEAU. 1987. Metabolism of monoterpenes: Demonstration of the hydroxylation of (+)-sabinene to (+)-*cis*-sabinol by an enzyme preparation from sage (*Salvia officinalis*) leaves. Arch. Biochem. Biophys. 256: 179-193.
14. LUCKNER, M. 1984. Secondary Metabolism in Microorganisms, Plants and Animals, 2nd edition. Springer-Verlag, Berlin, 576 pp.
15. TAKEUCHI, A., K. OBA, I. URITANI. 1977. Change in acetyl-CoA synthetase activity of sweet potato in response to infection by *Ceratocystis fimbriata* and injury. Agric. Biol. Chem. 41: 1141-1145.
16. TAKEUCHI, A., M. YAMAGUCHI, I. URITANI. 1981. ATP Citrate lyase from *Ipomoea batatas* root tissue infected with *Ceratocystis fimbriata*. Phytochemistry 20: 1235-1239.
17. SABINE, J.R., ed. 1983. 3-Hydroxy-3-methylglutaryl Coenzyme A Reductase. CRC Press, Boca Raton, Florida, 271 pp.
18. RUSSELL, D.W. 1985. 3-Hydroxy-3-methylglutaryl-CoA reductases from pea seedlings. Meth. Enzymol. 110: 26-40.
19. SUZUKI, H., K. OBA, I. URITANI. 1975. The occurrence and some properties of 3-hydroxy-3-methylglutaryl coenzyme A reductase in sweet potato roots infected by *Ceratocystis fimbriata*. Physiol. Plant Pathol. 7: 265-276.
20. OBA, K., I. URITANI. 1979. Biosynthesis of furano-terpenes by sweet potato cell culture. Plant Cell Physiol. 20: 819-826.
21. OBA, K., K. KONDO, N. DOKE, I. URITANI. 1985. Induction of 3-hydroxy-3-methylglutaryl CoA reductase in potato tubers after slicing, fungal infection or chemical treatment, and some properties of the enzyme. Plant Cell Physiol. 26: 873-880.
22. STERMER, B.A., R.M. BOSTOCK. 1987. Involvement of 3-hydroxy-3-methylglutaryl coenzyme A reductase in the regulation of sesquiterpenoid phytoalexin synthesis in potato. Plant Physiol. 84: 404-408.
23. VOGELI, U., J. CHAPPELL. 1988. Induction of sesquiterpene cyclase and suppression of squalene synthetase activities in plant cell cultures treated with fungal elicitor. Plant Physiol. 88: 1291-1296.
24. CHAPPELL, J., R. NABLE. 1987. Induction of sesquiterpenoid biosynthesis in tobacco cell suspension cultures by fungal elicitor. Plant Physiol. 85: 469-

473.
25. BACH, T.J., H.K. LICHTENTHALER. 1982. Mevinolin: A highly specific inhibitor of microsomal 3-hydroxy-3-methylglutaryl-coenzyme A reductase of radish plants. Z. Naturforsch. 37c: 46-50.
26. CHAPPELL, J., C. VON LANKEN, U. VOGELI, P. BHATT. 1989. Sterol and sesquiterpenoid biosynthesis during a growth cycle of tobacco cell suspension cultures. Plant Cell Reports 8: 48-52.
27. LYNEN, F. 1967. Biosynthetic pathways from acetate to natural products. Pure Appl. Chem. 14: 137-167.
28. HEPPER, C.M., B.G. AUDLEY. 1969. The biosynthesis of rubber from 3-hydroxy-3-methylglutaryl-coenzyme A in *Hevea brasiliensis* latex. Biochem. J. 114: 379-386.
29. WITITSUWANNAKUL, R. 1986. Diurnal variation of 3-hydroxy-3-methylglutaryl coenzyme A reductase activity in latex of *Hevea brasiliensis* and its relation to rubber content. Experientia 42: 44-45.
30. BACH, T.J. 1986. Hydroxymethylglutaryl-CoA reductase, a key enzyme in phytosterol synthesis? Lipids 21: 82-88.
31. NISHI, A., I. TSURITANI. 1983. Effect of auxin on the metabolism of mevalonic acid in suspension-cultured carrot cells. Phytochemistry 22: 399-401.
32. AREBALO, R.E., E.D. MITCHELL. 1984. Cellular distribution of 3-hydroxy-3-methylglutaryl coenzyme A reductase and mevalonate kinase in leaves of *Nepeta cataria*. Phytochemistry 23: 13-18.
33. BROOKER, J.D., D.W. RUSSELL. 1975. Subcellular localization of 3-hydroxy-3-methylglutaryl coenzyme A reductase in *Pisum sativum* seedlings. Arch. Biochem. Biophys. 167: 730-737.
34. CAMARA, B., F. BARDAT, O. DOGBO, J. BRANGEON, R. MONEGER. 1983. Terpenoid metabolism in plastids: Isolation and biochemical characteristics of *Capsicum annuum* chromoplasts. Plant Physiol. 73: 94-99.
35. RAMACHANDRA REDDY, A., V.S.R. DAS. 1986. Partial purification and characterization of 3-hydroxy-3-methylglutaryl coenzyme A reductase from the leaves of guayule (*Parthenium argentatum*). Phytochemistry 25: 2471-2474.
36. SUZUKI, H., I. URITANI. 1976. Subcellular localization of 3-hydroxy-3-methylglutaryl coenzyme A reductase and other membrane-bound enzymes in sweet potato roots. Plant Cell Physiol. 17: 691-700.
37. WONG, R.J., D.K. MC CORMACK, D.W. RUSSELL. 1982. Plastid 3-hydroxy-3-methylglutaryl coenzyme A reductase has

distinctive kinetic and regulatory features: Properties of the enzyme and positive phytochrome control of activity in pea seedlings. Arch. Biochem. Biophys. 216: 631-638.

38. BACH, T.J., H.K. LICHTENTHALER. 1982. Inhibition of mevalonate biosynthesis and of plant growth by the fungal metabolite mevinolin. In: Biochemistry and Metabolism of Plant Lipids. (J.F.G.M. Wintermans, P.J.C. Kuiper, eds.), Elsevier Biomedical Press, Amsterdam, pp. 515-521.
39. BACH, T.J. 1987. Synthesis and metabolism of mevalonic acid in plants. Plant Physiol. Biochem. 25: 163-178.
40. ITO, R., K. OBA, I. URITANI. 1979. Mechanism for the induction of 3-hydroxy-3-methylglutaryl coenzyme A reductase in $HgCl_2$-treated sweet potato root tissue. Plant Cell Physiol. 20: 867-874.
41. OBA, K., H. TATEMATSU, K. YAMASHITA, I. URITANI. 1976. Induction of furano-terpene production and formation of the enzyme system from mevalonate to isopentenyl pyrophosphate in sweet potato root tissue injured by *Ceratocystis fimbriata* and by toxic chemicals. Plant Physiol. 58: 51-56.
42. NARITA, J.O., W. GRUISSEM. 1989. Tomato hydroxymethyl-glutaryl-CoA reductase is required early in fruit development but not during ripening. Plant Cell 1: 181-190.
43. SIPAT, A.B. 1982. Hydroxymethylglutaryl CoA reductase (NADPH) in the latex of *Hevea brasiliensis*. Phytochemistry 21: 2613-2618.
44. SIPAT, A. 1985. 3-Hydroxy-3-methylglutaryl-CoA reductase in the latex of *Hevea brasiliensis*. Meth. Enzymol. 110: 40-51.
45. BACH, T.J., D.H. ROGERS, H. RUDNEY. 1986. Detergent-solubilization, purification, and characterization of membrane-bound 3-hydroxy-3-methylglutaryl-coenzyme A reductase from radish seedlings. Eur. J. Biochem. 154: 103-111.
46. BROOKER, J.D., D.W. RUSSELL. 1975. Properties of microsomal 3-hydroxy-3-methylglutaryl coenzyme A reductase from *Pisum sativum* seedlings. Arch. Biochem. Biophys. 167: 723-729.
47. BACH, T.J., H.K. LICHTENTHALER. 1984. Application of modified Lineweaver-Burk plots to studies of kinetics and regulation of radish 3-hydroxy-3-methylglutaryl-CoA reductase. Biochim. Biophys. Acta. 794: 152-161.
48. CLEGG, R.J., B. MIDDLETON, G.D. BELL, D.A. WHITE. 1980.

Inhibition of hepatic cholesterol synthesis and *S*-3-hydroxy-3-methylglutaryl-CoA reductase by mono and bicyclic monoterpenes administered *in vivo*. Biochem. Pharmacol. 29: 2125-2127.

49. QURESHI, A.A., W.R. MANGELS, Z.Z. DIN, C.E. ELSON. 1988. Inhibition of hepatic mevalonate biosynthesis by the monoterpene, *d*-limonene. J. Agric. Food Chem. 36: 1220-1224.
50. ELSON, C.E., G.L. UNDERBAKKE, P. HANSON, E. SHRAGO, R.H. WAINBERG, A.A. QURESHI. 1989. Impact of lemongrass oil, an essential oil, on serum cholesterol. Lipids 24: 677-679.
51. ISA, R.B.M., A.B. SIPAT. 1982. 3-Hydroxy-3-methylglutaryl CoA reductase of *Hevea* latex: The occurrence of a heat-stable activator in the C-serum. Biochem. Biophys. Res. Commun. 108: 206-212.
52. LEARNED, R.M., G.R. FINK. 1989. 3-Hydroxy-3-methylglutaryl-coenzyme A reductase from *Arabidopsis thaliana* is structurally distinct from the yeast and animal enzymes. Proc. Natl. Acad. Sci. USA 86: 2779-2783.
53. DUDLEY, M.W., M.T. DUEBER, C.A. WEST. 1986. Biosynthesis of the macrocyclic diterpene casbene in castor bean (*Ricinus communis* L.) seedlings: Changes in enzyme levels induced by fungal infection and intracellular localization of the pathway. Plant Physiol. 81: 335-342.
54. ATKINSON, D.E. 1969. Regulation of enzyme function. Ann. Rev. Microbiol. 23: 47-68.
55. KNOTZ, J., R.C. COOLBAUGH, C.A. WEST. 1977. Regulation of the biosynthesis of *ent*-kaurene from mevalonate in the endosperm of immature *Marah macrocarpus* seeds by adenylate energy charge. Plant Physiol. 60: 81-85.
56. SKILLETER, D.N., R.G.O. KEKWICK. 1971. The enzymes forming isopentenyl pyrophosphate from 5-phosphomevalonate (mevalonate 5-phosphate) in the latex of *Hevea brasiliensis*. Biochem. J. 124: 407-417.
57. OGURA, K., T. KOYAMA, T. SHIBUYA, T. NISHINO, S. SETO. 1969. Inhibitory effect of substrate analogs on isopentenyl pyrophosphate isomerase and prenyltransferase. J. Biochem. 66: 117-118.
58. LANDAU, B.R., H. BRUNEGRABER. 1985. Shunt pathway of mevalonate metabolism. Meth. Enzymol. 110: 100-114.
59. NES, W.D., T.J. BACH. 1985. Evidence for a mevalonate shunt in a tracheophyte. Proc. R. Soc. Lond. B 225: 425-444.

60. SKRUKRUD, C.L., S.E. TAYLOR, D.R. HAWKINS, E.K. NEMETHY, M. CALVIN. 1988. Subcellular fractionation of triterpenoid biosynthesis in *Euphorbia lathyris* latex. Physiol. Plant. 74: 306-316.
61. WEST, C.A., M.W. DUDLEY, M.T. DUEBER. 1979. Regulation of terpenoid biosynthesis in higher plants. Rec. Adv. Phytochem. 13: 163-198.
62. DUDLEY, M.W., T.R. GREEN, C.A. WEST. 1986. Biosynthesis of the macrocyclic diterpene casbene in castor bean (*Ricinus communis* L.) seedlings: The purification and properties of farnesyl transferase from elicited seedlings. Plant Physiol. 81: 343-348.
63. OGURA, K., T. SHINKA, S. SETO. 1972. The purification and properties of geranylgeranyl pyrophosphate synthetase from pumpkin fruit. J. Biochem. 72: 1101-1108.
64. DE LA FUENTE, M., L.M. PEREZ, U. HASHAGEN, L. CHAYET, C. ROJAS, G. PORTILLA, O. CORI. 1981. Prenyltransferases from the flavedo of *Citrus sinensis*. Phytochemistry 20: 1551-1557.
65. PEREZ, L.M., R. LOZADA, O. CORI. 1983. Biosynthesis of allylic isoprenoid pyrophosphates by an enzyme preparation from the flavedo of *Citrus paradisii*. Phytochemistry 22: 431-433.
66. DOGBO, O., B. CAMARA. 1987. Purification of isopentenyl pyrophosphate isomerase and geranylgeranyl pyrophosphate synthase from *Capsicum* chromoplasts by affinity chromatography. Biochim. Biophys. Acta 920: 140-148.
67. CROTEAU, R., P.T. PURKETT. 1989. Geranyl pyrophosphate synthase: Characterization of the enzyme and evidence that this chain-length specific prenyltransferase is associated with monoterpene biosynthesis in sage (*Salvia officinalis*). Arch. Biochem. Biophys. 271: 524-535.
68. CROTEAU, R., M.A. JOHNSON. 1984. Biosynthesis of terpenoids in glandular trichomes. In: Biology and Chemistry of Plant Trichomes. (E. Rodriguez, P.L. Healy, I. Mehta, eds.), Plenum Publishing Co., New York, pp. 133-185.
69. GERSHENZON, J., M. MAFFEI, R. CROTEAU. 1989. Biochemical and histochemical localization of monoterpene biosynthesis in the glandular trichomes of spearmint (*Mentha spicata*). Plant Physiol. 89: 1351-1357.
70. GERSHENZON, J., M.A. DUFFY, F. KARP, R. CROTEAU. 1987. Mechanized techniques for the selective extraction of

enzymes from plant epidermal glands. Anal. Biochem. 163: 159-164.
71. WIDMAIER, R., J. HOWE, P. HEINSTEIN. 1980. Prenyltransferase from *Gossypium hirsutum*. Arch. Biochem. Biophys. 200: 609-616.
72. SAGAMI, H., K. OGURA, S. SETO, T. KUROKAWA. 1978. A new prenyltransferase from *Micrococcus lysodeikticus*. Biochem. Biophys. Res. Comm. 85: 572-578.
73. SAGAMI, H., K. OGURA. 1981. Geranylgeranyl pyrophosphate synthetase lacking geranyl-transferring activity from *Micrococcus luteus*. J. Biochem. 89: 1573-1580.
74. HEIDE, L., U. BERGER. 1989. Partial purification and properties of geranyl pyrophosphate synthase from *Lithospermum erythrorhizon* cell cultures. Arch. Biochem. Biophys. 273: 331-338.
75. PAULY, B., L. BELINGHERI, A. MARPEAU, M. GLEIZES. 1986. Monoterpene formation by leucoplasts of *Citrofortunella mitis* and *Citrus unshiu*: Steps and conditions of biosynthesis. Plant Cell Rep. 5: 19-22.
76. GLEIZES, M., G. PAULY, J.-P. CARDE, A. MARPEAU, C. BERNARD-DAGAN. 1983. Monoterpene hydrocarbon biosynthesis by isolated leucoplasts of *Citrofortunella mitis*. Planta 159: 373-381.
77. HEIDE, L. 1988. Geranylpyrophosphate synthase from cell cultures of *Lithospermum erythrorhizon*. FEBS Lett. 237: 159-162.
78. BRUENGER, E., H.C. RILLING. 1988. Determination of isopentenyl diphosphate and farnesyl diphosphate in tissue samples with a comment on secondary regulation of polyisoprenoid biosynthesis. Anal. Biochem. 173: 321-327.
79. BANTHORPE, D.V., D.R.S. LONG, C.R. PINK. 1983. Biosynthesis of geraniol and related monoterpenes in *Pelargonium graveolens*. Phytochemistry 22: 2459-2463.
80. GREEN, T.R., D.J. BAISTED. 1971. Development of the squalene-synthesizing system during early stages of pea seed germination. Biochem. J. 125: 1145-1147.
81. GREEN, T.R., D.J. BAISTED. 1972. Development of the activities of enzymes of the isoprenoid pathway during early stages of pea-seed germination. Biochem. J. 130: 983-995.
82. BENEDICT, C.R. 1983. Biosynthesis of rubber. In: Biosynthesis of Isoprenoid Compounds, Vol. 2. (J.W. Porter, S.L. Spurgeon, eds.), John Wiley and Sons, New York, pp. 355-369.
83. RAMACHANDRA REDDY, A., V.S.R. DAS. 1988. Enhanced rubber

accumulation and rubber transferase activity in guayule under water stress. J. Plant Physiol. 133: 152-155.

84. MADHAVAN, S., G.A. GREENBLATT, M.A. FOSTER, C.R. BENEDICT. 1989. Stimulation of isopentenyl pyrophosphate incorporation into polyisoprene in extracts from guayule plants (*Parthenium argentatum* Gray) by low temperature and 2-(3,4-dichlorophenoxy)triethylamine. Plant Physiol. 89: 506-511.

85. ARCHER, B.L., B.G. AUDLEY. 1987. New aspects of rubber biosynthesis. Bot. J. Linn. Soc. 94: 181-196.

86. LEUBE, J., H. GRISEBACH. 1983. Further studies on induction of enzymes of phytoalexin synthesis in soybean and cultured soybean cells. Z. Naturforsch. 38c: 730-735.

87. TIETJEN, K.G., U. MATERN. 1983. Differential response of cultured parsley cells to elicitors from two non-pathogenic strains of fungi. 2. Effects on enzyme activities. Eur. J. Biochem. 131: 409-413.

88. HEIDE, L., M. TABATA. 1987. Geranylpyrophosphate:*p*-hydroxybenzoate geranyltransferase activity in extracts of *Lithospermum erythrorhizon* cell cultures. Phytochemistry 26: 1651-1655.

89. CROTEAU, R., A.J. BURBOTT, W.D. LOOMIS. 1972. Biosynthesis of mono- and sesquiterpenes in peppermint from glucose-^{14}C and $^{14}CO_2$. Phytochemistry 11: 2459-2467.

90. CHARLWOOD, B.V., D.V. BANTHORPE. 1978. The biosynthesis of monoterpenes. In: Progress in Phytochemistry, Vol. 5. (L. Reinhold, J.B. Harborne, T. Swain, eds.), Pergamon Press, Oxford, pp. 65-125.

91. SHAH, D.H., W.W. CLELAND, J.W. PORTER. 1965. The partial purification, properties, and mechanisms of action of pig liver isopentenyl pyrophosphate isomerase. J. Biol. Chem. 240: 1946-1956.

92. STEIGER, A., U. MITZKA-SCHNABEL, W. RAU, J. SOLL, W. RUDIGER. 1985. Inhibition of geranylgeranyl diphosphate synthesis in *in vitro* systems. Phytochemistry 24: 739-743.

93. CROTEAU, R., M. FELTON, F. KARP, R. KJONAAS. 1981. Relationship of camphor biosynthesis to leaf development in sage (*Salvia officinalis*). Plant Physiol. 67: 820-824.

94. RAFFA, K.F., A.A. BERRYMAN. 1983. Physiological aspects of lodgepole pine wound responses to a fungal symbiont of the mountain pine beetle, *Dendroctonus ponderosae* (Coleoptera: Scolytidae). Can. Ent. 115: 723-734.

95. MILLER, R.H., A.A. BERRYMAN, C.A. RYAN. 1986. Biotic elicitors of defense reactions in lodgepole pine. Phytochemistry 25: 611-612.
96. CROTEAU, R., S. GURKEWITZ, M.A. JOHNSON, H.J. FISK. 1987. Biochemistry of oleoresinosis: Monoterpene and diterpene biosynthesis in lodgepole pine saplings infected with *Ceratocystis clavigera* or treated with carbohydrate elicitors. Plant Physiol. 85: 1123-1128.
97. EL-KELTAWI, N.E., R. CROTEAU. 1987. Influence of foliar applied cytokinins on growth and essential oil content of several members of the Lamiaceae. Phytochemistry 26: 891-895.
98. MAURER, K.-H., D. MECKE. 1986. Regulation of enzymes involved in the biosynthesis of the sesquiterpene antibiotic pentalenolactone in *Streptomyces arenae*. J. Antibiotics 39: 266-271.
99. THRELFALL, D.R., I.M. WHITEHEAD. 1988. Co-ordinated inhibition of squalene synthetase and induction of enzymes of sesquiterpenoid phytoalexin biosynthesis in cultures of *Nicotiana tabacum*. Phytochemistry 27: 2567-2580.
100. BRINDLE, P.A., P.J. KUHN, D.R. THRELFALL. 1988. Biosynthesis and metabolism of sesquiterpenoid phytoalexins and triterpenoids in potato cell suspension cultures. Phytochemistry 27: 133-150.
101. MOESTA, P., C.A. WEST. 1985. Casbene synthetase: Regulation of phytoalexin biosynthesis in *Ricinus communis* L. seedlings. Arch. Biochem. Biophys. 238: 325-333.
102. CANE, D.E., C. PARGELLIS. 1987. Partial purification and characterization of pentalenene synthase. Arch. Biochem. Biophys. 254: 421-429.
103. CROTEAU, R., F. KARP. 1976. Biosynthesis of monoterpenes: Enzymatic conversion of neryl pyrophosphate to 1,8-cineole, a-terpineol, and cyclic monoterpene hydrocarbons by a cell-free preparation from sage (*Salvia officinalis*). Arch. Biochem. Biophys. 176: 734-746.
104. CROTEAU, R., F. KARP. 1977. Biosynthesis of monoterpenes: Partial purification and characterization of 1,8-cineole synthetase from *Salvia officinalis*. Arch. Biochem. Biophys. 179: 257-265.
105. POULOSE, A.J., R. CROTEAU. 1978. γ-Terpinene synthetase: A key enzyme in the biosynthesis of aromatic monoterpenes. Arch. Biochem. Biophys. 191: 400-411.

106. WHITEHEAD, I.M., D.R. THRELFALL, D.F. EWING. 1989. 5-*epi*-Aristolochene is a common precursor of the sesquiterpenoid phytoalexins capsidiol and debneyol. Phytochemistry 28: 775-779.

107. ROGERS, L.J., S.P.J. SHAH, T.W. GOODWIN. 1968. Compartmentation of biosynthesis of terpenoids in green plants. Photosynthetica 2: 184-207.

108. RAMACHANDRA REDDY, A., V.S.R. DAS. 1987. Chloroplast autonomy for the biosynthesis of isopentenyl diphosphate in guayule (*Parthenium argentatum* Gray). New Phytol. 106: 457-464.

109. SCHULZE-SIEBERT, D., A. HEINTZE, G. SCHULTZ. 1987. Substrate flow from photosynthetic carbon metabolism to chloroplast isoprenoid synthesis in spinach: Evidence for a plastidic phosphoglycerate mutase. Z. Naturforsch. 42c: 570-580.

110. KREUZ, K., H. KLEINIG. 1981. On the compartmentation of isopentenyl diphosphate synthesis and utilization in plant cells. Planta 153: 578-581.

111. KREUZ, K., H. KLEINIG. 1984. Synthesis of prenyl lipids in cells of spinach leaf: Compartmentation of enzymes for formation of isopentenyl diphosphate. Eur. J. Biochem. 141: 531-535.

112. CARDE, J.-P. 1984. Leucoplasts: A distinct kind of organelles lacking typical 70S ribosomes and free thylakoids. Eur. J. Cell Biol. 34: 18-26.

113. SCHNEPF, E. 1974. Gland cells. In: Dynamic Aspects of Plant Ultrastructure. (A.W. Roberts, ed.), McGraw-Hill, New York, pp. 331-357.

114. BOSABALIDIS, A., I. TSEKOS. 1982. Ultrastructural studies on the secretory cavities of *Citrus deliciosa* Ten. I. Early stages of gland cell differentiation. Protoplasma 112: 55-62.

115. CHENICLET, C., J.-P. CARDE. 1985. Presence of leucoplasts in secretory cells and of monoterpenes in the essential oil: A correlative study. Israel J. Bot. 34: 219-238.

116. METTAL, U., W. BOLAND, P. BEYER, H. KLEINIG. 1988. Biosynthesis of monoterpene hydrocarbons by isolated chromoplasts from daffodil flowers. Eur. J. Biochem. 170: 613-616.

117. FALK, H., B. LIEDVOGEL, P. SITTE. 1974. Circular DNA in isolated chromoplasts. Z. Naturforsch. 29c: 541-544.

118. LIEDVOGEL, B., P. SITTE, H. FALK. 1976. Chromoplasts in the daffodil: Fine structure and chemistry. Cytobiologie 12: 155-174.

119. CHAYET, L., C. ROJAS, E. CARDEMIL, A.M. JABALQUINTO, R. VICUNA, O. CORI. 1977. Biosynthesis of monoterpene hydrocarbons from [1-^{3}H]neryl pyrophosphate and [1-^{3}H]geranyl pyrophosphate by soluble enzymes from *Citrus limonum*. Arch. Biochem. Biophys. 180: 318-327.
120. AMELUNXEN, F. 1965. Electronenmikroskopisch Untersuchungen an den Drusenschuppen von *Mentha piperita* L. Planta Med. 13: 457-473.
121. WOODING, F.B.P., D.H. NORTHCOTE. 1965. The fine structure of the mature resin cells of *Pinus pinea*. J. Ultrastr. Res. 13: 233-244.
122. BOSABALIDIS, A., I. TSEKOS. 1982. Glandular scale development and essential oil secretion in *Origanum dictamnus* L. Planta 156: 496-504.
123. CROTEAU, R., K.V. VENKATACHALAM. 1986. Metabolism of monoterpenes: Demonstration that (+)-*cis*-isopulegone, not piperitenone, is the key intermediate in the conversion of (-)-isopiperitenone to (+)-pulegone in peppermint (*Mentha piperita*). Arch. Biochem. Biophys. 249: 306-315.
124. CROTEAU, R., C.L. HOOPER. 1978. Metabolism of monoterpenes: Acetylation of (-)-menthol by a soluble enzyme preparation from peppermint (*Mentha piperita*) leaves. Plant Physiol. 61: 737-742.
125. POTTY, V.H., J.H. BRUEMMER. 1970. Oxidation of geraniol by an enzyme system from orange. Phytochemistry 9: 1003-1007.
126. KJONAAS, R., C. MARTINKUS-TAYLOR, R. CROTEAU. 1982. Metabolism of monoterpenes: Conversion of *l*-menthone to *l*-menthol and *d*-neomenthol by stereospecific dehydrogenases from peppermint (*Mentha piperita*) leaves. Plant Physiol. 69: 1013-1017.
127. KJONAAS, R.B., K.V. VENKATACHALAM, R. CROTEAU. 1985. Metabolism of monoterpenes: Oxidation of isopiperitenol to isopiperitenone, and subsequent isomerization to piperitenone by soluble enzyme preparations from peppermint (*Mentha piperita*) leaves. Arch. Biochem. Biophys. 238: 49-60.
128. DEHAL, S.S., R. CROTEAU. 1987. Metabolism of monoterpenes: Specificity of the dehydrogenases responsible for the biosynthesis of camphor, 3-thujone, and 3-isothujone. Arch. Biochem. Biophys. 258: 287-291.
129. CROTEAU, R., C.L. HOOPER, M. FELTON. 1978. Biosynthesis of monoterpenes: Partial purification and characterization of bicyclic monoterpenol dehydrogenase

from sage (*Salvia officinalis*). Arch. Biochem. Biophys. 188: 182-193.

130. CROTEAU, R., N.M. FELTON. 1980. Substrate specificity of monoterpenol dehydrogenases from *Foeniculum vulgare* and *Tanacetum vulgare*. Phytochemistry 19: 1343-1347.

131. BERNARD-DAGAN, C., G. PAULY, A. MARPEAU, M. GLEIZES, J.-P. CARDE, P. BARADAT. 1982. Control and compartmentation of terpene biosynthesis in leaves of *Pinus pinaster*. Physiol. Veg. 20: 775-795.

132. BELINGHERI, L., G. PAULY, M. GLEIZES, A. MARPEAU. 1988. Isolation by an aqueous two-polymer phase system and identification of endomembranes from *Citrofortunella mitis* fruits for sesquiterpene hydrocarbon synthesis. J. Plant Physiol. 132: 80-85.

133. GLEIZES, M., J.-P. CARDE, G. PAULY, C. BERNARD-DAGAN. 1980. *In vivo* formation of sesquiterpene hydrocarbons in the endoplasmic reticulum of pine. Plant Sci. Lett. 20: 79-90.

134. CROTEAU, R., A. GUNDY. 1984. Cyclization of farnesyl pyrophosphate to the sesquiterpene olefins humulene and caryophyllene by an enzyme system from sage (*Salvia officinalis*). Arch. Biochem. Biophys. 233: 838-841.

135. DEHAL, S.S., R. CROTEAU. 1988. Partial purification and characterization of two sesquiterpene cyclases from sage (*Salvia officinalis*) which catalyze the respective conversion of farnesyl pyrophosphate to humulene and caryophyllene. Arch. Biochem. Biophys. 261: 346-356.

136. CROTEAU, R., S.L. MUNCK, C.C. AKOH, H.J. FISK, D.M. SATTERWHITE. 1987. Biosynthesis of the sesquiterpene patchoulol from farnesyl pyrophosphate in leaf extracts of *Pogostemon cablin* (patchouli): Mechanistic considerations. Arch. Biochem. Biophys. 256: 56-68.

137. KUDAKASSERIL, G.J., L. LAM, E.J. STABA. 1987. Effect of sterol inhibitors on the incorporation of ^{14}C-isopentenyl pyrophosphate into artemisinin by a cell-free system from *Artemisia annua* tissue cultures and plants. Planta Med. 280-284.

138. SIMCOX, P.D., D.T. DENNIS, C.A. WEST. 1975. Kaurene synthetase from plastids of developing plant tissues. Biochem. Biophys. Res. Comm. 66: 166-172.

139. RAILTON, I.D., B. FELLOWS, C.A. WEST. 1984. *ent*-Kaurene synthesis in chloroplasts from higher plants. Phytochemistry 23: 1261-1267.

140. LUCKNER, M., B. DIETTRICH, W. LERBS. 1980. Cellular compartmentation and channeling of secondary metabolism in microorganisms and higher plants. In: Progress in

Phytochemistry, Vol. 6. (L. Reinhold, J.B. Harborne, T. Swain, eds.), Pergamon Press, Oxford, pp. 103-142.
141. SCHNEIDER, M.M., R. HAMPP, H. ZIEGLER. 1977. Envelope permeability to possible precursors of carotenoid biosynthesis during chloroplast-chromoplast transformation. Plant Physiol. 60: 518-520.
142. GLEIZES, M., B. CAMARA, J. WALTER. 1987. Some characteristics of terpenoid biosynthesis by leucoplasts of *Citrofortunella mitis*. Planta 170: 138-140.
143. REED, L.J., D.J. COX. 1970. Multienzyme complexes. In: The Enzymes: Structure and Control, Vol. 1. (P.D. Boyer, ed.), Academic Press, New York, pp. 213-240.
144. STAFFORD, H.A. 1981. Compartmentation in natural product biosynthesis by multienzyme complexes. In: The Biochemistry of Plants, Vol. 7, Secondary Plant Products. (E.E. Conn, ed.), Academic Press, New York, pp. 117-137.
145. MAUDINAS, B., M.L. BUCHOLTZ, C. PAPASTEPHANOU, S.S. KATIYAR, A.V. BRIEDIS, J.W. PORTER. 1977. The partial purification and properties of a phytoene synthesizing enzyme system. Arch. Biochem. Biophys. 180: 354-362.
146. LUTZOW, M., P. BEYER. 1988. The isopentenyl-diphosphate Δ-isomerase and its relation to the phytoene synthase complex in daffodil chromoplasts. Biochim. Biophys. Acta 959: 118-126.
147. LOOMIS, W.D., R. CROTEAU. 1973. Biochemistry and physiology of lower terpenoids. Rec. Adv. Phytochem. 6: 147-186.
148. CROTEAU, R., W.D. LOOMIS. 1972. Biosynthesis of mono- and sesquiterpenes in peppermint from mevalonate-2-^{14}C. Phytochemistry 11: 1055-1066.
149. KEENE, C.K., G.J. WAGNER. 1985. Direct demonstration of duvatrienediol biosynthesis in glandular heads of tobacco trichomes. Plant Physiol. 79: 1026-1032.
150. CROTEAU, R., W.D. LOOMIS. 1973. Biosynthesis of squalene and other triterpenes in *Mentha piperita* from mevalonate-2-^{14}C. Phytochemistry 12: 1957-1965.
151. BERNARD-DAGAN, C., J.P. CARDE, M. GLEIZES. 1979. Etude des composes terpiniques au cours de la croissance des aiguilles du Pin maritime: comparaison de donnees biochimiques et ultrastructurales. Can. J. Bot. 57: 255-263.
152. BANTHORPE, D.V., O. EKUNDAYO. 1976. Biosynthesis of (+)-car-3-ene in *Pinus* species. Phytochemistry 15: 109-112.

153. HEINRICH, G., W. SCHULTZE, I. PFAB, M. BOTTGER. 1983. The site of essential oil biosynthesis in *Poncirus trifoliata* and *Monarda fistulosa*. Physiol. Veg. 21: 257-268.
154. FRANCIS, M.J.O., M. O'CONNELL. 1969. The incorporation of mevalonic acid into rose petal monoterpenes. Phytochemistry 8: 1705-1708.
155. STUBBS, J.M., M.J.O. FRANCIS. 1971. Electron microscopical studies of rose petal cells during flower maturation. Planta Med. 20: 211-218.
156. CROTEAU, R., A.J. BURBOTT, W.D. LOOMIS. 1972. Apparent energy deficiency in mono- and sesquiterpene biosynthesis in peppermint. Phytochemistry 11: 2937-2948.
157. MIHALIAK, C.A., D.E. LINCOLN. 1989. Changes in leaf mono- and sesquiterpene metabolism with nitrate availability and leaf age in *Heterotheca subaxillaris*. J. Chem. Ecol. 15: 1579-1588.
158. BURBOTT, A.J., W.D. LOOMIS. 1969. Evidence for metabolic turnover of monoterpenes in peppermint. Plant Physiol. 44: 173-179.
159. GLEIZES, M., G. PAULY, C. BERNARD-DAGAN, R. JACQUES. 1980. Effects of light on terpene hydrocarbon synthesis in *Pinus pinaster*. Physiol. Plant. 50: 16-20.
160. YAMAURA, T., S. TANAKA, M. TABATA. 1989. Light-dependent formation of glandular trichomes and monoterpenes in thyme seedlings. Phytochemistry 28: 741-744.
161. FIRMAGE, D.H. 1981. Environmental influences on the monoterpene variation in *Hedeoma drummondii*. Biochem. Syst. Ecol. 9: 53-58.
162. BURBOTT, A.J., W.D. LOOMIS. 1967. Effects of light and temperature on the monoterpenes of peppermint. Plant Physiol. 42: 20-28.
163. HORNOK, L. 1978. [cited in BERNATH, J. 1986. Production ecology of secondary plant products. In: Herbs, Spices and Medicinal Plants: Recent Advances in Botany, Horticulture and Pharmacology, Vol. 1. (L.E. Craker, J.E. Simon, eds.), Oryx Press, Phoenix, Arizona, pp. 185-234.]
164. LINCOLN, D.E., J.H. LANGENHEIM. 1978. Effect of light and temperature on monoterpenoid yield and composition in *Satureja douglasii*. Biochem. Syst. Ecol. 6: 21-32.
165. TANAKA, S., T. YAMAURA, M. TABATA. 1987. Localization and biosynthetic regulation of monoterpenoids in thyme

seedlings. Abstract, Bioflavour '87, International Conference, University of Wurzburg, West Germany, September 29-30, 1987.

166. TANAKA, S., T. YAMAURA, M. TABATA. 1988. Localization and photoregulation of monoterpenoid biosynthesis in thyme seedlings. In: Bioflavour '87: Analysis, Biochemistry, Biotechnology. (P. Schreier, ed.), Walter de Gruyter, Berlin, pp. 237-241.

167. LINCOLN, D.E., D. COUVET. 1989. The effect of carbon supply on allocation to allelochemicals and caterpillar consumption of peppermint. Oecologia 78: 112-114.

168. FAJER, E.D., M.D. BOWERS, F.A. BAZZAZ. 1989. The effects of enriched carbon dioxide atmospheres on plant-insect herbivore interactions. Science 243: 1198-1200.

169. CROTEAU, R. 1977. Site of monoterpene biosynthesis in *Majorana hortensis* leaves. Plant Physiol. 59: 519-520.

170. KJONAAS, R., R. CROTEAU. 1983. Demonstration that limonene is the first cyclic intermediate in the biosynthesis of oxygenated *p*-menthane monoterpenes in *Mentha piperita* and other *Mentha* species. Arch. Biochem. Biophys. 220: 79-89.

171. KARP, F., C.A. MIHALIAK, J.L. HARRIS, R. CROTEAU. 1990. Monoterpene biosynthesis: Specificity of the hydroxylations of (-)-limonene by enzyme preparations from peppermint (*Mentha piperita*), spearmint (*Mentha spicata*) and perilla (*Perilla frutescens*) leaves. Arch. Biochem. Biophys. (in press).

172. BANTHORPE, D.V. 1988. Monoterpenes and sesquiterpenes. In: Cell Culture and Somatic Cell Genetics, Vol. 5, Phytochemicals in Plant Cell Cultures. (F. Constabel, I.K. Vasil, eds.), Academic Press, San Diego, pp. 143-157.

173. CHARLWOOD, B.V., J.T. BROWN, C. MOUSTOU, G.S. MORRIS, K.A. CHARLWOOD. 1988. The accumulation of isoprenoid flavour compounds in plant cell cultures. In: Bioflavour '87: Analysis, Biochemistry, Biotechnology. (P. Schreier, ed.), Walter de Gruyter, Berlin, pp. 303-314.

174. MULDER-KRIEGER, T., R. VERPOORTE, A. BAERHEIM SVENDSEN, J.J.C. SCHEFFER. 1988. Production of essential oils and flavours in plant cell and tissue cultures. A review. Plant Cell, Tissue and Organ Culture 13: 85-154.

175. BECKER, H. 1970. Studies on the formation of volatile substances in plant tissue cultures. Biochem. Physiol.

Pflanzen 161: 425-441.

176. BRICOUT, J., C. PAUPARDIN. 1975. Sur la composition de l'huile essentielle de *Mentha piperita* L. cultivee *in vitro*: influence de quelques facteurs sur sa synthese. C. R. Acad. Sci. Paris Ser. D 281: 383-386.

177. BRICOUT, J., M.-J. GARCIA-RODRIGUEZ, C. PAUPARDIN, R. SAUSSAY. 1978. Biosynthese de composes monoterpeniques par les tissus de quelques especes de Menthes cultivees *in vitro*. C. R. Acad. Sci. Paris Ser. D 287: 611-613.

178. WINK, M. 1984. Evidence for an extracellular lytic compartment of plant cell suspension cultures: The cell culture medium. Naturwiss. 71: 635-637.

179. BROWN, J.T., P.K. HEGARTY, B.V. CHARLWOOD. 1987. The toxicity of monoterpenes to plant cell cultures. Plant Sci. 48: 195-201.

180. CARRIERE, F., G. GIL, P. TAPIE, P. CHAGVARDIEFF. 1989. Biotransformation of geraniol by photoautotrophic, photomixotrophic and heterotrophic plant cell suspensions. Phytochemistry 28: 1087-1090.

181. BERGER, R.G., F. DRAWERT. 1988. Glycosylation of terpenols and aromatic alcohols by cell suspension cultures of peppermint (*Mentha piperita* L.). Z. Naturforsch. 43c: 485-490.

182. CORMIER, F., C. AMBID. 1987. Extractive bioconversion of geraniol by a *Vitis vinifera* cell suspension employing a two-phase system. Plant Cell Rep. 6: 427-430.

Chapter Four

SESQUITERPENE LACTONES: BIOGENESIS AND BIOMIMETIC TRANSFORMATIONS

NIKOLAUS H. FISCHER

Department of Chemistry
Louisiana State University
Baton Rouge, LA 70803, U.S.A.

INTRODUCTION

The different skeletal types of sesquiterpene lactones (SLs) are classified on the basis of their carbocyclic ring skeleton, in which the suffix "olide" is used to indicate the presence of a lactone group. The majority of SLs from higher plants contain α-methylene-γ-lactone functions (**1**) in which H-

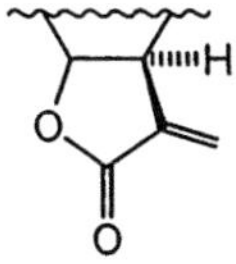

(*1*)

Biochemistry of the Mevalonic Acid Pathway to Terpenoids
Edited by G.H.N. Towers and H. A. Stafford
Plenum Press, New York

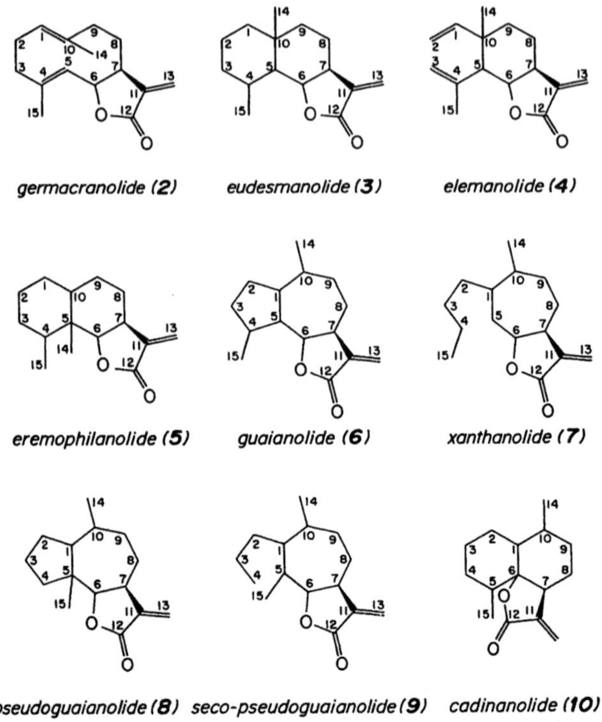

Fig. 1. Major skeletal types of sesquiterpene lactones shown as 12,6-lactonized representatives with the common numbering of the carbocyclic rings.

7 is, without exception, α-oriented.[1-4] Certain liverworts, however, produce SLs of the enantiomeric series.[5] In this review the biogenesis of only those common SLs will be considered in which one of the isopropyl methyls of the sesquiterpene ring is oxidized. The major skeletal types of SLs with the generally accepted numbering are outlined in Figure 1. The lactone rings are shown as 12,6-lactonized structures, although many SLs contain 12,8-lactones. The by far largest number of SLs has been isolated from the Asteraceae (Compositae),[1-4] but they are also found in other angiosperm families including the Acanthaceae, Amaranthaceae, Apiaceae,[6] Aristolochiaceae, Burseraceae, Bombacaceae, Coriariaceae,

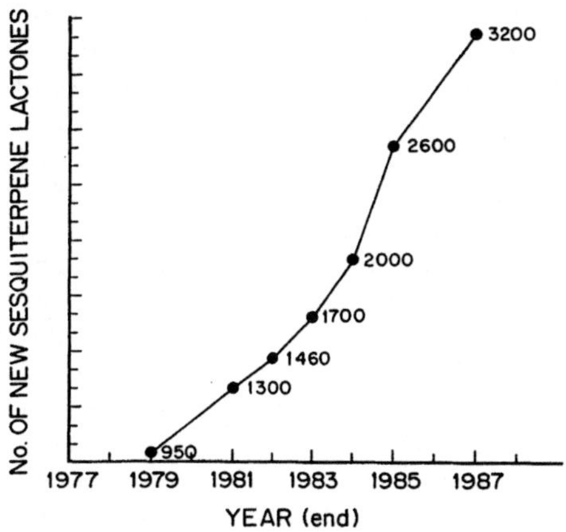

Fig. 2. The number of naturally occurring sesquiterpene lactones identified since 1979.

Illiciaceae, Magnoliaceae, Menispermaceae, Lamiaceae, Lauraceae, Polygonaceae and Winteraceae. In addition, SLs have been found in the gymnosperm family Cupressaceae[1] as well as in fungi[1] and liverworts.[5] Except for some earlier work,[7] studies on the isolation, structure and chemistry of SLs have only been significant during the last three decades. Several reviews have appeared on this subject since the late 1950s.[1-6] Additional reviews on selected skeletal types of SLs covered the germacranolides,[8-10] guaianolides,[9] pseudoguaianolides[11-16] as well as eremophilanolides and bakkenolides.[17] More recent general reviews on sesquiterpenes, including lactones can be found in the annual Natural Products Reports.[18-22]

During the past 30 years, a constantly increasing number of SLs have been reported in the literature as shown in Figure 2 (M. Vasquez, unpublished). Presently, well over 3200 naturally occurring SLs are known and there is a steady flow of papers describing more new compounds. This fact places the SLs among the largest classes of plant natural products.

In the early 1960s, major emphasis was on the use of SLs as taxonomic markers in biochemical systematic studies within the Asteraceae.[2,3,12-16] This provided the impetus for many detailed structural studies of SLs in this plant family. In the following decade, an intense search for cytotoxic and anti-cancer active SLs followed. In spite of extensive anti-tumor tests of several hundred SLs under the auspices of the U.S. National Cancer Institute, high toxicities of several promising compounds prevented clinical testing.[23-25] The first report of anti-feedant properties of SLs in *Vernonia* species (Asteraceae) upon herbivorous insects[26] created an awareness of a possible ecological role of SLs as natural deterrents against insects and mammals, as toxins against plant pathogens and as allelopathic agents in plant-plant interactions. In recent years, an increasing number of publications describing the ecological functions of SLs has been reported.[25] This will be without doubt the major area of study during the 1990s.

The wealth of structural information on SLs is contrasted by a nearly complete absence of biosynthetic data.[27] Therefore, biogenetic theory of SLs is based on biosynthetic pathways of analogous terpenoids. It is reasonable to assume that SL biosynthesis proceeds via the farnesyl/nerolidyl pyrophosphate route. Biogenetic relationships between different skeletal types of SLs are frequently based on organic-mechanistic considerations which include stereochemical and conformational considerations. In addition, biomimetic transformations of natural products and their derivatives into other skeletal types are frequently used as models to strengthen biogenetic proposals. Biogenetic relationships of the major skeletal types of SLs have been studied and several reviews on this topic have previously appeared.[1,28-30] It is the main purpose of this review to present an update on biogenetic proposals. In addition, significant new biomimetic studies, which have appeared in the literature during the past decade, will be discussed. However, this review is not meant to be a comprehensive treatement of this subject.

GERMACRANOLIDES

Based on the available structural data, it is apparent that many biomodifications in the advanced biosynethetic stages of SLs involve oxidative reactions. The isolation of a large number of SLs containing epoxide groups and the natural occurrence of hydroperoxide-bearing SLs suggests the

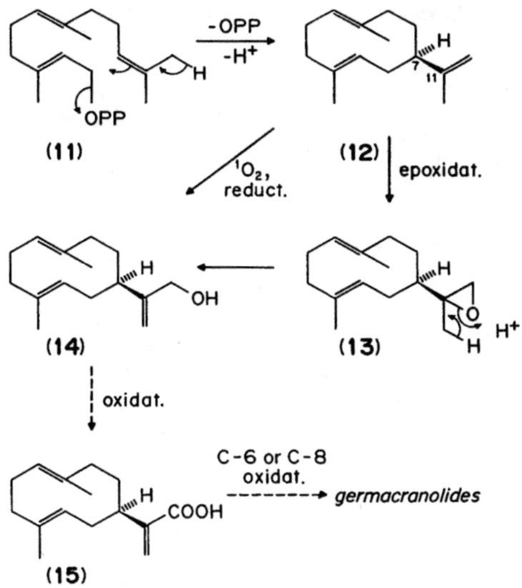

Fig. 3. The biogenesis of the α-methylene-γ-lactone ring and the germacranolide skeleton.

involvement of epoxides in sesquiterpene cyclizations and singlet oxygen type reactions in allylic oxidations.[1]

One hypothetical route in the biogenesis of the germacranolides (**2**) involves initial cyclization of *trans, trans*-farnesyl pyrophosphate (**11**) to give the naturally occurring germacrene A (**12**). In triene (**12**), all non-olefinic carbons are activated for allylic oxidations, except C-8. In its 7,11-double bond isomer, germacrene B, all non-olefinic carbons are allylicly activated, which suggests germacrene B as possible intermediate, in particular, in the formation of 8-oxygenated SLs.[1] As outlined in Figure 3, oxidative biomodification of the isopropenyl group in 12 could either proceed via initial formation of epoxide (**13**) or via a singlet oxygen process ($^1O_2$) followed by reduction of the intermediate hydroperoxide to provide the allylic alcohol (**14**) which upon further oxidation would give acid 15. Hydroxylations at C-6 or C-8 and subsequent lactonization would provide the 12,6- or 12,8-lactonized germacranolides. Questions regarding the biochemical steps remain open, including the sequence of oxidations, that is, medium ring oxidation at C-6 and/or C-8 before formation of the carboxylic acid moiety, at C-12 or *vice versa*.

Configuration *Conformation*

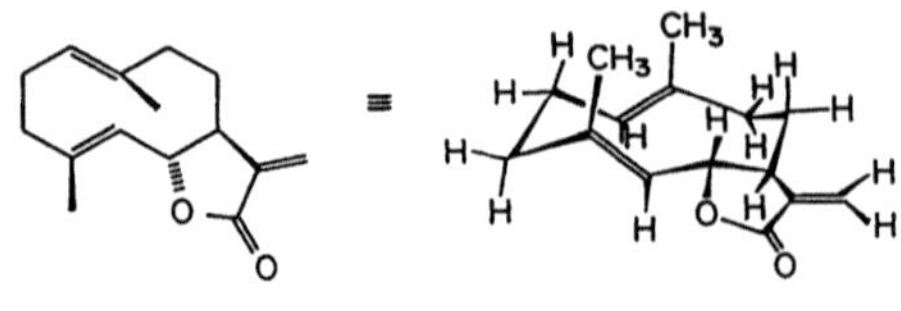

germacrolide (16)

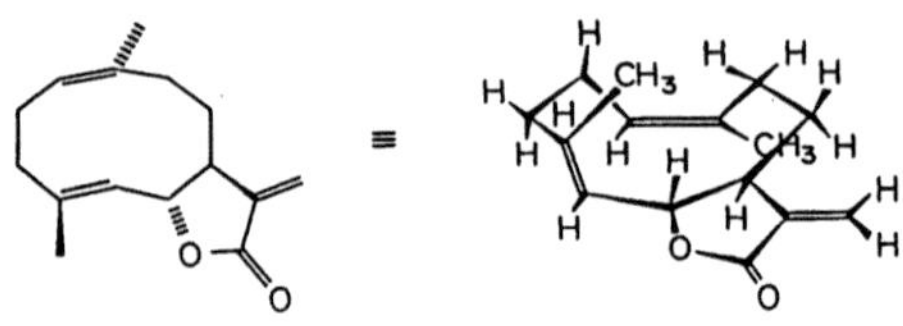

melampolide (17)

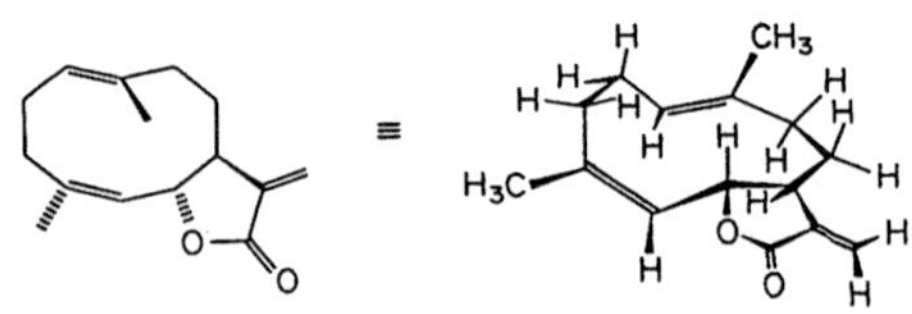

heliangolide (18)

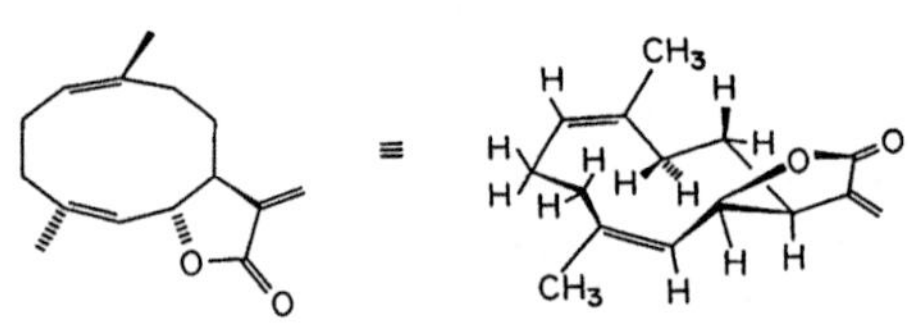

cis, cis-germacranolide (19)

Fig. 4. Configurational types of 12,6βH-germacranolides and their conformational representations.

By far the largest group of naturally occurring SLs are the germacranolides, which exhibit considerable structural variety because of the unique configurational and conformational features of the cyclodecadiene skeleton. Four germacranolide subgroups are found as natural products. They are classified on the basis of the configuration of the cyclodecadiene double bonds.[1,31] The germacrolides contain a *trans, trans*-cyclodecadiene ring, the melampolides[32] have a 1(10)-*cis*, 4-*trans*-cyclodecadiene skeleton, the heliangolides[1,33] possess a 1(10)-*trans*, 4-*cis*-germacradiene ring and the *cis*,

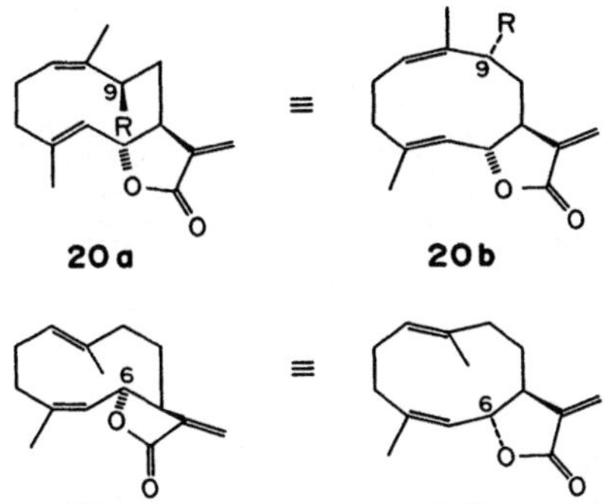

20a 20b 21a 21b

Fig. 5. Stereochemical ambiguities in common representations of melampolides (20a and 20b) and heliangolides (21a and 21b).

cis-germacranolides[34] contain 1(10)-*cis*, 4-*cis* medium ring double bonds (Fig 4).

The two-dimensional representation of the stereochemistry of germacranolides has in the past caused considerable confusion.[1] The reader is referred to page 59 in reference 1 for further details on this subject. Figure 5 presents two examples for common representations of melampolides (**20a** or **20b**) and heliangolides (**21a** and **21b**). In the melampolide structure **20b** the substituent R with broken lines at the apical C-9 correctly indicates its α-configuration. However, representation **20a** with a reentrant angle at C-9 should by common convention be correctly presented as 9β substituent, indicated by a wedge, but was in the past frequently drawn incorrectly with a broken line as in **20b**. Therefore, structure **20b** should be used instead of **20a**. Similarly, in the representation of heliangolides, structure **21a** with a reentrant C-6 should be avoided in presenting a 12,6α-lactonized structure; instead, **21b** with an apical C-6α-lactonic oxygen represents the correct stereochemistry. These confusing representations, which often cause incorrect configurational assignments of reentrant angles at chiral centers, should be avoided.

The stereoscopic representations of the molecular structures of four representative configurational types of germacranolides are shown in Figure 6. The structures represent the germacrolide tamaulipin A (**22**),[35] from *Ambrosia confertiflora*, the melampolide melampodinin A (**23**)[36] from *Melampodium*

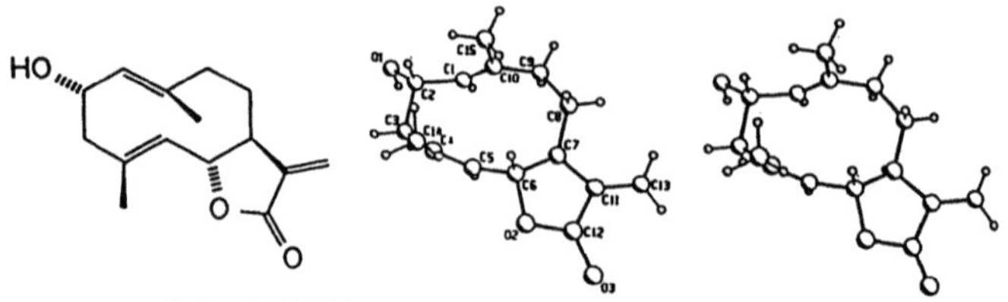

tamaulipin A (22)

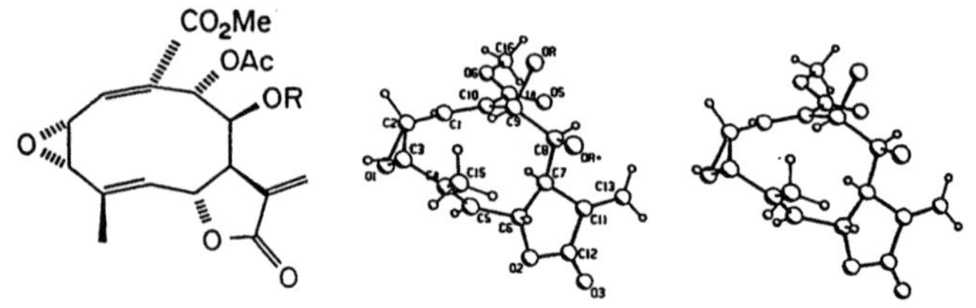

melampodinin A (23)

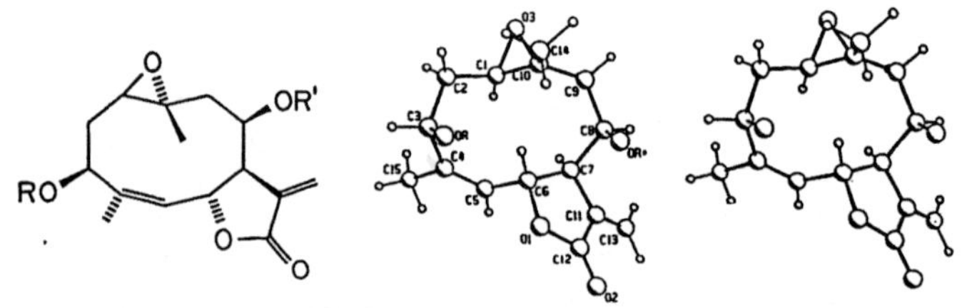

leptocarpin acetate (24)

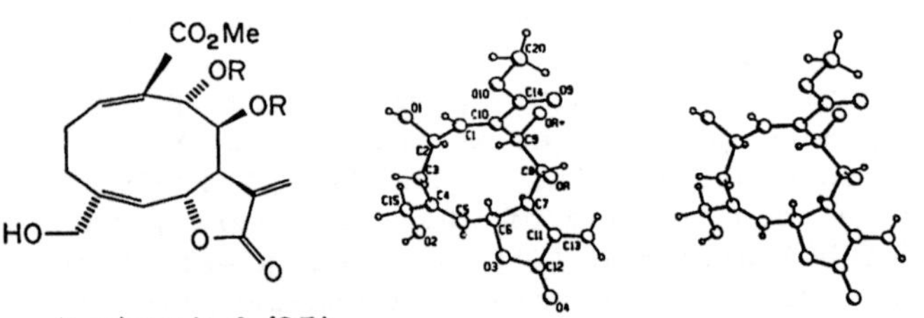

longicornin A (25)

Fig. 6. The stereoscopic representations of the molecular structures of the four major configurational types of germacranolides: The germacrolide tamaulipin A (**22**), the melampolide melampodinin A (**23**), the 1(10)-epoxyheliangolide leptocarpin acetate (**24**) and the *cis, cis*-germacranolide longicornin A (**25**). Ester substituents were deleted for clarity.

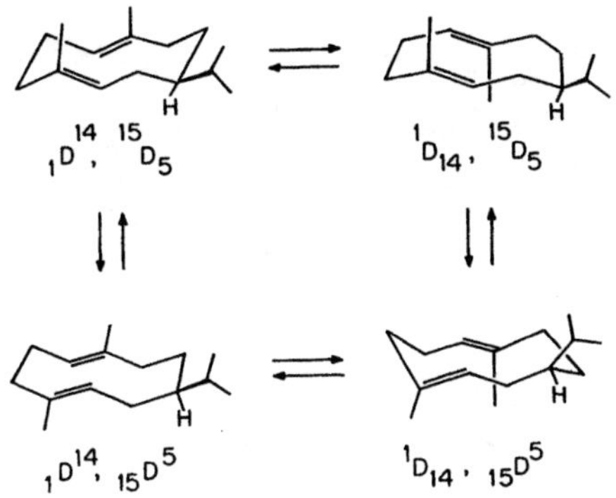

Fig. 7. Four major conformations of the (E,E)-germacradiene skeleton, using the conformational symbolism proposed by Samek and Harmatha.[38]

americanum, the heliangolide-1(10)-epoxide leptocarpin acetate (**24**)[37] from *Calea divaricata* and the *cis, cis*-germacranolide longicornin A (**25**)[34] from *Melampodium longicorne*.

Each of the four configurational types of germacradienolides shown in Figure 4 have four possible major conformations. In Figure 7, the four predominant conformers of the germacradiene skeleton (**12**) are shown, using the conformational symbolism proposed by Samek and Harmatha[1,38] and the structures given in Figures 6 and 8 show the conventions of Rogers, Moss and Neidle,[31] in which wedges and broken lines on double bonds indicate conformational orientations of substituents above or below the plane of the medium ring, respectively. In structures (**26**) to (**29**), the double bonds are nearly perpendicular to the medium ring plane with two crossed (**26** and **29**) and two parallel (**27** and **28**) double bond orientations.

Among the 12,6-lactonized germacrolides, the very common 12,6βH-lactonic compounds, which are based on the costunolide skeleton (**16**), exhibit a "frozen" [$_1D^{14,15}D_5$]-conformation. For instance, in the solid state[35] as well as in solution,[39] tamaulipin A (**22**) and its derivatives exhibit the [$_1D^{14,15}D_5$]-conformation. In contrast, montafrusin A (**30**) from *Montanoa frutescens*[40], which bears a 9β-hydroxyl group, adopts a [$^1D_{14},{}^{15}D_5$]-conformation which resembles conformer (**27**) in Figure 7. A change in the stereochemistry at C-6 from a germacra-12,6βH-olide (*trans*-lactone) to the less common

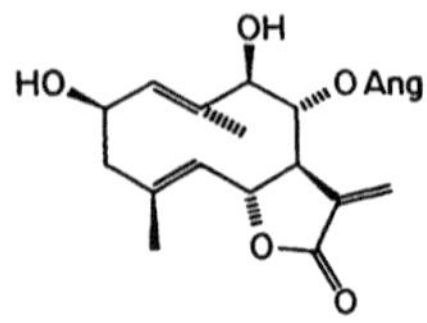

montafrusin A (**30**)
[$^{1}D_{14}, {}^{15}D_{5}$]

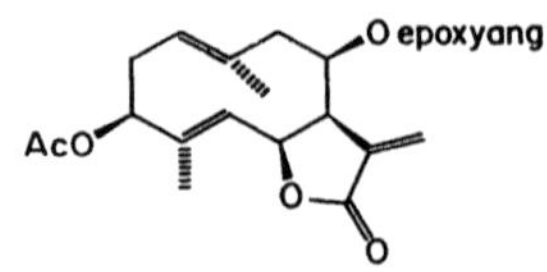

ursiniolide A (**31**)
[$^{1}D_{14}, {}_{15}D^{5}$]

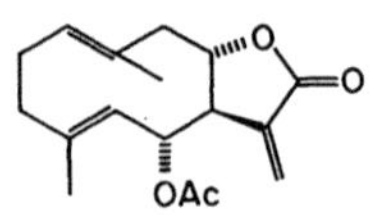

laurenobiolide (**32**)
(4 solute comformers)

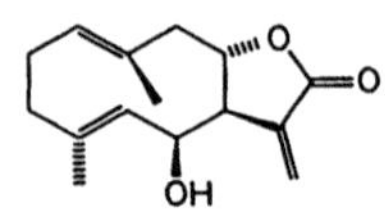

6-epi-desacetyl laurenobiolide (**33**)
[${}_{1}D^{14}, {}_{15}D^{5}$]

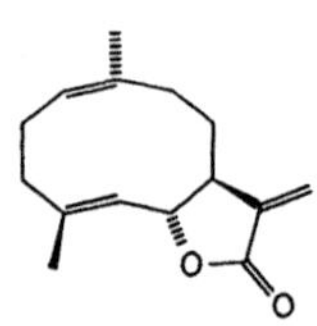

1(10)-cis-costunolide (**34**)
[${}_{1}D_{14}, {}^{15}D_{5}$]

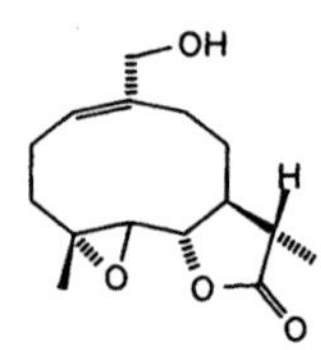

14-hydroxy-1(10)-cis-11βH(13)-dihydroparthenolide (**35**)

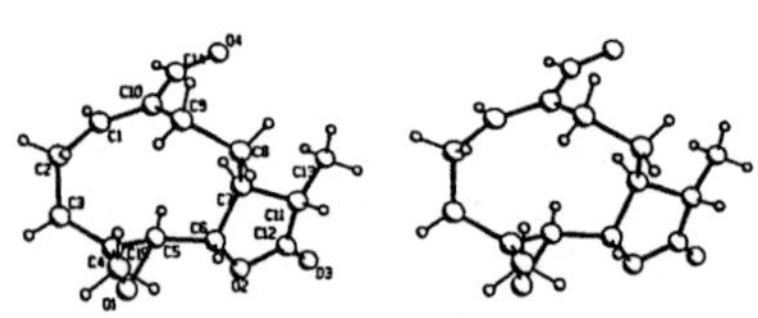

(**35a**)

Fig. 8. Conformational representatives of 12,6βH-, 12,6αH- and 12,8βH- germacra-1(10),4-dienolides using the conformational symbolism proposed by Rogers, Moss and Neidle.[31]

germacra-12,6αH-olide (*cis*-lactone) changes the medium ring conformation from [${}_{1}D^{14,15}D_{5}$], as in tamaulipin A (**22**), to [$^{1}D_{14}, {}_{15}D^{5}$] as exemplified by ursiniolide A (**31**)[41] from *Ursinia anthemoides* (Anthemideae) (Fig. 8). [$^{1}D_{14}, {}_{15}D^{5}$]-Germacra-12,6αH-olides have also been found in several genera including *Montanoa* (Asteraceae)[42] and they are common in the family Apiaceae (Umbelliferae).[6]

The "frozen" conformations found in the 12,6-lactonized germacranolides are contrasted by considerably greater medium ring flexibility in the germacra-12,8βH-olides. In the case of laurenobiolide[32] in solution all four conformational analogs (**26-29**) represented in Figure 7 could be detected.[43,44] In contrast, 6-epidesacetyl-laurenobiolide (**33**) shows a single conformation in its molecular structure.[45]

The question remains open whether the melampolides, heliangolides and *cis, cis*-germacranolides are derived from their respective (Z,E)-, (E,Z)- and (Z,Z)-farnesyl pyrophosphates, or whether the enzymatic and/or possibly photochemical isomerizations of the respective double bonds of the *trans, trans*-germacra-1(10), 4-diene skeleton (**12**) occur at a later biosynthic stage. This applies, for instance, to the biosynthesis of the simplest melampolide, 1(10)-*cis*-costunolide (**34**) from *Aristolochia yunnanensis* (Aristolochiaceae), which exhibits a solid-state conformation typical for melampolides.[46]

A distinct feature of most melampolides as well as *cis, cis*-germacranolides isolated from members of the subtribe Melampodiinae[32] and other taxa of the Asteraceae is the oxidized nature of C-14, which commonly represent alcohols, aldehydes or carbomethoxy groups.[32] The biosynthesis of C-14-oxidized melampolide seems to follow biological oxidation processes which finds its biomimetic analogy in a highly specific oxidative transformation of germacrolides into melampolides.[47-50] Reactions of germacrolides with selenium dioxide (SeO_2)[47] and SeO_2/t-butyl hydroperoxide (Bu^t-O-O-H) provide in good yields melampolides as well as *cis, cis*-germacramolides.[48-50] Figure 9 outlines the highly regio-and stereospecific transformations of the germacrolides epitulipinolide (**36**) with *SeO_2/Bu^t-O-O-H* in excellent total yield. These reactions not only involve double bond isomerizations but also cause conformational changes which lead to conformations typically found in natural melampolides. The stereopair (**35a**) of the molecular structure of melampolide **35**, derived from 11,13-dihydroparthenolide,[50] is shown on the bottom of Figure 8. Using the SeO_2/Bu^t-O-O-H methodology, *cis, cis*-germacranolides can be obtained from melampolides, as shown on the bottom of Figure 9 for the conversion of melampolide **38** into alcohol **39**.[48] The above *in vitro* transformations represent excellent models for the possible biosynthetic routes for C-14 hydroxylated melampolides and their derivatives, e.g. melampodinin A (**23**)[32,36] and also the *cis, cis*-germacranolides represented by longicornin A (**25**).[34]

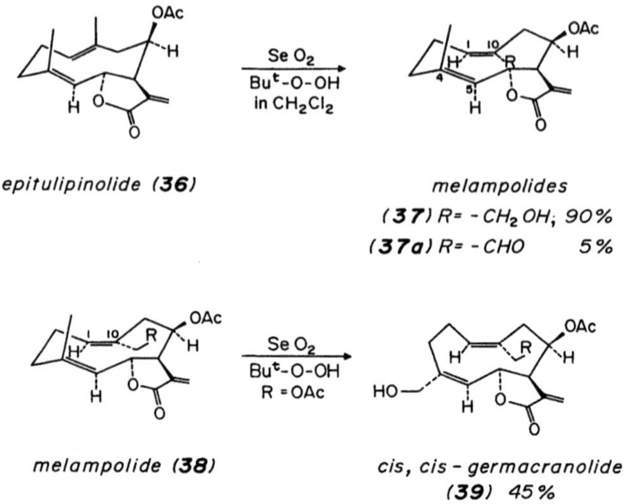

Fig. 9. Biomimetic transformations of germacrolides into melampolides and *cis, cis*-germacranolides.

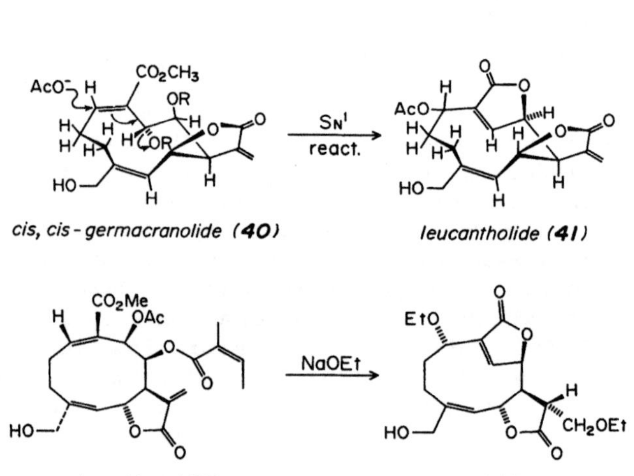

Fig. 10. Biogenesis of leucantholides from *cis, cis*-germacranolides and the biomimetic transformation of melcanthin A into a leucantholide dilactone.

In Melampodium, the *cis, cis*-germacranolides are often accompanied by leucantholides which appear to be biogenetically closely related.[52] As outlined in Figure 10, nucleophilic attack at C-1 of the *cis, cis*-germacranolide (**40**) and release of the C-9 acyl moiety followed by lactonization to C-8β-oxygen function provides the leucantholide skeleton (**41**). This type of transformation, which resembles an S_N1 reaction, was successfully accomplished in a biomimetic conversion of melcanthin A (**42**) to the dilactone (**43**), which was chemically correlated with cinerenin from *M. cinerium.*[53]

It is possible that formation of C-15-hydroxylated heliangolides involves similar oxidative double bond isomerizations at the 4-double bond of their hypothetical germacrolide precursors. The conformation of the heliangolide skeleton (**18**) support this biogenetic assumption.

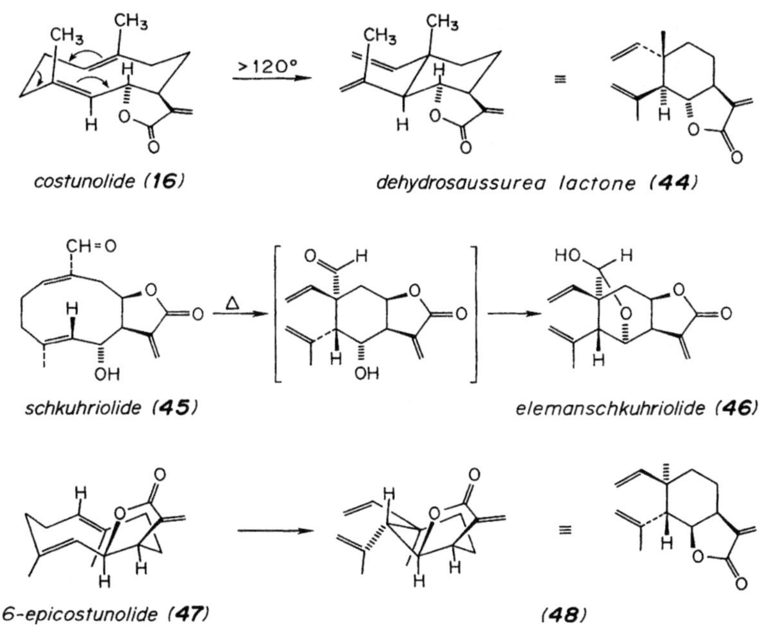

Fig. 11. Biogenetic relationships of germacra-1(10),4-dien-12,6βH-olides, germacra-1(10),4-dien-12,6αH-olides, and the stereochemistry of the resulting elemanolides.

ELEMANOLIDES

The biogenesis of the elemanolide skeleton (**3**) can be formulated as a Cope rearrangement of the $[_1D^{14,15}D_5]$-germacradienolide ring. This process is highly stereospecific in laboratory experiments and it generally requires a chair-like transition state of the 1,5-diene system, but exceptions are known.[1] In Figure 11, the thermal conversion of costunolide (**16**) into dehydrosaussurea lactone () is shown.[54] If the medium ring conformation does not adopt a chair-like transition state as in 6-epidesacetyllaurenobiolide (**33**), the Cope rearrangement is not favored.[44] Although many 12,6-lactonized melampolides, heliangolides and *cis, cis*-germacranolides possess the 1,5-diene moiety, no reports on Cope rearrangements of these skeletal types exist. This inherent difference between the 7,6-lactonized germacrolides and the other three skeletal types of 7,6-lactonic germacranolides is possibly due to the greater transannular distance between the medium ring double bonds.[55] The Cope rearrangement of the conformationally more flexible 7,8-lactonized melampolide schkuhriolide (**45**) to elemanschkuhriolide (**46**) with the opposite stereochemistry at C-5 and C-10 has been reported.[56] Similarly, one would predict from the $[^1D_{14,15}D^5]$-conformation of the germacra-12,6αH-olides with a *cis*-lactone moiety, the transition state conformation for the Cope rearrangements of germacrolide (**47**) *in vivo* or *in vitro* would dictate a stereochemistry of its derived elemanolide skeleton as represented by structure (**48**), which with reference to H-7 has also opposite stereochemistries at C-5 and C-10. Indeed, elemanolides isolated from plants that commonly produce germacra-12,6αH-olides of type 47 typically contain stereoisomeric elemanolides of type 48.[6,59] For instance, isolaserolide (**49**) from *Laser trilobum* (Apiaceae) is also obtained from its isomeric $[^1D_{14,15}D^5]$-germacrolide by thermal rearrangement (Fig. 12).[6,59] From the South African species *Pegolettia senegalensis* (Asteraceae, tribe Inuleae) the elemanolides of structural type (**50**) were isolated.[58] Thermolysis of an ursiniolide (**31**) analog afforded the elemanolide (**51**).[60]

In earlier studies some elemanolides claimed to be natural products were later shown to be artifacts formed from the germacrolide precursors during the preparative procedures at elevated temperature.[1] Saussurea lactone (**52**)[61] from *Saussurea lappa* is an example. However, vernolepin (**53**),[62] zinaflorin I (**54**),[63] and a series that represent biosynthetically modified elemanolides, are formed *in vivo*. Whether for-

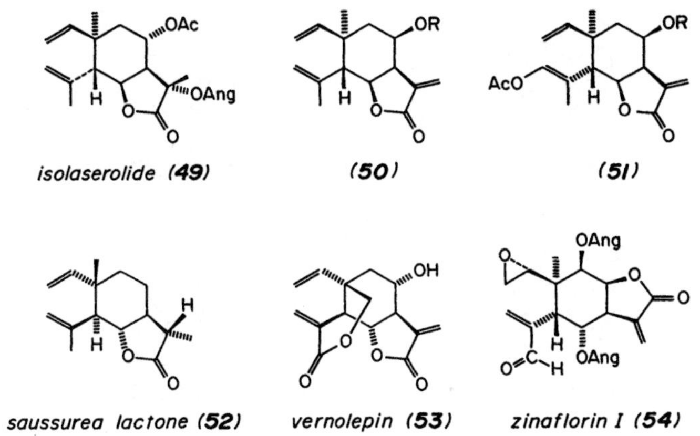

Fig. 12. Structural representatives of the two stereochemical types of elemanolides.

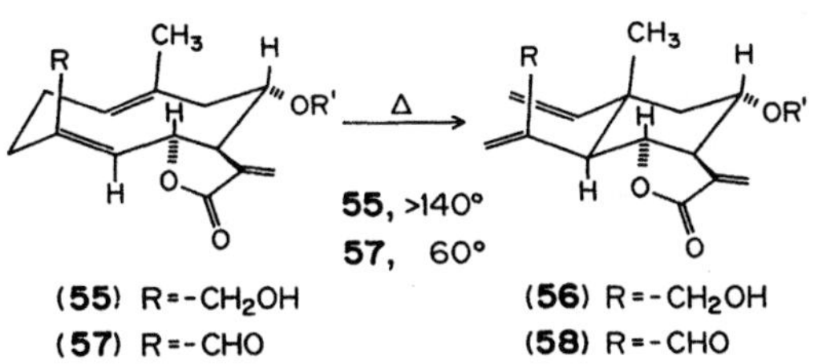

Fig. 13. Difference in temperature requirements for the Cope rearrangements of 15-hydroxygermacranolides and 15-oxagermacranolides.

mation of the elemanolide skeleton is an enzyme controlled process or whether this occurs spontaneously, is not known. It is of interest to note, however, that conversion of germacrolide (**55**) to elemanolide (**56**) required elevated temperatures (> 140^0), whereas the aldehyde derivative 57 was transformed into the elemanolide within 30 minutes at 60^0. (Fig. 13).[64] These mild thermal transformations of germacranolides with electron-withdrawing substituents, mainly aldehyde and/or carboxyl groups at C-4 and C-10, support the notion that spontaneous instead of enzyme-mediated Cope rearrangements may occur *in vivo*.

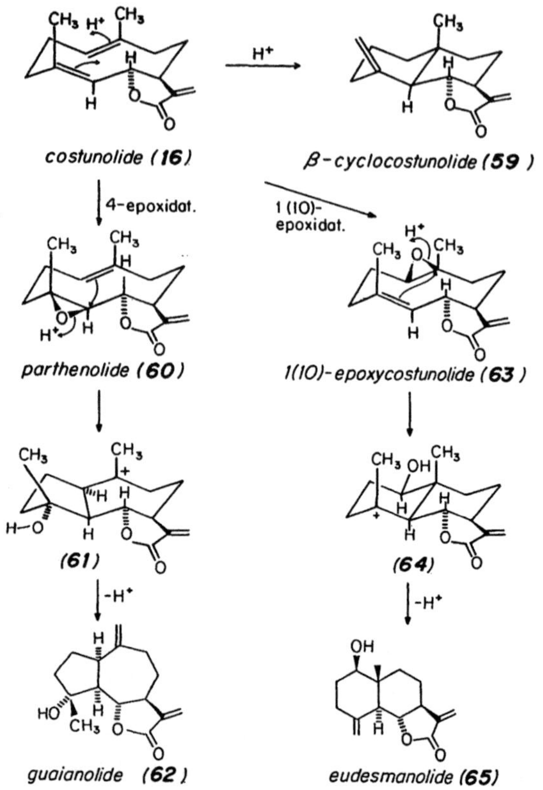

Fig. 14. The biogenesis of 1αH, 5αH-guaianolides and 5αH, 10β-methyleudesmanolides from $[_1D^{14,15}D_4]$-germacrolides and their epoxide precursors.

EUDESMANOLIDES, GUAIANOLIDES AND 8,10-CYCLOPROPANE-TYPE GUAIANOLIDES

A large number of stereospecific transformations of germacradienolides and their derivatives into 5αH,10β-methyleudesmanolides and 1αH, 5αH- guaianolides have been reported.[1,28,65] These biomimetic reactions provide strong indirect evidence for the involvement of germacradienolides, their monoepoxide intermediates and photo-oxygenation products in the biosynthesis of eudesmanolides and guaianolides.[43,58,65-73] In general, acid-mediated cyclizations of germacrolides lead to 5αH, 10β-methyleudesmanolides as shown in Figure 14 for

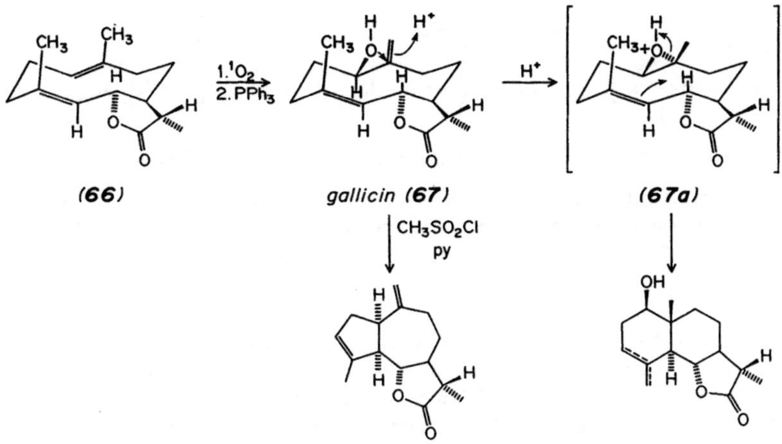

Fig. 15. The biogenesis of 1αH, 5αH-guaianolides and 5αH, 10β-methyleudesmanolides via the singlet oxygen 1(10)-epoxygermacrolide route.

the conversion of costunolide (**16**) into β-cyclocostunolide (**59**).[69]

Bio-epoxidations of one of the medium ring double bonds in (**16**) provide either parthenolide (**60**) or 1(10)-epoxycostunolide (**63**). In epoxides (**60**) and (**63**) the nucleophilicity of one of the double bonds is retained and the newly formed epoxide groups represent reactive receptor sites for nucleophilic attack. Thus, formation of the monoepoxides provides increased reactivity, as well as regio- and stereo-specificity of subsequent intramolecular cyclizations.[55] In parthenolide (**60**) and other germacra-1(10)-en-4-epoxy-olides, the epoxide nucleophile receptor, activated by protonation, undergoes a Markovnikov-type transannular cylization to give cation (**61**). Loss of a proton from C-14 of (**61**) provides the 1αH, 5αH-guaianolides, as shown for the formation of the guaianolide (**62**).[51] Alternatively, 1(10)-epoxycostunolide (**63**) and 1(10)-epoxide analogs in general cyclize to give the 5αH, 10β-methyleudesmanolide cation (**64**), which by proton loss from C-15 gives the 5αH, 10β-methyleudesmanolide skeleton (**65**).[70] In biomimetic experiments, epoxidation of costunolide (**16**) with

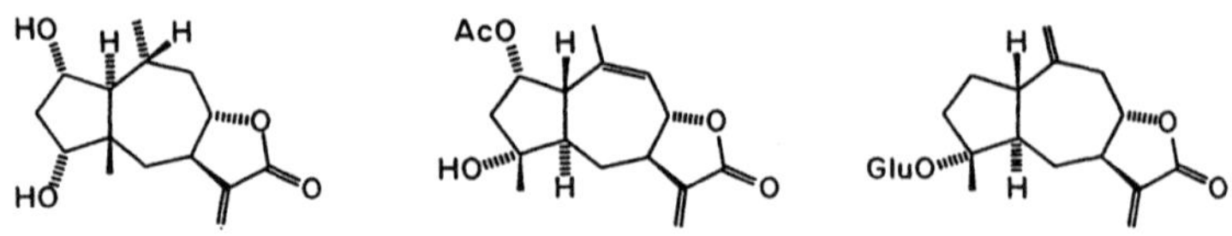

pulchellin (70) *gaillardin (71)* *(72)*

Fig. 16. Structural representatives of the *trans*-fused 1βH, 5αH-guaianolides and the structure of the helenanolide pulchellin (**70**).

peracids results in the formation of reynosin (**65**) and its 4-double bond isomer santamarine.[70,71] Another highly specific transformation of germacrolides into 1αH, 5αH-guaianolides and 5α,H 10β-methyleudesmanolides involves initial $^1O_2$-photo-oxygenation of the germacrolide skeleton as shown in Figure 15 for the laboratory conversion of 11βH,13-dihydrocostunolide (**66**) into gallicin (**67**) followed by further cyclizations.[72] Singlet oxygen reaction of (**66**) occurs from the outer face of the 1(10)-double bond, resulting in a C-1β-hydroperoxide which upon reduction with triphenylphosphine gives gallicin (**67**). Treatment with acid causes protonation at C-14 initiating transannular cyclization at C-10 via cation intermediate (**67a**) to give the 5αH, 10β-methyleudesmanolide (**68**).[73] Alternatively, conversion of the allylic alcohol at C-1 of **67** into a good leaving group (mesylate) initiates an intramolecular S_N2 reaction providing the 1αH, 5βH-guaianolide skeleton (**69**).[72,73]

In contrast to the *cis*-fused guaianolides of type **62** and **69**, which are very common in higher plants,[1] *trans*-fused guaianolides are rare in nature and generally occur in the genera *Helenium* and *Gaillardia* together with pseudoguaianolides[13] of the helenanolide type represented by pulchellin (**70**) from *Gaillardia pulchella*.[74] A typical representative of the 1βH, 5αH-guaianolide group is gaillardin (**71**) from *G. pulchella*.[75] More recently, two glycosides of *trans*-fused guaianolides (**72**) were isolated from *Helenium donianum* from Argentina.[76]

It has been postulated that the biogenesis of *trans*-fused guaianolides proceeds via melampolides,[13] as shown in Figure 17 for the acid-induced cyclization of the 4-epoxymelampolide (**73**) to give the 1βH, 5αH-guaianolide (**74**). Alternatively, the *trans*-fused lactone (**74**) could be derived

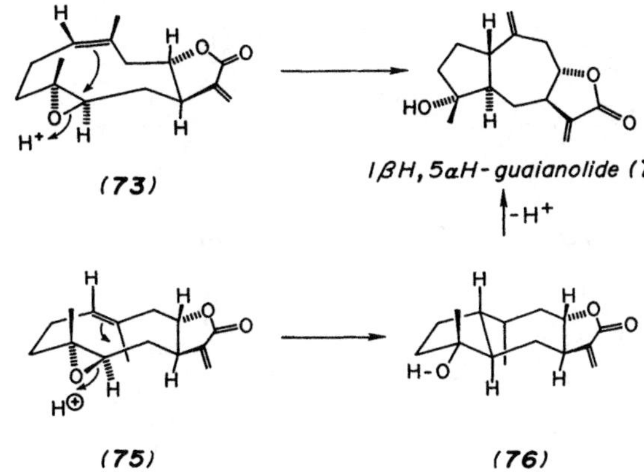

Fig. 17. The biogenesis of the 1βH, 5αH-guaianolide skeleton by the [$^1D^{14}$,$^{15}D_5$]-melampolide and [$^1D_{14}$,$^{15}D_5$]-germacrolide routes.

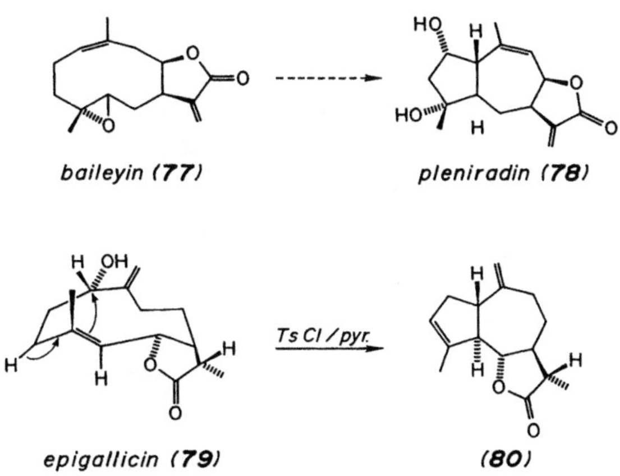

Fig. 18. The biogenetic relationship of baileyin (**77**) and pleniradin (**78**) and the biomimetic transformation of epigallicin (**79**) into a 1βH, 5αH- guaianolide.

from the [$^1D_{14}$,$^{15}E_5$]-germacrolide (**75**) via the cationic intermediate **76**, which upon loss of a proton from C-14 would also give **74**.[55] Although on paper both biogenetic routes are equal, the melampolide route is strongly supported by the fact that the melampolide-4-epoxide baileyin (**77**) and the 1βH, 5βH-guaianolide pleniradin (**78**) co-occur in *Baileya pleniradiata*.[77]

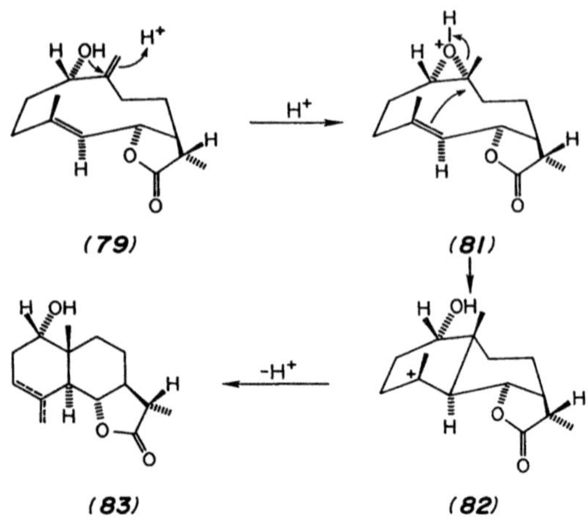

Fig. 19. The biomimetic transformation of epigallicin (**79**) into a 1α-hydroxylated 5αH, 10β-methyleudesmanolide via a [$^{1}D^{14}$,$^{15}D_{5}$]-1(10)-epoxymelampolide intermediate.

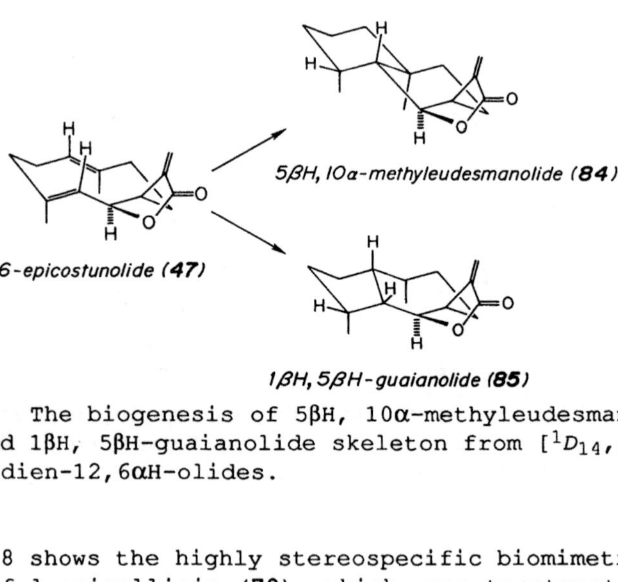

Fig. 20. The biogenesis of 5βH, 10α-methyleudesman-12,6αH-olide and 1βH, 5βH-guaianolide skeleton from [$^{1}D_{14}$,$_{15}D^{5}$]-germacradien-12,6αH-olides.

Figure 18 shows the highly stereospecific biomimetic transformation of 1-epigallicin (**79**), which upon treatment with tosyl chloride (TsCl)/pyridine gives the *trans*-fused guaianolide (**80**). This reaction was interpreted by the authors as a process that may proceed via the melampolide route with initial photo-oxygenation of the melampolide-1(10)-double bond in

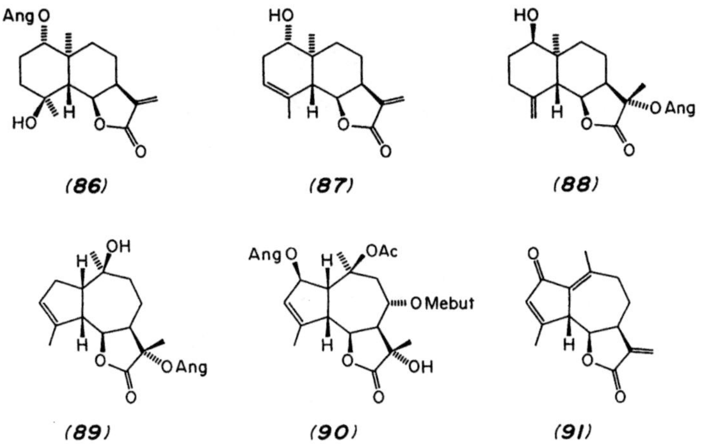

Fig. 21. Structural representatives of naturally occuring 5βH, 10α-methyleudesmanolides and 1βH, 5βH-guaianolides.

analogy to the reaction shown in Figure 15.[68] In contrast, treatment of 1-epigallicin (**79**) with HCl in dichloromethane provided a mixture of two eudesmanolides (**83**) with a 1α-hydroxyl group (Fig. 19).[78]

Figure 20 outlines the cyclization processes of germacra-12,6αH-olides that adopt the $[^{1}D_{14,15}D^{5}]$ conformation as shown for 6-epicostunolide (**47**). This conformation would dictate the formation of 5βH, 10α-methyleudesmanolides (**84**) and 5βH-guaianolides (**85**).[6,28] Indeed, these stereochemical types of eudesmanolides and guaianolides are found in plant taxa that also produce **12**,6αH-lactonic germacranolides. Representative eudesmanolides are compound **86** from *Montanoa dumicola*,[79] lactone **87** from *Pegolettia senegalensis* [58] and lasolide (**88**) from *Laserpitium trilobum*.[6,80] Lactones **89**[81] and **90**[82] from *L. trilobum* as well as compound (**91**) from *Montanoa imbricata* [83] exhibit the typical "anti"-stereo-chemistry of the guaianolides skeleton (Fig. 21).

The hypothetical 8α-hydroxyguaianolide precursor (**92**), which must be formed in analogy to guaianolide (**62**) given in Figure 14, could be the biogenetic precursor for the 8,10-cyclopropane-type guaianolides represented by axivalin (**95**).[84] As shown in Figure 22, intramolecular cyclopropane formation involving the 1(10)-double bond and the homoallylic C-8α-hydroxyl group in (**92**), as indicated by the arrows, would give

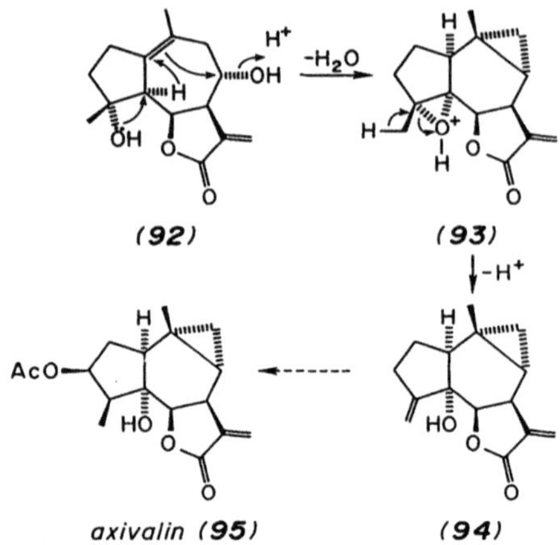

Fig. 22. The biogenesis of 8,10-cyclopropane-type guaianolides represented by axivalin (**95**).

intermediate (**93**). This involves a C-5αH to C-1α hydride shift followed by attack at C-5 of the C-4α-hydroxyl group to give the α-epoxide intermediate (**93**). Eliminative opening of the 4α-epoxide would result in the 4α-hydroxy-8(10)-cyclopropane-type guaianolide (**94**) and further biomodification would provide axivalin (**95**) from *Iva axillaris*.[84]

XANTHANOLIDES (SECO-GUAIANOLIDES) AND CYCLOPROPANE-TYPE XANTHANOLIDES

As outlined in Figure 23 (top), cyclization of $[^1D_{14,15}D_5]$-germacrolide-4-epoxide (**96**) in the Markovnikov fashion would first provide the *cis*-fused guaianolide cation (**97**). Subsequent hydride shift of the C-1αH to C-10α could initiate fragmentation of the 4,5-carbon-carbon bond as indicated by the arrows, resulting in the seco-guaianolide skeleton (**98**). Most naturally occuring xanthanolides from the Asteraceae genera *Iva, Xanthium, Parthenice* and others represent skeletal analogs of (**98**).[1,2] The above biogenetic pathway is supported by the recent biomimetic conversions of lipiferolide (**96**)[66] and the 11βH,13-dihydroderivative of parthenolide (**60**)[67] into the related natural xanthanolides.

Also shown in Figure 23 is the biogenesis of the seco-guaianolide carabrone (**101**) from Helenium and related genera.

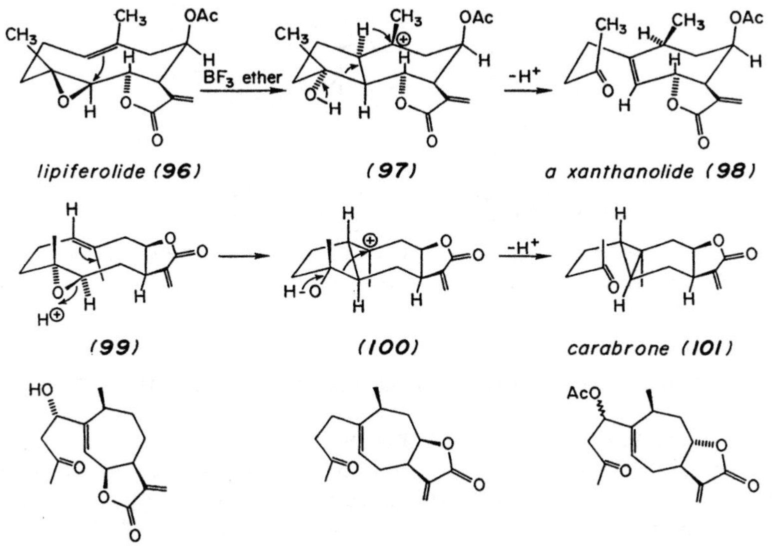

Fig. 23. The biomimetic transformation of [$_1E^{14,15}D_5$]-germacrolides into xanthanolides and the biogenesis of the cyclopropane-type xanthanolide carabrone (**101**). Structural representatives of the xanthanolides.

It represents a cyclopropane-type xanthanolide, the biogenesis of which can be formulated via a 1βH,5αH-guaianolide cation (**100**) which could be formed by cyclization via the melampolide route[13] or the [$^1D_{14}$,$^{15}D_5$]-germacrolide pathway (shown),[55] similar to the initial cyclization given in Figure 17. The intermediate *trans*-fused guaianolide cation (**100**) is stereoelectronically suited for a fragmentation type rearrangement as indicated by the arrows in structure (**100**) to give carabrone (**101**).[85] Structure (**101**) shows the stereochemistry dictated by the initial germacranolide (germacrolide and/or melampolide) conformation.[1] Parthemollin (**102**) from *Parthenice mollis*,[86] tomentosin (**103**)[87] and xanthinin (**104**)[1,2] are typical representatives of naturally occuring xanthanolides (Fig. 23).

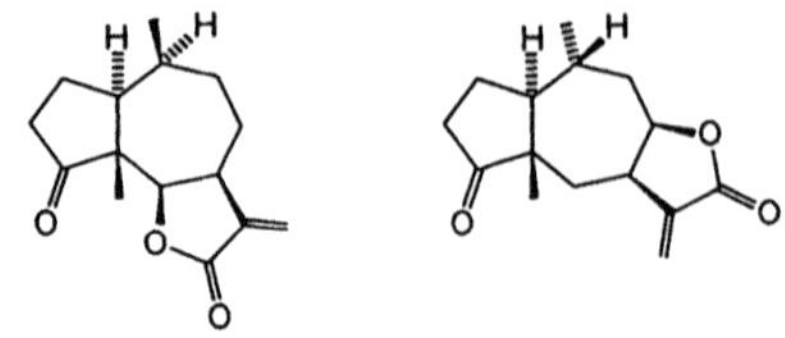

damsin (105)
an ambrosanolide

aromatin (106)
a helenanolide

Fig. 24. Structural representatives of the pseudoguaianolides: ambrosanolides and helenanolides.

PSEUDOGUAIANOLIDES AND SECO-PSEUDOGUAIANOLIDES

The pseudoguaianolides, which typically contain a C-5β-methyl group at the 5,7-ring junction, are represented by two stereochemically distinctly different skeletal types (Fig. 24). The ambrosanolides, represented by damsin (**105**), which occur in the subtribe Ambrosiinae, contain a C-10β-methyl group with the lactone ring generally being closed toward C-6, and then exclusively as 12,6αH-lactones (*cis*-lactones).[1,2] In addition, 12,8αH- as well as 12,8βH- lactones are found as natural products.

The helenanolides, represented in Figure 24 by aromatin (**106**), commonly occur in the tribe Heleniae of the Asteraceae. In contrast to the ambrosanolides, the most distinct feature of the helenanolides is the α-orientation of the C-10 methyl group and the lactone ring closure toward C-8 with *cis*- and *trans*-lactones commonly observed. The biogenetic aspects of these two types of lactones have over the years received considerable attention.[1,12-14,28,55] The biogenesis of the ambrosanolide skeleton and its seco-derivatives can be formulated as outlined in Figure 25. Cyclization of the hypothetical $[^{1}D^{14,15}D_{5}]$-germacrolide-4α,5β-epoxide (**107**), a conformational isomer of 6-epicostunolide (**47**) which is not yet represented by a natural product, or possibly its nonlactonized acid form would initially give the guaianolide cation (**108**). As indicated by arrows, the cation (**108**) could undergo double hydride shift followed by C-4 to C-5-methyl migration providing the ambrosanolide skeleton as represented by damsin (**105**), a compound commonly found in the genera *Ambrosia, Parthenium* and related taxa.[1,2] Biological Baeyer-Villiger oxidation, most likely via hydroperoxide intermediates, provides the dilactone psilostachyin C (**109**), which can also be prepared

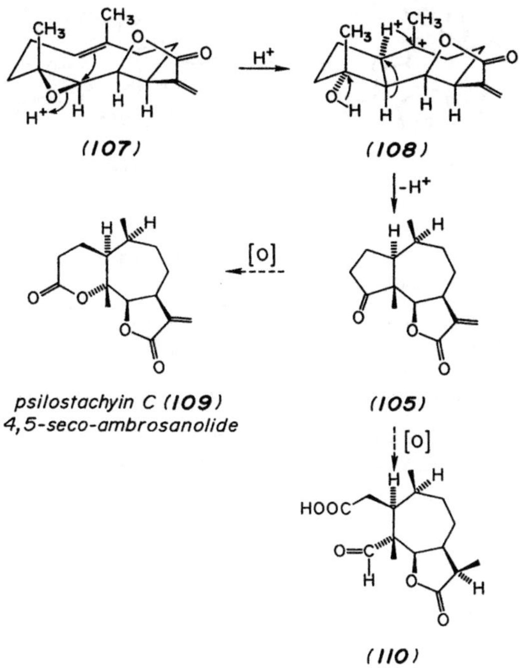

Fig. 25. The biogenesis of the ambrosanolide skeleton and the 4,5- and 3,4-seco-ambrosanolides.

from damsin with high stereospecificity by peracetic acid oxidation.[88] An analog oxidative cleavage would provide the 3,4-seco-ambrosanolide (**110**) which was recently isolated from Yugoslavian *Ambrosia artemisiifolia*.

The biomimetic conversion of germacrolides and other possible precursors into ambrosanolides have in the past been attempted, but with limited success.[1,12-14] Recently, two biomimetic type syntheses of pseudoguaianolides have been reported (Fig. 26). One involves the acid-mediated rearrangement of the 4α,5α-epoxyguaianolide (**111**) into the pseudoguaianolide (**112**).[90] The other example represents the first one-step conversion of a germacrolide-4α,5β-epoxide into a pseudoguaianolide, a process which strongly supports the proposed biogenetic pathway formulated in Figure 25.[91] Treatment of a chloroform solution of 4α, 5β-epoxyinunolide (**113**) from *Stevia tephrophylla* [92] by absorption on bentonitic earth provided the pseudoguaianolide 8-epiconfertin (**114**) plus two guaianolides.[91]

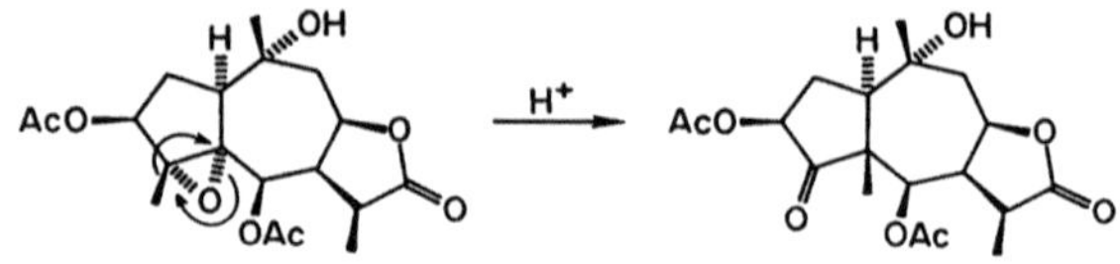

(111) *(112)*

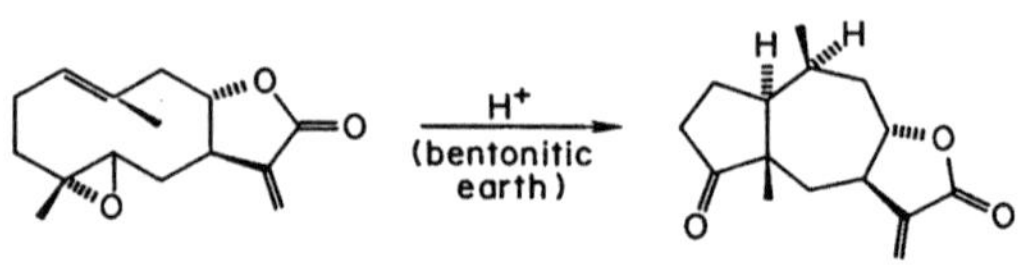

4α,5β-epoxyinunolide (113) *8-epiconfertin (114)*

Fig. 26. Biomimetic transformations of an 4α,5α-epoxyguaianolide and a germacrolide-4α, 5β-epoxide into ambrosanolides.

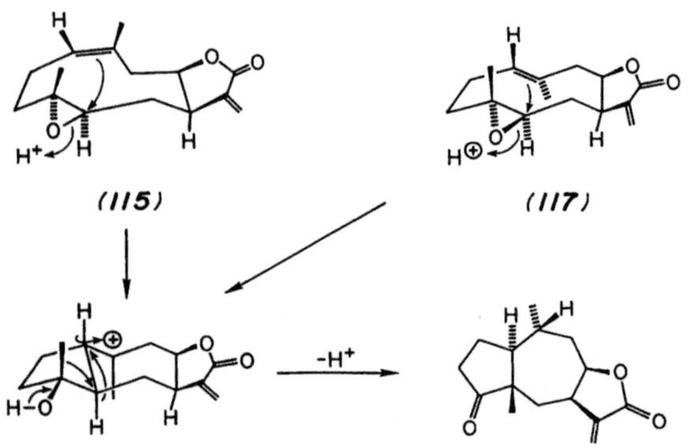

(116) *aromatin (106)*

Fig. 27. The biogenesis of the helenanolide aromatin **(106)** *via* the melampolide and germacrolide route.

The bentonitic earth appears to mimic the enzymatic process by fulfilling a two-fold function in this highly specific cyclization-rearrangement reaction. Firstly, its cavities enforce conformational rigidity during the cyclization process. Secondly, stabilization of the catonic intermediate by Lewis base sites of the catalist surface increases the life span of the cation, intermediate, thus allowing the consecutive hydride shifts and methyl migration to occur.[55,91]

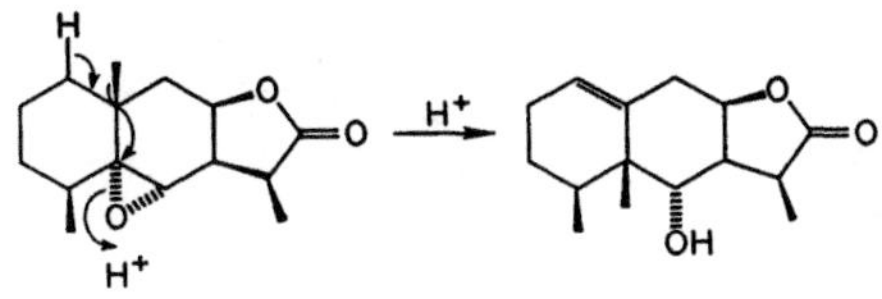

(118) *(119)*

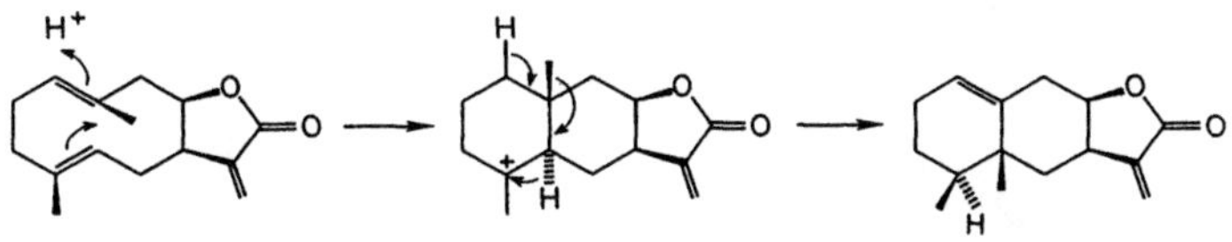

(120) *(121)* *xanthanodiene (122)*

Fig. 28. Biogenetic proposal for the formation of eremophilanolides from eudesmanolides and germacrolides.

The biogenesis of the helenanolide skeleton is shown in Figure 27. As previously proposed by Herz,[12-14,30] acid-induced cyclization of the melampolide-4α,5β-epoxide **(115)** provides the guaianolide cation **(116)** from which by the indicated hydride shifts and methyl migration the helenanolide skeleton **(106)** is formed. Alternatively, the intermediate cation **(116)** could be derived from the [$^{1}D_{14},{}^{15}D_{1}$]-germacrolide skeleton **(117)**. Since the helenanolides are generally represented by 12,8-lactonized compounds, the conformational flexibility of the two possible germacranolide precursors **(115)** and **(117)** would allow adoption of the shown conformations which are essential requirements for the stereochemistry found in aromatin **(106)**.

EREMOPHILANOLIDES

An early example of a biomimetic transformation of a eudesmanolide into an eremophilanolide involved the acid-mediated conversion of 11αH-dihydroalantolactone,5α,6α-epoxide **(118)** into the rearranged eremophilanolide **(119)**, by a route indicated by the arrows (Fig. 28).[93]

Alternatively, acid-mediated cyclization of germacrolide **(120)** would form the eudesmanolide cation **(121)** which by indicated hydride and methyl shifts, would provide the eremophilanolide skeleton, as represented by xanthanodiene **(122)** from

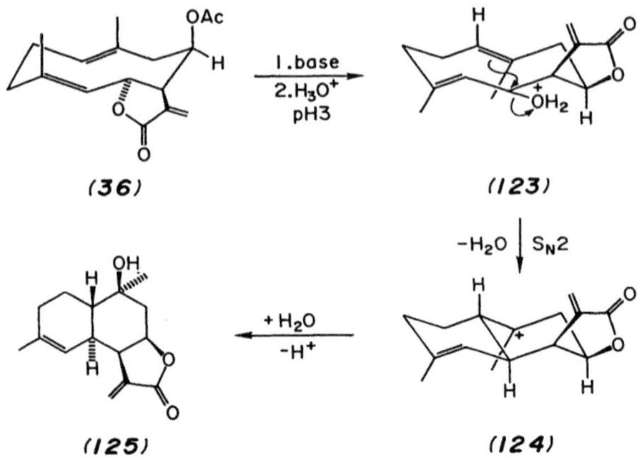

Fig. 29. Biomimetic transformation of epitulipinolide into a cadinanolide.

Xanthium canadense.[94] More recently, a series of eremophilanolide analogs of xanthanodiene were isolated from *Stevia achalensis.*[95]

CADINANOLIDES

The biogenesis of the cadinanolide skeleton (**10**) with its stereochemical details is best demonstrated by a biomimetic transformation of a germacrolide, epitulipinolide (**36**) into the cadinanolide (**125**).[96] This process is detailed in Figure 29. Hydrolysis of epitulipinolide (**36**) with dilute aqueous base and subsequent controlled acid treatment at pH 3 proceeds via the carboxylate ion which will allow rotational change from the [$_1D^{14}$,$^{15}D_5$]- to the [$^1D_{14,15}D^5$] conformation of the medium ring. Acid-mediated relactonization would give the more favored 12,8αH-lactonic intermediate with its typical conformation (**123**). Protonation of the 6α-hydroxyl group in (**123**) would initiate an intramolecular Markovnikov-type substitution to give cation (**124**). Uptake of H_2O from the outer face of the cation leads to the cadinanolide (**125**), the stereochemistry of which is again dictated by the conformation of its germacrolide precursor (**123**).

In Figure 30 the biogenesis of arteannuin B (**128**)[97] and C (**129**)[98] derived from a conformational analog of cation (**124**) in Figure 29 is shown. If cation (**126**) undergoes consecutive

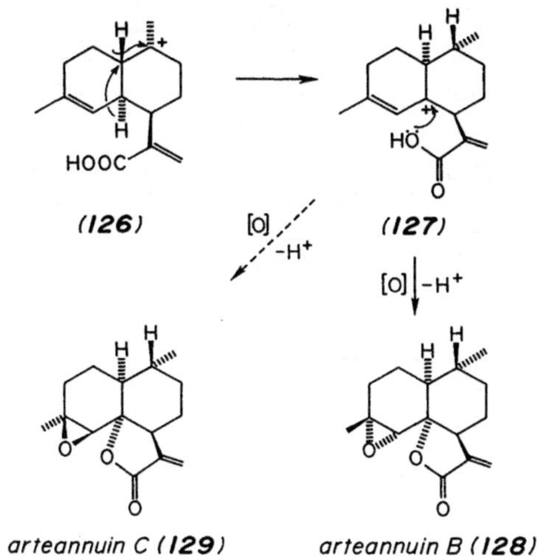

Fig. 30. Biogenesis of the cadinanolides arteannuin B and C.

hydride shifts (1βH to C-10 and 6αH to C-1), formation of the more stable allylic cation (**127**) can be expected. Attack of the carboxyl group at C-6 followed by bioepoxidations would result in the correct stereochemistry for arteannuin B (**128**)[97] and C (**129**) from *Artemisia annua*.[98]

MISCELLANEOUS SECO-SESQUITERPENE LACTONES

Three common types of terpenoid oxidations exist, which most likely involve $^1O_2$ or radical type processes with electron-rich double bonds (Fig. 31). Type 1 represents a cycloaddition of an electron-rich double bond (**130**) with singlet oxygen to form a highly unstable 1,2-dioxetane (**131**) which fragments to two carbonyls (**132**). Another frequently observed process (type 2) represents an ene-reaction of the alkene with oxygen. This is indicated by the transformation of alkene (**133**) into a hydroperoxide (**134**). A number of SLs either contain the allylic hydroperoxide function (**134**) or the reductively modified allylic alcohols (**135**). This process finds its analogy in a number of biomimetic $^1O_2$ reactions.[43,72] The type 3 reactions involving 1,3-dienes (**136**) and oxygen representing Diels-Alder type transformations to give endoperoxides (**137**), structures which are not generally seen in SLs but are frequently found in other terpenoids.

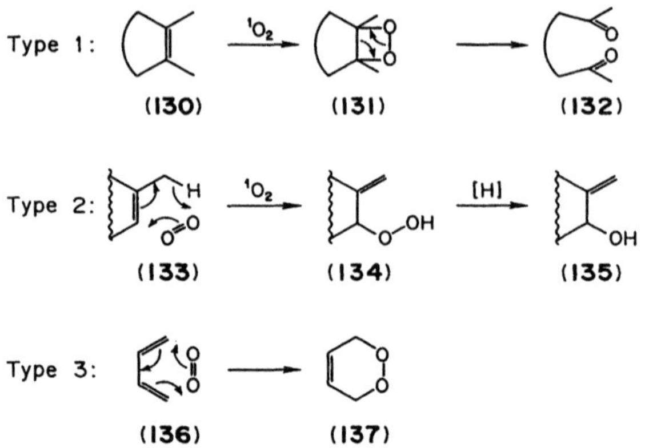

Fig. 31. Types of singlet oxygen oxidations involving electron-rich alkenes.

Examples of the type 1 reaction forming seco-derivatives **(139)** and **(141)** from the irrespective hypothetical guaianolides **(138)** and **(140)** are listed in Figure 32. Lactones **(139)** and **(141)** were isolated from *Artemisia rutifolia* [99] and *Dittrichia graveolus*,[100] respectively. Umbellifolide **(143)** from *Artemisia umbelliformis* appears to be a 4,5-seco-derivative of the alantolactone isomer **(142)**.[101] It can be envisioned that the three oxidative modifications described above represent $^1O_2$- quenching steps to protect essential biomolecules within the plant against oxidative destruction.

Also sketched in Figure 32 is a possible biogenetic route from precursor **(144)** and **(145)** into the anti-malarial SL artemisinin (quinhaosu) **(146)** from *Artemisia annua*.[102] This pathway is supported by the fact that arteannuin B **(128)** is a biosynthetic precursor for artemisinin **(146)** in cell-free leave extracts of *A. annua*[103] and hydroperoxide **(144)** is a likely precursor for arteannuin B **(128)** as demonstrated by biomimetic transformations.

Angular hydroxylations of eudesmanolides most likely involve biological oxidation reactions of type 2 which can be easily performed in the laboratory. For instance, magnoliolide **(147)** from *Magnolia grandiflora* was transformed by methylene blue-sensitized photo-oxygenation into the hydroperoxide **(148)** and upon reduction with triphenylphosphine gave alcohol **(149)** that was identical with natural artemin (Fig. 33).[43]

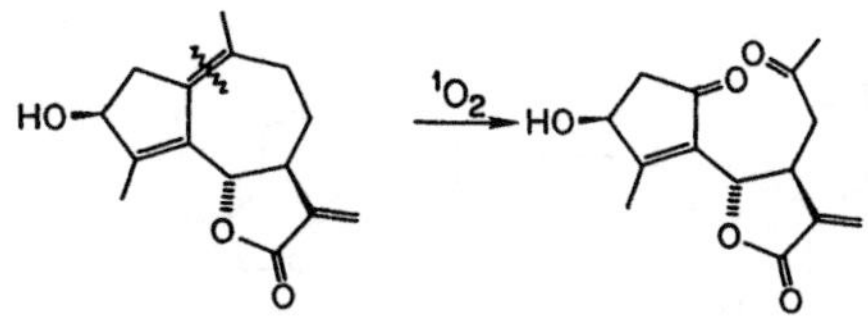

(138) *(139)*

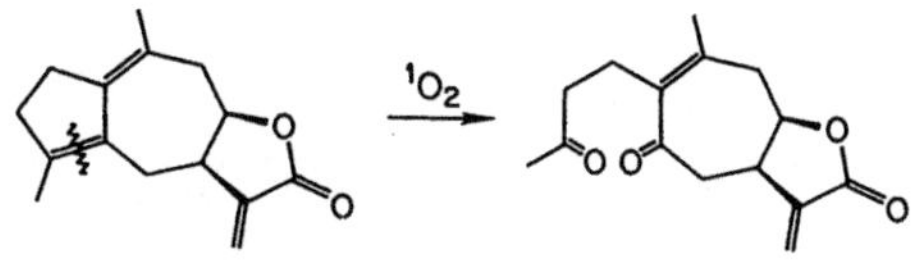

(140) *(141)*

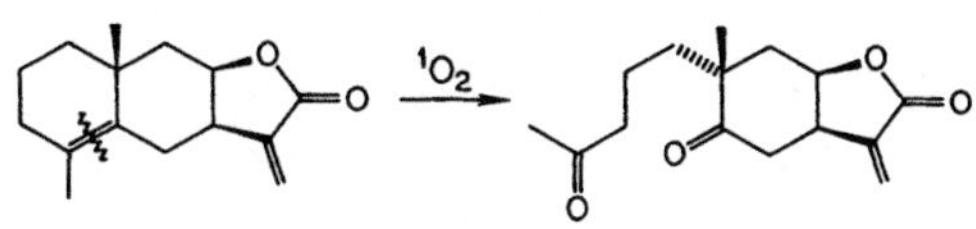

(142) *(143)*

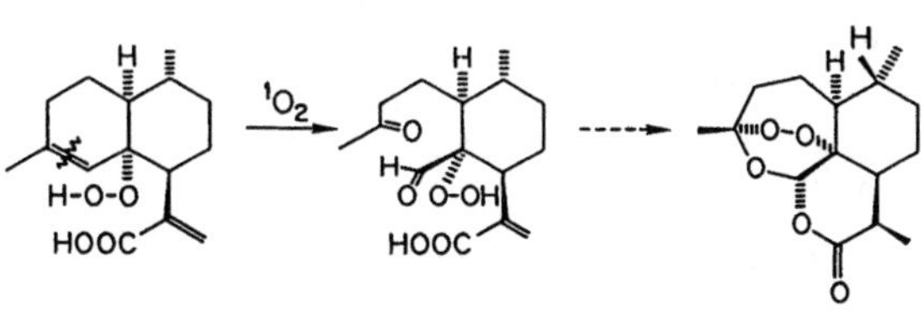

(144) *(145)* *artemisinin (146)*

Fig. 32. Biogenetic examples of oxidations leading to seco-sesquiterpene lactones.

An analog of the intermediate hydroperoxide **(148)** was proposed to be the precursor for the 4,5-seco-eudesmanolide skeleton represented by structure **(143)** in Figure 32.[28,101] The biogenesis of the 1(10)-seco-eudesmanolide ivangulin **(153)** was earlier formulated to proceed via hydroperoxide intermediates, as indicated by the arrows in structures **(151)** and **(152)**, leading to the ivangulin skeleton **(153)**.[1] Alternatively, a pathway via the 5αH,10β-methyleudesmanolide **(154)** and its hydroperoxide equivalent **(155)** was proposed.[28]

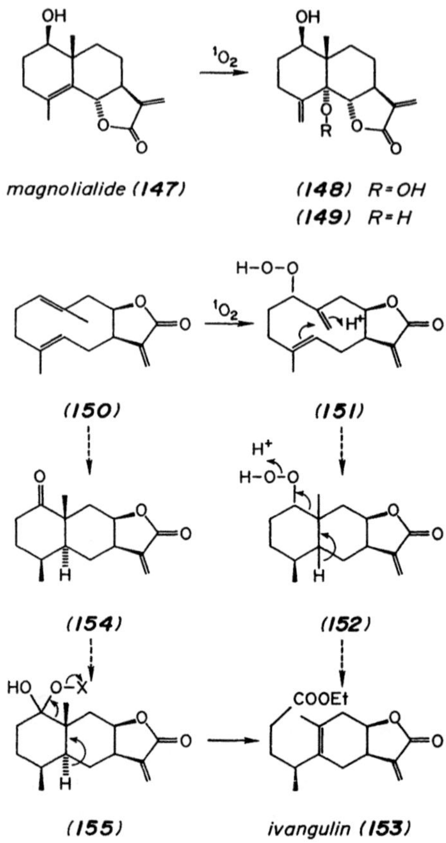

Fig. 33. Biogenesis of angular hydroperoxide eudesmanolides and 1,10-seco-eudesmanolides.

CONCLUSION

Although major progress has been made in the isolation and structural aspects of a large number of sesquiterpene lactones, the biosynthesis and regulation of the biosynthetic pathways is still not understood. Therefore, chemists and biologists still have to rely on biogenetic proposals supported by biomimetic experiments. In spite of the speculative nature of many biogenetic pathways presented in this review, it is important to note that in many instances stereochemical and mechanistic organic chemistry concepts can be applied as predictive tools in the study of biosynthetic routes. A solid understanding of the influences of configurations and conformations of the germacranolide precursors permits insight into the stereochemical requirements of hypothetical biosynthetic processes and allows stereochemical predictions in naturally occuring lactones. In return, the information gained from biosynthetic modeling and biomimetic transformations can assist in structural studies of new sesquiterpene lactones.

Considerable knowledge related to the biosynthetic pathways and the enzymes involved in the regulation of other natural products such as flavonoids and certain alkaloids has accumulated. The time has now come to direct our attention toward biosynthetic studies of one of the major groups of natural products, the sesquiterpene lactones. Perhaps, past difficulties encountered in attempts to incorporate biosynthetic precursors into sesquiterpene lactones will be overcome by the use of cell and/or organ cultures. In particular, "hairy root" cultures show considerable potential.[105] These differentiated organs generally express the same pattern of secondary metabolites which are characteristic of the species from which they are derived.

ACKNOWLEDGEMENTS

My sincere thanks are extended to my former and present graduate students, postdoctorals and visiting scientist for their major contributions to gain insight into the structural and biogenetic aspects of sesquiterpene lactone chemistry. Many thanks to Frank Fronczek for providing the stereopairs of several molecular structures.

I thank Ms. Linda Gauthier for typing the original manuscript and Ms. Luz Barona and Mr. Henry Hutado for drawing all figures for this review. Many thanks to my wife Helga for her patience and encouragement and apologies to our grandson Christopher for not keeping my promise to visit him more frequently.

Support by the Louisiana Education Quality Support Fund (86-89)-RD-A-13 and the National Science Foundation Biotechnology Program (Project No. EET-8713078) is thankfully acknowledged.

REFERENCES

1. FISCHER, N.H., E.J. OLIVIER, H.D. FISCHER. 1979. The biogenesis and chemistry of sesquiterpene lactones. In: Prog. Chem. Org. Nat. Prod. (W. Herz, H. Grisebach, G.W. Kirby, eds.) Springer, Wien, New York, Vol. 38, pp. 47-390.
2. SEAMAN, F.C. 1982. Sesquiterpene lactones as characters in the Asteraceae. Bot. Review 48: 121-592.
3. YOSHIOKA, H., T.J. MABRY, B.N. TIMMERMANN. 1973. Sesquiterpene Lactones: Chemistry, NMR, and Plant Distribution. Univ. of Tokyo Press, Tokyo, 544 pp.
4. BRETON FUNES, J.L. 1974. Lactonas sesquiterpenicas. In: Servicio de Publicaciones de la Caja General de Ahorros de Santa Cruz de Tenerife. Santa Cruz de Tenerife, pp. 12-119.
5. ASAKAWA, Y. 1982. Chemical constituents of the Hepaticae. In: Prog. Chem. Org. Nat. Prod. (W. Herz, H. Grisebach, G. W. Kirby, eds.) Springer, Wien, New York, Vol. 42, 1-285.
6. HOLUB, M., M. BUDESINSKY. 1986. Sesquiterpene lactones of the Umbelliferae. Phytochemistry 25: 2015-2026.
7. KORTE, F., H. BARKEMEYER, I. KORTE. 1959. Neue Ergebnisse der Chemie Pflanzlicher Bitterstoffe. In: Fortschr. Chem. Org. Naturst. (L. Zechmeister, ed.), Springer, Wien, Vol. 17, pp. 124-182.
8. SORM, F. 1961. Medium ring terpenes. In: Fortschr. Chem. Org. Naturst. (L. Zechmeister, ed.), Springer, Wien, Vol. 19, pp. 1-31.
9. SORM, F., L. DOLEJS. 1966. Guaianolides and germacranolides. In: Chimie de Substances Naturelles (E. Lederer, ed.) Herman, Paris, 153 pp.
10. SORM, R. 1970. Advances in terpene chemistry, Pure Appl. Chem. 21: 263-283.
11. ROMO, J., A. ROMO DE VIVAR. 1967. The Pseudoguaianolides. In: Fortschr. Chem. Org. Naturst. (L. Zechmeister, ed.), Springer, Wien, Vol. 25, 90-130.
12. HERZ, W. 1968. Pseudoguaianolides in Compositae. In: Recent Advances in Phytochemistry. (T.J. Mabry, R.E. Alston, V.C. Runeckles, eds.), Appleton-Century-Crofts, New York, Vol. 1, pp. 229-269.
13. HERZ, W. 1973. Pseudoguaianolides in Compositae. In: Chemistry in Botanical Classification. Nobel Symposium 25: Academic Press, New York, pp. 153-172.
14. HERZ, W. 1978. Sesquiterpene lactones in the

Compositae. In: Biology and chemistry of the Compositae (V.H. Heywood, J.B. Harborne, B.L. Turner, eds.), Academic Press, London, New York, San Francisco, Vol. 1, pp. 337-357.
15. MABRY, T.J. 1970. Intraspecific variation of sesquiterpene lactones in *Ambrosia* (Compositae). Application to evolutionary problems at the populational level. In: Phytochemical Phylogeny (J. B. Harborne, ed.), Academic Press, London, pp. 269-300.
16. MABRY, T.J. 1973. Chemistry of geographical races. Pure Appl. Chem. 34: 377-400.
17. PINDER, A.R. 1977. The chemistry of the eremophilanes and related sesquiterpenes. In: Prog. Chem. Org. Nat. Prod. (W. Herz, H. Grisebach, G.W. Kirby, eds.), Springer, Wien, Vol. 34, pp. 81-186.
18. ROBERTS, J.S., I. BRYSON. 1984. Sesquiterpenoids. Nat. Prod. Rep. 1: 105-169.
19. FRAGA, B. M. 1985. Natural sesquiterpenoids. Nat. Prod. Rep. 2: 147-161.
20. FRAGA, B. M. 1986. Natural sesquiterpenoids. Nat. Prod. Rep. 3: 273-296.
21. FRAGA, B. M. 1987. Natural sesquiterpenoids. Nat. Prod. Rep. 4: 473-498.
22. FRAGA, B. M. 1988. Natural sesquiterpenoids. Nat. Prod. Rep. 5: 497-521.
23. CASSADY, J.M., M. SUFFNESS. 1980. Terpenoid antitumor agents. In: Anticancer Agents Based on Natural Products Models (J. M. Cassady, J.D. Douros, eds.) Academic Press, London, pp. 201-269.
24. MISRA, R., R.C. PANDEY. 1981. Cytotoxic and antitumor terpenoids. In: Antitumor Compounds of Natural Origin: Chemistry and Biochemistry (A. Aszalos, ed.), CRC Press, Boca Raton, Vol. 2, pp. 145-192.
25. PICMAN, A.K. 1986. Biological activities of sesquiterpene lactones. Biochem. System. Ecol. 14: 255-281.
26. BURNETT, W.C., JR., S.B. JONES, JR., T.J. MABRY, W.G. PADOLINA. 1974. Sesquiterpene lactones--insect feeding deterrents in Vernonia. Biochem. System. Ecol. 2: 25-29.
27. CORDELL, G.A. 1976. Biosynthesis of sesquiterpenes. Chem. Rev. 76:425-460.
28. HERZ, W. 1986. Biogenetic aspects of sesquiterpene lactone chemistry. In: Natural Product Chemistry (Atta-ur-Rahman, ed.) Springer, Berlin, Heidelberg, pp. 154-174.
29. GEISSMAN, T.A. 1973. The biogenesis of sesquiterpene lactones of the Compositae. In: Recent Advances in

Phytochemistry (V. C. Runeckles, T. J. Mabry, eds.) Academic Press, New York and London, Vol. 6, pp. 65-95.

30. HERZ, W. 1977. Biogenetic aspects of sesquiterpene lactone chemistry. Israel J. Chem. 16: 32-44.

31. ROGERS, D., G.P. MOSS, S. NEIDLE. 1972. Proposed convention for describing germacranolide sesquiterpenes. Chem. Commun. 1972: 142-143.

32. SEAMAN, F.C., N.H. FISCHER, T.F. STUESSY. 1980. Systematic implications of sesquiterpene lactones in the subtribe Melampodiinae. Biochem. Syst. Ecol. 8: 263-271.

33. NEIDLE, S., D. ROGERS. 1972. X-ray determination of the structure and absolute configuration of a novel sesquiterpene melampodin. Chem. Commun. 1972: 140-141.

34. MALCOLM, A.J., J.F. CARPENTER, F.R. FRONCZEK, N.H. FISCHER. 1983. Longicornins A to D, four *cis*-1(10), *cis*-4-germacradienolides from *Melampodium longicorne*. Phytochemistry 22: 2759-2766.

35. WITT, M.E., S.F. WATKINS. 1978. Crystal structure of tamaulipin A, a 1(10)-*trans*, 4-*trans*-germacranolide sesquiterpene lactone. J. Chem. Soc. Perkin Trans. 2, 1978: 204-208.

36. FRONCZEK, F.R., A. MALCOLM, N.H. FISCHER. 1983. The molecular structure of melampodinin-A. J. Nat. Prod. 46: 170-173.

37. OBER, A.G., F.R. FRONCZEK, N.H. FISCHER. 1985. Sesquiterpene lactones of *Calea divaricata* and the molecular structure of leptocarpin acetate. J. Nat. Prod. 48: 302-306.

38. SAMEK, Z., J. HARMATHA. 1978. Use of structural changes for stereochemical assignments of natural α-exomethylene γ-lactones of the germacra-1(10),4-dienolide type on the basis of allylic and vicinal couplings of bridgehead protons. Hydrogenation of endocyclic double bonds. Collect. Czech. Chem. Commun. 43: 2779-2799.

39. BHACCA, N.S., N.H. FISCHER. 1969. The determination of the conformation of a germacranolide (dihydrotamaulipin A acetate) with the aid of nuclear Overhauser effects. Chem. Commun. 1969: 68-69.

40. QUIJANO, L., J.S. CALDERON, F. GOMEZ-GARIBAY, S. BAUTISTA, T. RIOS, F.R. FRONCZEK. 1986. Montafrusin B, a germacrolide from *Montanoa frutescens* and the molecular strucure of montafrusin A. Phytochemistry 25: 695-697.

41. RYCHLEWSKA, U. 1981. Stereochemistry of a novel sesquiterpene lactone. X-ray determination of the

structure of ursiniolide A monohydrate. J. Chem. Soc. Perkin Trans. 2, 1981: 660-663.

42. SEAMAN, F.C., A.J. MALCOLM, N.H. FISCHER. 1983. Germacra-12,6β-olides from *Montanoa revealii* and *M. mollissima*. Phytochemistry 23:1063-1066.

43. EL-FERALY, F.S., D.A. BENIGNI, A.T. MCPHAIL. 1983. Biogenetic-type synthesis of santonin, chrysanolide, dihydrochrysanolide, tulirinol, arbusculin C, tanacetin, and artemin. J. Chem. Soc. Perkin Trans. 1, 1983: 355-364.

44. TORI, K., I. HORIBE, Y. TAMURA, K. KURIYAMA, H. TADA, K. TAKEDA. 1976. Re-investigation of the conformation of laurenobiolide, a ten-membered ring sesquiterpene lactone by variable-temperature carbon-13 nmr spectroscopy. Evidence for the presence of four conformational isomers in solution. Tetrahedron Letters 1976: 387-390.

45. QUIJANO, L., J.S. CALDERON, G.F. GOMEZ, P.J. LOPEZ, T. RIOS, F.R. FRONCZEK. 1984. The crystal structure of 6-epi-desacetyllaurenobiolide, a germacra-a(10), 4-diene-12,8α-olide from *Montanoa grandiflora*. Phytochemistry 23: 1971-1974.

46. MING, C.W., R. MAYER, H. ZIMMERMANN, G. RUECKER. 1989. A non-oxidized melampolide and other germacranolides from *Aristolochia yunnanensis*. Phytochemistry 28: 3233-3234.

47. KRISHNAN, S., S.K. PAKNIKAR, S.C. BHATTACHARYYA, A.L. HALL, W. HERZ. 1978. Biogenetic-type transformation of germacranolide to melampolides. J. Indian Chem. Soc. 55: 1142-1147.

48. HARUNA, M., K. ITO. 1981. Regio- and stereo-specific allylic oxidation of germacrane-type sesquiterpene lactones with selenium dioxide and t-butyl hydroperoxide. J. Chem. Soc. Chem. Comm. 1981:483-485.

49. EL-FERALY, F.S. Melampolides from *Magnolia grandiflora*. 1984. Phytochemistry 23: 2372-2374.

50. MALCOLM, A.J. 1983. Ph.D. dissertation, Louisiana State University, Baton Rouge, Louisiana, U.S.A., pp. 215.

51. GOVINDACHARI, T.R., B.S. JOSHI, V.N. KAMAT. 1965. Structure of parthenolide. Tetrahedron 21: 1509-1519.

52. FISCHER, N.H., R.A. WILEY, D.L. PERRY. 1976. Sesquiterpene lactones from *Melampodium* (Compositae, Heliantheae); Structural and biosynthetic considerations. Rev. Latinoamer. Quim. 7: 87-93.

53. OLIVIER, E.J., A.J. MALCOLM, D.V. ALLIN, N.H. FISCHER. 1983. Melrosin A, B, and C, three *cis*-1(10)-*cis*-4-

germacradienolides from *Melampodium rosei*. Phytochemistry 22: 1453-1456.
54. JAIN, T.C., C.M. BANKS, J.E. MCCLOSKEY. 1976. Reversible dimethylamine addition as a protecting reaction for α,β-unsaturated methylene groups of γ-lactones and its regeneration by basic elimination of quaternary ammonium salts. Tetrahedron 32: 765-768.
55. FISCHER, N.H. 1978. On the biogenesis of pseudoguaianolides. Rev. Latinoamer. Quim. 9: 41-46.
56. DELGADO, G., H. HERNANDEZ, A. ROMO DE VIVAR. 1984. Structure of elemanschkuhriolide. Melampolides as possible biogenetic precursors of $C_{14}\alpha, C_5\beta$-elemanolides. J. Org. Chem. 49: 2994-2996.
57. BRECKNELL, D.J., R.M. CARMAN. 1979. The interconversion of two elemadienolides through two consecutive Cope rearrangements. Aust. J. Chem. 32: 2097-2102.
58. BOHLMANN, F., J. JAKUPOVIC, A. SCHUSTER. 1983. 8-Hydroxypegolettiolide, a sesquiterpene lactone with a new carbon skeleton and further constituents from *Pegolettia senegalensis*. Phytochemistry 22: 1637-1644.
59. HOLUB, M., Z. SAMEK, V. HEROUT. 1972. Structure of isolaserolide from *Laser trilobum*. Phytochemistry 11: 3053-3055.
60. SAMEK, Z., M. HOLUB, U. RYCHLEWSKA, H. GRABARCZYK, B. DROZDZ. 1979. Germacra-1(10),4-dien-*cis*-6,12-olides: a novel stereochemical group of natural sesquiterpenic lactones from *Ursinia anthemoides* (L.) Poiret. Tetrahedron Letters 29: 2691-2694.
61. RAO, A.S., A.P. SADGOPAL, S.C. BHATTACHARYYA. 1961. Structure of suassurea lactone. Tetrahedron 13: 319-323.
62. KUPCHAN, S.M., R.J. HEMINGWAY, D. WERNER, A. KARIM. 1969. Vernolepin, a novel sesquiterpene dilactone tumor inhibitor from *Vernonia hymenolepis* A. Rich. J. Org. Chem. 34: 3903-3908.
63. MALDONADO, E., M. SORIANO-GARCIA, C. GUERRERO, A. ROMO DE VIVAR, A. ORTEGA. 1985. The structures of 9-hydroxyzinnolides and their rearranged acetates. Phytochemistry 24: 991-994.
64. BOHLMANN, F., C. ZDERO. 1979. Neue Germacranolide and andere Inhaltsstoffe aus Vertretern der Subtribus Gochnatiinae. Phytochemistry 18: 95-98.
65. SUTHERLAND, J.K. 1974. Regio- and stereo-specificity in the cyclization of medium ring 1,5-dienes. Tetrahedron 30: 1651-1660.
66. WILTON, J.H., R.W. DOSKOTCH. 1983. Acid cyclization and

other products of germacranolide epoxide lipiferolide. J. Org. Chem. 48: 4251-4256.
67. PARODI, F.J., F.R. FRONCZEK, N.H. FISCHER. 1989. Biomimetic transformations of 11,13-dihydroparthenolide and oxidative rearrangements of a guai-1(10)-en-6,12-olide. J. Nat. Prod. 52: 554-566.
68. GONZALEZ, A.G., A. GALINDO, H. MANSILLA, J.A. PALENZUELA. 1983. Evidence for the biogenesis of *trans*-(1β-H; 5α-H)-guaianolides. Tetrahedron Letters 24: 969-972.
69. KULKARNI, G.H., G.R. KELKAR, S.C. BHATTACHARYYA. 1964. Cyclocostunolides. Tetrahedron 20: 2639-2645.
70. RODRIGUES, A.A.S., M. GARCIA, J.A. RABI. 1978. Facile biomimetic synthesis of costunolide-1(10)-epoxide, santamarin and reynosin. Phytochemistry 17: 953-945.
71. PATHAK, S.P., B.V. BAPAT, G.H. KULKARNI. 1970. Conversion of costunolide into santamarine and reynosin. Indian J. Chem. 8: 471-472.
72. GONZALEZ, A.G., A. GALINDO, H. MANSILLA. 1980. Biomimetic cyclization of gallicin to form guaianolides. Tetrahedron 36: 2015-2017.
73. GONZALEZ, A.G., A. GALINDO, H. MANSILLA, A. ALEMANY. 1979. Conformation of gallicin, a ten-membered ring sesquiterpene lactone. Tetrahedron Letters 39: 3769-3772.
74. AOTA, K., C.N. CAUGHLAN, M.T. EMERSON, W. HERZ, S. INAYAMA, M. UL-HAQUE. 1970. Structure and absolute configuration of pulchellin. Crystal and molecular structure of 3-bromoanhydrodehydrodihydropulchellin. J. Org. Chem. 35: 1448-1452.
75. DULLFORCE, T.A., G.A. SIM, D.N.J. WHITE, J.E. KELSEY, S.M. KUPCHAN. 1969. The stereochemistry of gaillardin. Tetrahedron Letters 1969: 973-976.
76. ZDERO, C., F. BOHLMANN, R.M. KING, H. ROBINSON. 1986. Guaianolide glucosides from *Helenium donianum*. Planta Medica 1986: 22-24.
77. HERZ, W., R. MURARI, J.F. BLOUNT. 1979. Revised structures of pleniradin and baileyin and their bearing on the biogenesis of helenanolides. J. Org. Chem. 44: 1873-1876.
78. GONZALEZ, A.G., A. GALINDO, H. MANSILLA, A. GUTIERREZ. 1982. Evidence for the biogenesis of 1a-hydroxy-*trans*-eudesmanolides. J. Chem. Soc. Perkin 1, 1982: 881-884.
79. BOHLMANN, F., G. SCHMEDA-HIRSCHMANN, J. JAKUPOVIC. 1984. Further 6,12-*cis*-germacranolides and eudesmanolides from *Montanoa* species. J. Nat. Prod. 47:663-672.
80. HOLUB, M., M. BUDESINSKY, Z. SMITALOVA, D. SAMAN, U.

RYCHLEWSKA. 1985. Structure of Isosilerolide, relative and absolute configuration of silerolide and lasolide--sesquiterpenic lactones of new stereoisometric type of eudesmanolides. Collect. Czech. Chem. Commun. 51: 903-929.
81. HOLUB, M., Z. SAMEK, S. VASICKOVA, M. MASOJIDKOVA. 1978. 11-Hydroxy-1βH, 5βH, 6αH, 7αH-guaian-6,12-olides: relative and absolute configuration of the sesquiterpenic lactones montanolide, isomontanolide, acetylisomontanolide and related substances. Collec. Czech. Chem. Commun. 43: 2444-2470.
82. SMITALOVA, Z., M. BUDESINSKY, D. SAMAN, M. HOLUB. 1986. Minor sesquiterpenic lactones of *Laser trilobum* (L.) Borkh. Species. Collect. Czech. Chem. Commun. 51: 1323-1339.
83. SEAMAN, F.C., N.H. FISCHER, T.J. MABRY. 1986. Isodehydroleucodin and another novel *cis*-lactonized guaianolide from *Montanoa imbricata*. Phytochemistry 25: 2663-2664.
84. ANDERSON, G.D., R.S. MCEWEN, W. HERZ. 1972. Relative and absolute configuration of axivalin and its congeners. Tetrahedron Letters 1972: 4423-4426.
85. MINATO, H., S. NOSAKA, I. HORIBE. 1964. The structure of carabrone, a new component of *Carpesium abrotanoides*, Linn. J. Chem. Soc. 1964: 5503-5510.
86. HERZ, W., S.V. BHAT, A.L. HALL. 1970. Parthemollin, a new xanthanolide from *Parthenice mollis* Gray. J. Org. Chem. 35: 1110-1114.
87. RODRIGUEZ, E., H. YOSHIOKA, T.J. MABRY. 1971. The sesquiterpene lactone chemistry of the genus *Parthenium*. Phytochemistry 10: 1145-1154.
88. KAGAN, H.B., H.E. MILLER, W. RENOLD, M.V. LAKSHMIKANTHAM, L.R. TETHER, W. HERZ, T.J. MABRY. 1966. The structure of psilostachyin C, a new sesquiterpene lactone from *Ambrosia psilostachya* D.C. J. Org. Chem. 31: 1629-1632.
89. STEFANOVIC, M., I. ALJANCIC-SOLAJA, S. MILOSAVLJEVIC. 1987. A 3,4-seco-ambrosanolide from *Ambrosia artemisiifolia*. Phytochemistry 26: 850-852.
90. BORDOLOI, M.J., R.P. SHARMA, J.C. SARMA. 1986. Biomimetic transformation of a guaianolide to a pseudoguaianolide. Tetrahedron Letters 27: 4633-4634.
91. ORTEGA, A., E. MALDONADO. 1986. A one-step transformation of 4α,β-epoxygermacrolides into pseudoguaianolides. First Symposium of the Latin American Phytochemical Academy. Mexico City, March 3-6, 1986.
92. BOHLMANN, F., J. JAKUPOVIC, M. AHMED, A. SCHUSTER. 1983.

Sesquiterpene lactones and other constituents from *Schistostephium* species. Phytochemistry 22: 1623-1636.

93. KITAGAWA, I., Y. YAMAZOE, H. SHIBUYA, R. TAKEDA, H. TAKENO, I. YOSIOKA. 1974. Biogenetically patterned transformation of eudesmanolide to eremophilanolide I. Angular methyl migration of 5α,6α-epoxydihydroalantolactone. Chem. Pharm. Bull. 22: 2662-2674.
94. TANAKA, N., T. YAZAWA, K. AOYAMA, T. MURAKAMI. 1976. Chemical studies on the constituents of *Xanthium canadense* Mill. Chem. Pharm. Bull. 24: 1419-1421.
95. BOHLMANN, F., C. ZDERO, R.M. KING, H. ROBINSON. 1986. Neue Sesquiterpenlactone und andere Inhaltsstoffe aus *Stevia mercedensis* und *Stevia achalensis*. Liebigs Ann. Chem. 1986: 799-813.
96. DOSKOTCH, R.W., C.D. HUFFORD, F.S. EL-FERALY. 1972. Further studies on the sesquiterpene lactones tulipinolide and epitulipinolide from *Liriodendron tulipifera* L. J. Org. Chem. 37: 2740-2744.
97. USKOKOVIC, M.R., T.H. WILLIAMS, J.F. BLOUNT. 1974. The structure and absolute configuration of arteannuin B. Helv. Chim. Acta 57: 600-602.
98. MISRA, L.N. 1986. Arteannuin-C, a sesquiterpene lactone from *Artemisia annua*. Phytochemistry 25: 2892-2893.
99. HUNECK, S., C. ZDERO, F. BOHLMANN. 1986. Seco-guaianolides and other constituents from *Artemisia* species. Phytochemistry 25: 883-889.
100. RUSTAIYAN, A., A. BAMONIERI, M. RAFFATRAD, J. JAKUPOVIC, F. BOHLMANN. 1987. Eudesmane derivatives and highly oxygenated monoterpenes from Iranian *Artemisia* species. Phytochemistry 26: 2307-2310.
101. APPENDINO, G., P. GARIBOLDI, M. CALLERI, G. CHIARI, D. VITERBO. 1983. The structure and conformation of umbellifolide, a 4,5-seco-eudesmane derivative. J. Chem. Soc. Perkin Trans. 1. 1983: 2705-2709.
102. KLAYMAN, D.L. 1985. Quinghaosu (Artemisinin): an antimalarial drug from China. Science 228: 1049-1055.
103. NAIR, M.S.R., N. ACTON, D.L. KLAYMAN, K. KENDRICK, H.H. LEHMAN, S. MANTE. 1985. Int. Research Congress on Nat. Prod., Chapel Hill, NC, U.S.A., Abstract No. 102.
104. EL-FERALY, F.S., I.A. AL-MESHAL, M.A. AL-YAHYA, M.S. HIFNAWY. 1986. On the possible role of quinghao acid in the biosynthesis of artemisinin. Phytochemistry 25: 2777-2778.
105. FLORES, H.E. 1987. Use of plant cells and organ culture in the production of biological chemicals. ACS Symposium Series 334, ACS, Washington DC, pp. 66-86.

Chapter Five

PLANT HORMONES AND THE BIOSYNTHESIS OF GIBBERELLINS: THE EARLY-13-HYDROXYLATION PATHWAY LEADING TO GA_1

BERNARD O. PHINNEY AND CLIVE R. SPRAY

Department of Biology
University of California
Los Angeles, CA 90024-1606, U.S.A.

INTRODUCTION

This report will briefly review the gibberellin (GA) biosynthetic pathway, from mevalonic acid (MVA) to *ent*-kaurene, to GA_{12}-aldehyde, to the C_{19} gibberellins. Emphasis will be given to the early-13-hydroxylation pathway, a branch pathway from GA_{12}-aldehyde that leads to the bioactive gibberellin, GA_1. The use of GA mutants in the analysis of GA biosynthesis will be discussed. Thus it is not the purpose of this paper to give a detailed analysis of the GA biosynthetic pathway, nor to give an analysis of the chemistry of this interesting class of natural products, nor to review the physiology of the gibberellins.

Specific and detailed information on the chemistry and biosynthesis of gibberellins is available in chapters from a number of books (e.g. Chapter 3, by Graebe and Ropers, in Vol. 1, Phytohormones and Related Compounds: A Comprehensive Treatise:[1] Chapter 1, by Bearder, in Hormonal Regulation of

Biochemistry of the Mevalonic Acid Pathway to Terpenoids
Edited by G.H.N. Towers and H. A. Stafford
Plenum Press, New York

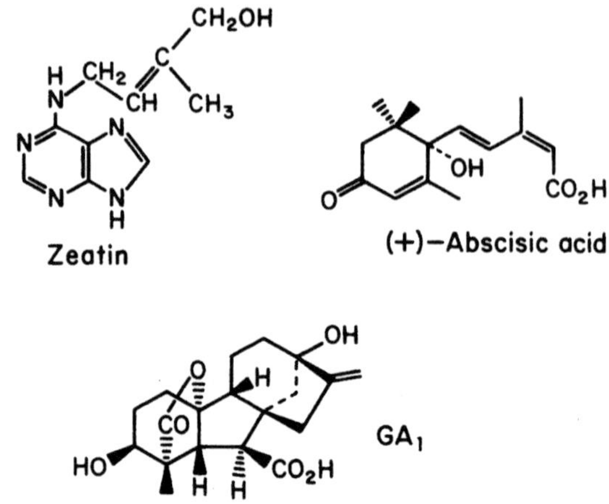

Fig. 1. Specific examples (zeatin, abscisic acid, and GA_1) of the three classes of plant hormones, the *cytokinins*, the *abscisins*, and the *gibberellins*. They originate, all or part, from the MVA biosynthetic pathway.

Development, Molecular Aspects of Plant Hormones:[2] and all chapters in The Biochemistry and Physiology of Gibberellins, Vol. 1, edited by A. Crozier.[3] Additional details are available from symposia publications by Graebe *et al.*,[4] Graebe,[5] MacMillan and Phinney,[6] and Phinney.[7] The physiological importance of GAs in higher plants has been analyzed and discussed in reviews,[8-11] and the biosynthesis of GAs has recently been reviewed by Graebe.[12] Information on the genetics and physiology of GA mutants (dwarf mutants) in higher plants has been elegantly analyzed in a recent mini-review by Reid.[13]

GENERAL BACKGROUND

The three classes of plant hormones (Fig. 1), *abscisins*, *cytokinins*, and *gibberellins*, originate, all or in part, from the mevalonic acid (MVA) biosynthetic pathway. The *abscisins* are sesquiterpenes that branch from farnesylpyrophosphate (FPP) (see Fig. 4). They promote leaf abscission (hence their name); they often play an inhibitory role in growth and apparently control dormancy; they are involved in water stress and in the opening and closing of stomata. The isoprenoid sidechains of *cytokinins* come from isopentenylpyrophosphate (IPP). Cytokinins, in conjunction with auxin, control many

Fig. 2. Growth response of the single gene dwarf mutant, *dwarf-3* (*d3*) to exogenously applied GA_1. The gibberellin was applied in acetone:water (1:10) to the uppermost unfolding leaves in amounts and frequencies required to give a "normal type" growth response to the treated dwarf; one normal plant was given a similar treatment. Total treatment was 150μg/plant. Plants read from left to right: non-treated dwarf control (*d3/d3*), treated dwarf (*d3/d3*), treated normal (+/+), and non-treated normal control (+/+): *dwarf-3* is a simple recessive mutation that controls a step after GA_{12}-aldehyde and before GA_{53}.

aspects of cell division and plant differentiation; cytokinins also delay senescence. The *gibberellins* (GAs) branch from copalylpyrophosphate (CPP). They also control a variety of growth responses, e.g. the release of α-amylase during seed germination, fruit size, and shoot elongation, especially the reversal of genetic dwarfism (Fig. 2) and the bolting of many rosetted biennials. Thus the MVA pathway plays an important role in the control of the growth of higher plants.

Gibberellins have been identified from many species of flowering plants; they are present in conifers, ferns, and fern allies. Gibberellins are also produced by two species of fungi (*Ascomycetes*), and there is recent evidence for their occurrence in bacteria.[14,15] Gibberellins are present in animals as a part of their ingested food (plants). Gibberellins apparently play no role in animal nutrition. While the number of identified GAs is currently 79, only a limited number are found in any one species, with even fewer being present in a particular organ of the plant. Of these GAs, most are either dead-end branch metabolites, or intermediates to a very limited number of GAs that have *per se* bioactivity.

Gibberellins were first identified as secondary metabolites from the fungus, *Gibberella fujikuroi*. The history of the discovery of the GAs has been reviewed by Phinney[16]. While the function of GAs in the fungus is still unknown, the spectacular responses of higher plants to the fungal GAs gave the original impetus that led to their isolation as natural products from higher plants (albeit some 20 years after the original discovery by the Japanese of GAs in the fungus). Until recently, biosynthetic studies with GAs and their precursors were limited to the fungus for at least two reasons: first, the fungus produces relatively large amounts of GAs (mainly GA_3) over a relatively short period of time (e.g. 6 g. L^{-1} of culture media, over a period of 2 weeks); by contrast, the levels of endogenous GAs found in the green tissues of plants is in the order of 10^{-9}g. g^{-1} fresh weight. Second, while it is relatively easy to purify fungal extracts for analytical studies, this is not the case for vegetative tissues of higher plants. As a result, there were few reports on biosynthetic studies using higher plants prior to 1980. A number of technical advances led to the studies with higher plants. One was the development of highly sensitive gas chromatography - mass spectrometry (GC-MS) instrumentation for the identification of GAs. Another was the development of

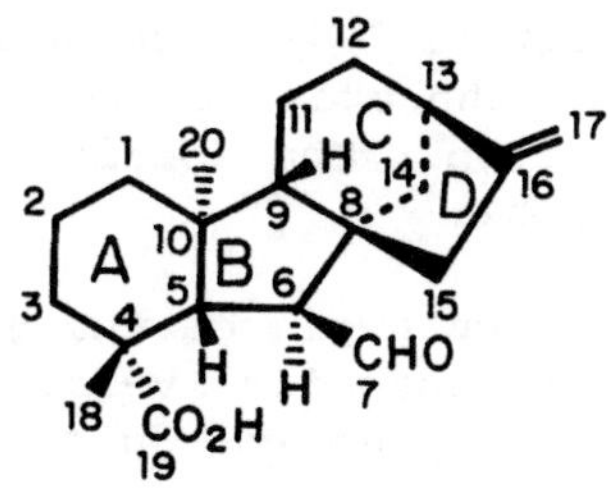

Fig. 3. Numbering system for the *ent*-gibberellane carbon skeleton. The example is GA_{12}-aldehyde, the precursor to all the GAs.

rigorous purification steps such as HPLC (and currently immunoaffinity chromatography[17,18]) that allowed the preparation of plant extracts having the purity necessary for GC-MS analysis. In addition, there have been rapid developments in chemical methods which allow for the preparation of GAs doubly-labeled with a radioactive and a stable isotope,[19] and also GAs stereospecifically labeled with deuterium.[20] An additional and crucial feature of the studies on the GA biosynthetic pathway in maize has been the use of single gene GA-mutants,[21,22] mutants that control specific and different steps in the pathway.[23,24] Gibberellin-mutants have also been used to study GA-dependent shoot elongation in pea,[25-27] rice,[28,29] tomato,[30] *Arabidopsis*,[31] and *Brassica*.[32]

CHEMISTRY

All gibberellins (GAs) originate from the *ent*-gibberellane, GA_{12}-aldehyde; all GAs have either the *ent*-gibberellane (C_{20} GAs) or the *ent*-20-norgibberellane (C_{19} GAs) carbon skeleton. The systematic nomenclature for the gibberellins follows the accepted rules for diterpenes (Fig. 3). The gibberellins fall into two groups, the C_{20} GAs and the C_{19} GAs. The C_{20} GAs are the precursors to the C_{19} GAs, the latter originating through the loss of carbon-20. Thus the GAs differ from each other in the presence or absence of carbon-20; also in the position and number of hydroxyl groups, the oxidation state of carbon-18 and carbon-20, lactone formation, the presence and position of double bonds in ring A, epoxide formation, the presence of a carbonyl group, and hydration of the 16,17-double bond. The trivial name, gibberellin, is abbreviated, GA. The gibberellins are numbered chronologically, GA_1-GA_n, based on the order of isolation (discovery) and structure determination. The specific number is assigned by MacMillan and Takahashi.[33]

Studies on the biosynthesis of the gibberellins have been based on the use of isotopically-labeled GAs and their precursors. There are many reports on the preparation of labeled GAs either chemically or biologically (c.f. 19,20,34). An isotopic label is required for the unequivocal identification (by MS) of metabolites produced on administration of a precursor. Both non-radioactive (e.g. ^2H, ^{13}C, or ^{18}O) and radioactive (e.g. ^3H or ^{14}C) isotopes have been used.

The general procedure for studies on the biosynthesis of the GAs is as follows: the labeled compound is fed to the tissue or organ under investigation. After incubation, the metabolites from the feed are purified by a combination of solvent partitioning and chromatography. The fractions containing radioactive metabolites are detected with a scintillation spectrometer. After final purification, radioactive fractions are derivatized for analysis by GC-MS. The specific gibberellins are identified by comparison to authentic standards of GC retention times, Kovats retention indicies, and mass spectra. Specific GAs are identified as metabolites of the feed by the appearence of a specific isotope peak in the MS.

Isotopically labeled GAs are also used to determine the levels of endogenous GAs present in the plant (for example, see reference 19). For this purpose, a known amount of labeled GA is added to the plant extract, and the extract is purified and analyzed by GC-MS. The level of the specific endogenous gibberellin is then calculated from the amount of dilution of label observed from the MS data.

THE GIBBERELLIN BIOSYNTHETIC PATHWAY

All metabolic steps from MVA to the diterpene GGPP have been shown to be present in a variety of organs from many higher plants as well as in the fungus, *Gibberella fujikuroi*. Details of this early portion of the pathway (MVA to GGPP) are the same in plants, micro-organisms, and animals. It is the cyclization of GGPP to *ent*-kaurene that is unique to higher plants and to the fungus *Gibberella fujikuroi*. In both kinds of organisms, the cyclization of GGPP to *ent*-kaurene is catalyzed by the enzyme *ent*-kaurene synthetase. This enzyme consists of activity A, which catalyses the step GGPP to CPP, and activity B which catalyses the step CPP to *ent*-kaurene.[35] The enzyme complex has been extensively studied by West and coworkers and it is currently being purified by the same group. Interestingly, the *dwarf-5* (*d5*) mutant of maize blocks the step CPP to *ent*-kaurene through an altered B-activity of *ent*-kaurene synthetase.[36] Following cyclization, carbon-19 of *ent*-kaurene is stepwise oxidized to *ent*-kaurenol, *ent*-kaurenal, and *ent*-kaurenoic acid; carbon-7 is hydroxylated to give *ent*-7α-hydroxykaurenoic acid. Recent studies have conclusively demonstrated these metabolic steps.[37] Contraction of ring B then results in the formation of the *ent*-gibberellane, GA_{12}-aldehyde (Fig. 4).

Several pathways diverge from GA_{12}-aldehyde to give the known 79 GAs (Figs. 5,6). The kinds and numbers of branch pathways differ depending on the organism and organ under study. The position of initial hydroxylation is the basis for the initiation of a specific pathway; there is evidence for the presence of at least five different branch pathways, some present in both the fungus and higher plants, others present in higher plants only. Hydroxylation does not occur prior to the biosynthesis of GA_{12}-aldehyde. There are hydroxylated derivatives of *ent*-kaurene, e.g. stevioside; however, they are branch products unrelated to the pathway leading to GA_{12}-aldehyde. Interestingly, the sequence of oxidation steps in the branch pathways parallel each other. Several of the enzymes catalyzing these steps show substrate non-specificity, especially in the fungus, *Gibberella fujikuroi*.[38] Radiolabeled feeding studies with intact plants suggest that the pathways branching from GA_{12}-aldehyde are separate and not interrelated. However, studies with cell-free systems from higher plants (primarily from developing embryos and the surrounding parts) clearly show the presence of a *grid* that interrelates members of the branch pathways.[39,40] This grid could arise from the break-down of intracellular *compartmentation* due to the use of cell-free systems.

The early-13-hydroxylation pathway found in maize is initiated from GA_{12}-aldehyde by the hydroxylation of carbon-13 (Fig. 5). However, the specific step in the pathway at which initial hydroxylation occurs is still ambiguous. It is not known whether GA_{12}-aldehyde is first oxidized to the acid, GA_{12}, followed by 13-hydroxylation to GA_{53}, or whether GA_{12}-

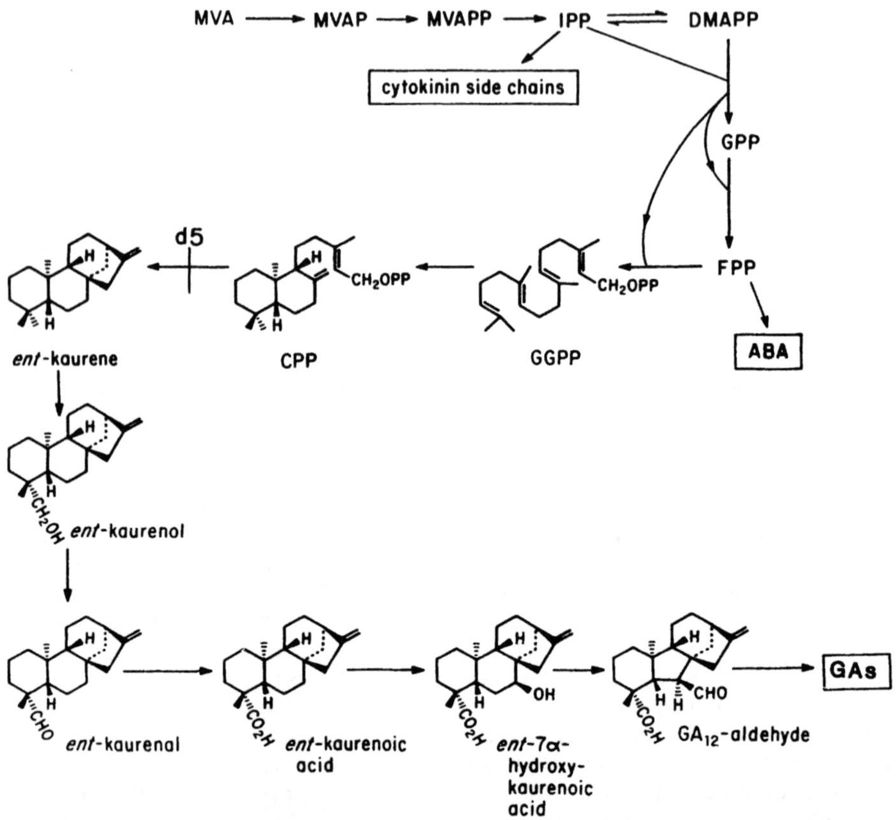

Fig. 4. Gibberellin biosynthetic pathway from MVA to GA_{12}-aldehyde. Note the biosynthetic origin of the three classes of plant hormones, the *cytokinins*, the *abscisins* (ABA), and the *gibberellins* (GAs). The *d5* mutant of maize blocks the step CPP to *ent*-kaurene.

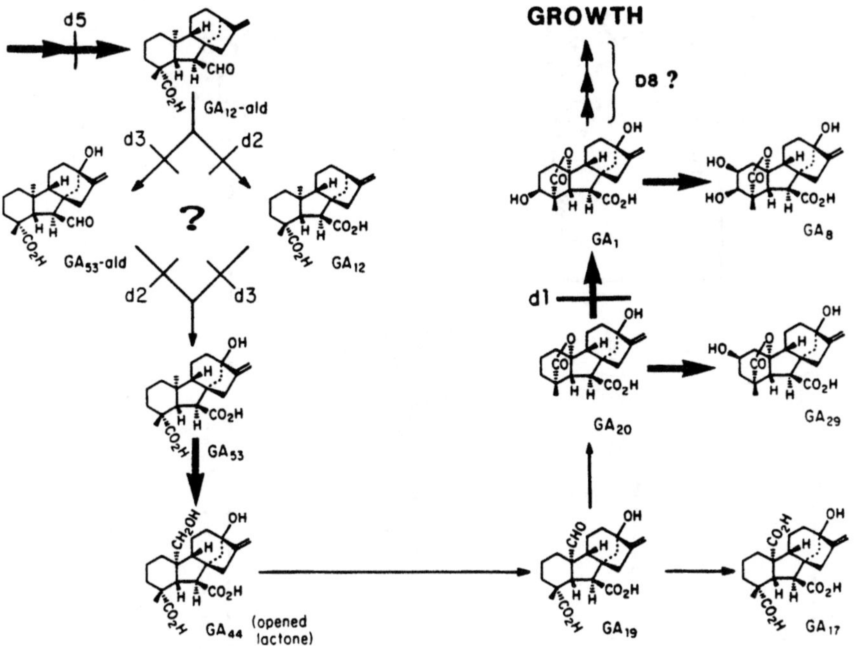

Fig. 5. Gibberellin biosynthetic pathway from GA_{12}-aldehyde to GA_1 *via* the early-13-hydroxylation pathway. Bold arrows indicate steps defined in maize by feeding studies. The single gene, non-allelic, dwarf mutants (GA mutants), *d1*, *d2*, *d3*, and *d5*, block specific and different steps as shown (see also figures 4 and 7). (The positions of the *d2* and *d3* mutants are still ambiguous).

aldehyde is first 13-hydroxylated to GA_{53}-aldehyde, and then oxidized at carbon-7 to GA_{53} (Fig. 5). The subsequent steps are well defined. Thus carbon-20 of GA_{53} is oxidized stepwise to the alcohol, GA_{44} (opened lactone), then to the aldehyde, GA_{19}, followed by the loss of carbon-20, with the formation of a γ-lactone to give GA_{20}, the first C_{19} gibberellin in the pathway. Finally, GA_{20} is 3β-hydroxylated to GA_1.

There are three additional gibberellins in the pathway: GA_{29} and GA_8, 2β-hydroxylated metabolites of GA_{20} and GA_1, respectively; and GA_{17}, the carbon-20 oxidation product of GA_{19}. Both GA_{29} and GA_8 are biologically inactive and not metabolized to bioactive gibberellins; also GA_{17} is not further metabolized to the C_{19}-GAs. The evidence for this

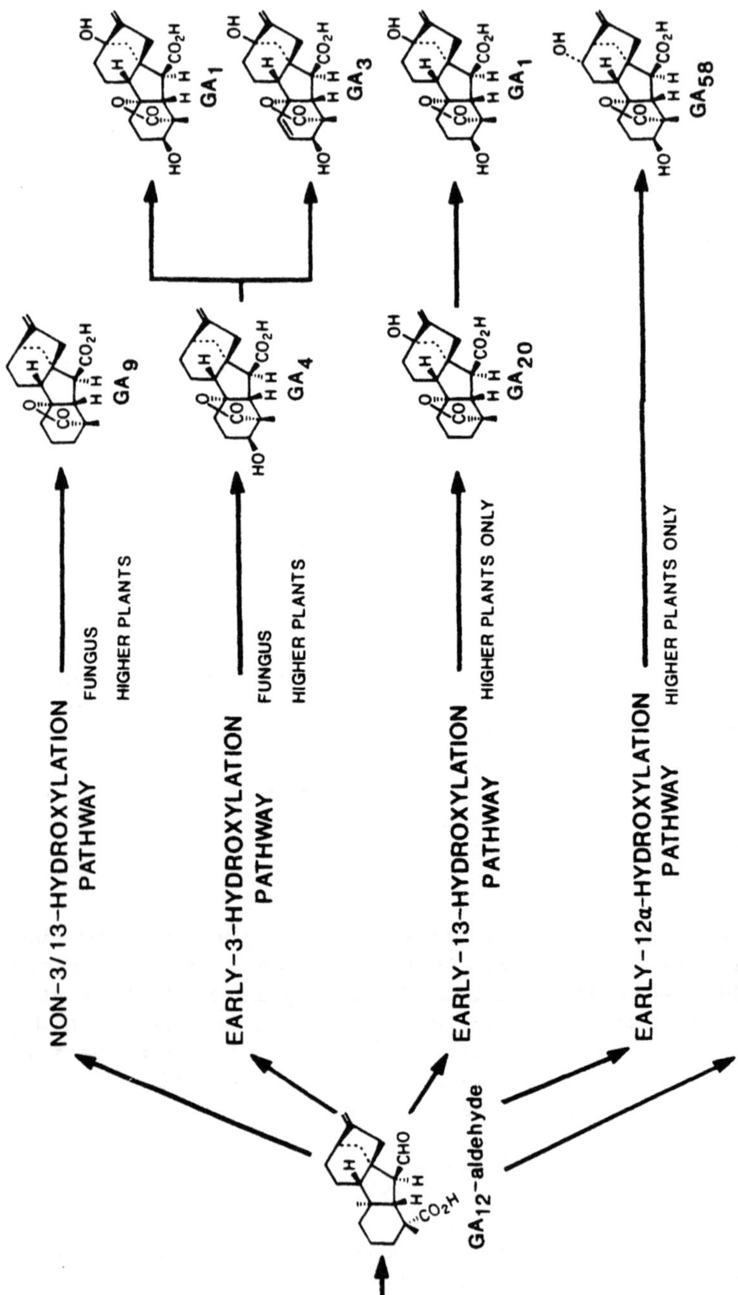

Fig. 6. Branch pathways from GA_{12}-aldehyde in *G. fujikuroi* and in higher plants.

comes from feeding studies in which the carbon-20 acids, GA_{13} and GA_{25}, were not converted into C_{19} GAs in cell-free preparations from *Cucurbita maxima* and *Pisum sativum*, respectively.[4,39] Thus the three gibberellins, GA_8, GA_{17}, and GA_{29}, are considered to be members of dead-end branches from the main pathway.

All of the gibberellins in the early-13-hydroxylation pathway for maize have been identified from vegetative shoots.[19] They are GA_{53}, GA_{44}, GA_{19}, GA_{17}, GA_{20}, GA_{29}, GA_1, and GA_8. Radio-labeled feeds have defined the specific steps; GA_{53} to GA_{44},[41] GA_{20} to GA_{29},[24,41] GA_{20} to GA_1,[24] and GA_1 to GA_8 (B.O. Phinney *et al.*, unpublished). The significance of this pathway in the control of GA-dependent shoot elongation in maize will be considered in the next section.

GIBBERELLIN MUTANTS (DWARF MUTANTS) AND SHOOT ELONGATION IN MAIZE

The dwarf mutants of maize have provided critical evidence for the evaluation of the biological significance of the shoot GAs of maize. The maize gibberellins, GA_{53}, GA_{44}, GA_{19}, GA_{20}, and GA_1, are each biologically active in stimulating growth when added to shoots of the *dwarf-5* mutant of maize. The question to be asked is why are there so many biologically active gibberellins in one organ of one species of plant? The answer comes from the consideration of the alternative questions, (1) is each gibberellin active *per se*?, or, (2) is there only one gibberellin active *per se*? i.e. the other GAs are biologically active only through their metabolism to the one active gibberellin. Clearly the five bioactive gibberellins are related to each other metabolically, and GA_1 is the terminal active gibberellin in the pathway. Studies with the *dwarf-5* (*d5*) and *dwarf-1* (*d1*) mutants have resolved the questions as follows: all five gibberellins are bioactive when added to the *d5* mutant, while only GA_1 is active when added to the *d1* mutant (i.e. all other maize GAs are inactive on *d1* or have less than 1% the activity of GA_1).[23] Such growth response data suggest that the *d5* mutant could be blocking an early step in the pathway, the *d1* mutant a late step in the pathway. Radio-labeled feeding experiments and analyses of the endogenous levels of GAs support the interpretation of blockage at two specific steps, CPP to *ent*-kaurene for the *d5* mutant[36] and GA_{20} to GA_1 for the *d1* mutant.[24] It follows that GA_{53}, GA_{44}, GA_{19}, and GA_{20}, are

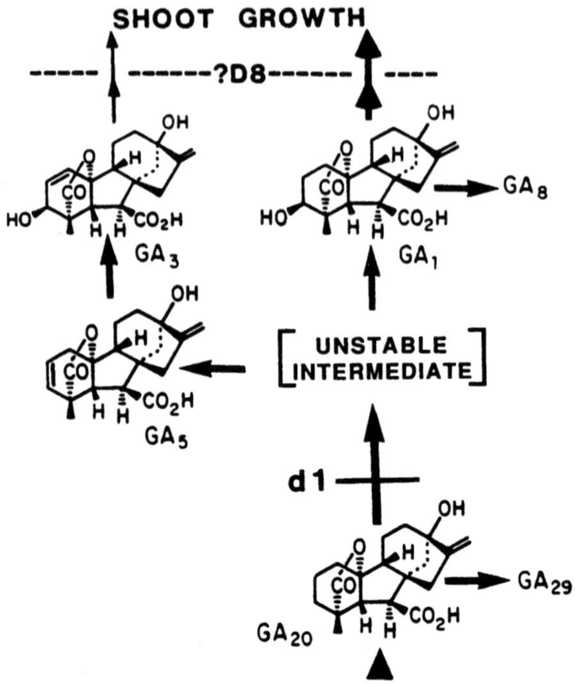

Fig. 7. Metabolism of GA_{20} in maize seedlings (B.O. Phinney *et al.*, unpublished data.)

bioactive on *d5* only through their metabolism to GA_1. The presence of only one active gibberellin, GA_1, is strongly supported by the observation that GA_{20} accumulates in the *d1* mutant to a level ten times that found in normals, yet *d1* has a phenotype indistin-guishable from the *d5* mutant, which lacks endogenous GAs.[19] We have concluded that GA_1 is the only (or the primary) gibberellin active *per se* in the control of GA-dependent shoot growth in maize.

We have recently identified two additional gibberellins in maize shoots, albeit in trace amounts.[42] They are GA_3 and GA_5 (Fig. 7.). These two GAs are active in stimulating shoot growth in both *d5* and *d1* mutants (B.O. Phinney, unpublished observation). We have also shown that GA_3 originates from GA_{20} *via* GA_5 (B.O. Phinney *et al.*, manuscript submitted). GA_3 is not metabolized to GA_1, and *vice versa* (B.O. Phinney *et al.*, unpublished observation). GA_1 and GA_5 are probably related to each other through a common unstable intermediate (see Fig. 7); if so the *d1* mutant may block an oxidation step, possibly the formation of an unstable intermediate having a free radical at carbon-3. Electron rearrangement and the loss of a hydrogen from ring A could give GA_5; hydroxylation at carbon-3 could give GA_1 (for a proposed intermediate in this type of enzymatic reaction, see reference 43). (The step GA_5 to GA_3 may be controlled by a different enzyme). The relative roles of GA_1 and GA_3 will probably not be resolved until information becomes available on the properties of the 3ß-hydroxylase together with information on the hypothetical receptor presumably involved in the action of the GAs.

REFERENCES

1. GRAEBE, J.E., and J. ROPERS. 1978. Gibberellins In: Phytohormones and related compounds: A comprehensive treatise. (D.S. Letham, P.B. Goodwin and T.J. Higgins, eds.), Vol. 1, Elsevier/North Holland Medical Press, Amsterdam, pp 107-204.
2. BEARDER, J.R. 1980. Plant hormones and other growth substances - their background, structures and occurrence In: Encyclopedia of plant physiology, New Series, (J. MacMillan, ed.), Vol. 9, Springer-Verlag, Berlin, pp 9-112.
3. CROZIER, A., ed. 1983. The biochemistry and physiology of gibberellins, Vol. 1, Praeger, New York. 568 pp.
4. GRAEBE, J.E., P. HEDDEN and W. RADEMACHER. 1980. Gibberellin biosynthesis In: Gibberellins - chemistry, physiology and use, Monograph 5. (J.R. Lenton, ed.), British Plant Growth Regulator Group, Wantage, U.K., pp 31-47.
5. GRAEBE, J.E. 1986. Gibberellin biosynthesis from gibberellin A_{12}-aldehyde In: Plant growth substances 1985. (M. Bopp, ed.), Springer-Verlag, Berlin, pp. 74-82.
6. MACMILLAN, J., and B.O. PHINNEY. 1987. Biochemical genetics and the regulation of stem elongation by gibberellins In: Physiology of cell expansion during growth. (D.J. Cosgrove and D.P. Knievel, eds.), American Society of Plant Pysiologists, Rockville, pp. 156-171.
7. PHINNEY, B.O. 1985. Gibberellin A_1, dwarfism and shoot elongation in higher plants. Biologia Plantarum (Prague) 27: 47-53.

8. JONES, R.L., and J.L. STODDART. 1977. Gibberellins and seed germination In: The physiology and biochemistry of seed dormancy and germination. (A.A. Khan, ed.), Elsevier/North Holland Biomedical Press, Amsterdam, pp. 77-109.
9. PHARIS, R.P., and R.W. KING. 1985. Gibberellins and reproductive development in seed plants. Ann. Rev. Plant Physiol. 36: 517-568.
10. REID, J.B. 1986. Gibberellin mutants In: A genetic approach to plant biochemistry. (A.D. Blonstein and P.J. King, eds.), Springer-Verlag, New York, pp. 1-34.
11. REID, J.B. 1987. The genetic control of growth *via* hormones In: Plant hormones and their role in plant growth and development. (P.J. Davies, ed.), Martinus Nijhoff, Dordrecht, pp. 318-340.
12. GRAEBE, J.E. 1987. Gibberellin biosynthesis and control. Ann. Rev. Plant Physiol. 38: 419-465.
13. REID, J.B. 1990. Phytohormone mutants in plant research. J. Plant Growth Regul. 9: (in press).
14. ATZORN, R., A. CROZIER, C.T. WHEELER and G. SANDBERG. 1988. Production of gibberellins and indole-3-acetic acid by *Rhizobium phaseoli* in relation to nodulation of *Phaseolus vulgaris* roots. Planta 175: 532-538.
15. BOTTINI R., M. FULCHIERI, D. PEARCE and R.P. PHARIS. 1989. Identification of gibberellins A_1, A_3, and iso-A_3 in cultures of *Azospirillum lipoferum*. Plant Physiol. 90: 45-47.
16. PHINNEY, B.O. 1983. The history of the gibberellins. In: The biochemistry and physiology of gibberellins. (A. Crozier, ed.), Vol. 1, Praeger, New York. pp. 19-52.
17. DURLEY, R.C., C.R. SHARP, S.L. MAKI, M.L. BRENNER and M.G. CARNES. 1989. Immunoaffinity techniques applied to the purification of gibberellins from plant extracts. Plant Physiol. 90: 445-451.
18. SMITH, V.A., and J. MACMILLAN. 1989. An immunological approach to gibberellin purification and quantification. Plant Physiol. 90: 1148-1155.
19. FUJIOKA, S., H. YAMANE, C.R. SPRAY, P. GASKIN, J. MACMILLAN, B.O. PHINNEY and N. TAKAHASHI. 1988. Qualitative and quantitative analyses of gibberellins in vegetative shoots of normal, *dwarf*-1, *dwarf*-2, *dwarf*-3, and *dwarf*-5 seedlings of *Zea mays* L. Plant Physiol. 88: 1367-1372.
20. ALBONE, K.S., J. MACMILLAN, A.R. PITT and C.L. WILLIS. 1986. Isotope labelling in ring A of gibberellin A_{20}.

Tetrahedron 42: 3203-3214.

21. PHINNEY, B.O. 1956. Growth response of single-gene dwarf mutants in maize to gibberellic acid. Proc. Natl. Acad. Sci. USA 42: 185-189.
22. COE, E.H., and M.G. NEUFFER. 1977. The genetics of corn In: Corn and corn improvement. (G.F. Sprague, ed.), Agronomy 18: 111-223.
23. PHINNEY, B.O., and C.R. SPRAY. 1982. Chemical genetics and the gibberellin pathway in *Zea mays* L. In: Plant growth substances 1982. (P.F. Wareing, ed.), Academic Press, London, pp. 101-110.
24. SPRAY, C.R., B.O. PHINNEY, P. GASKIN, S.J. GILMOUR and J. MACMILLAN. 1984. Internode length in *Zea mays* L. The *dwarf-1* mutation controls the 3β-hydroxylation of gibberellin A_{20} to gibberellin A_1. Planta 160: 464-468.
25. INGRAM, T.J., J.B. REID, W.C. POTTS and I.C. MURFET. 1983. Internode length in *Pisum*. IV. The effect of the *Le* gene on gibberellin metabolism. Physiol. Plant. 59: 607-616.
26. INGRAM, T.J., J.B. REID, I.C. MURFET, P. GASKIN, C.L. WILLIS and J. MACMILLAN. 1984. Internode length in *Pisum*. The *le*-gene controls the 3β-hydroxylation of gibberellin A_{20} to gibberellin A_1. Planta 160: 455-463.
27. INGRAM, T.J., J.B. REID and J. MACMILLAN. 1986. The quantitative relationship between gibberellin A_1 and internode elongation in *Pisum sativum* L. Planta 168: 414-420.
28. MURAKAMI, Y. 1972. Dwarfing genes in rice and their relation to gibberellin biosynthesis In: Plant growth substances 1970. (D.J. Carr, ed.), Springer-Verlag, Berlin, pp. 166-174.
29. SUZUKI, Y., S. KUROGOCHI, N. MUROFUSHI, Y. OTA and N. TAKAHASHI. 1981. Seasonal changes of GA_1, GA_{19} and abscisic acid in three rice cultivars. Plant Cell Physiol. 22: 1085-1093.
30. ZEEVAART, J.A.D. 1984. Gibberellins in single gene dwarf mutants of tomato. Plant Physiol. (Suppl.) 75: 186.
31. KOORNNEEF, M., and J.H. VAN DER VEEN. 1980. Induction and analysis of gibberellin sensitive mutants in *Arabidopsis thaliana* (L.) Heynh. Theor. Appl. Genet. 58: 257-263.
32. ROOD, S.B., D. PEARCE, P.H. WILLIAMS and R.P. PHARIS. 1989. A gibberellin-deficient *Brassica* mutant - *rosette*. Plant Physiol. 89: 482-487.
33. MACMILLAN, J., and N. TAKAHASHI. 1968. Proposed proce-

dure for the allocation of trivial names to the gibberellins. Nature (London) 217: 170-171.
34. BIRNBERG, P.R., S.L. MAKI, M.L. BRENNER, G.C. DAVIES and M.G. CARNES. 1986. An improved enzymatic synthesis of labeled gibberellin A_{12}-aldehyde and gibberellin A_{12}. Anal. Biochem. 153: 1-8.
35. DUNCAN, J.D., and C.A. WEST. 1981. Properties of kaurene synthetase from *Marah macrocarpus* endosperm: evidence for the participation of separate but interacting enzymes. Plant Physiol. 68: 1128-1134.
36. HEDDEN, P., and B.O. PHINNEY. 1979. Comparison of *ent*-kaurene and *ent*-isokaurene synthesis in cell-free systems from etiolated shoots of normal and *dwarf*-5 maize seedlings. Phytochemistry 18: 1475-1479.
37. SUZUKI, Y., B.O. PHINNEY, P. GASKIN and J. MACMILLAN. 1989. Elongating internodes of *Zea mays* (maize): early steps in the GA biosynthetic pathway. Plant Physiol. (Suppl.) 89: 107.
38. BEARDER, J.R., J. MACMILLAN, C.M. WELS and B.O. PHINNEY. 1973. Metabolism of steviol and its derivatives by *Gibberella fujikuroi*, mutant B1-41a. J. Chem. Soc. Chem. Commun. 778-779.
39. KAMIYA, Y., and J.E. GRAEBE. 1983. The biosynthesis of all major pea gibberellins in a cell-free system from *Pisum sativum*. Phytochemistry 22: 681-689.
40. TAKAHASHI, M., Y. KAMIYA, N. TAKAHASHI and J.E. GRAEBE. 1986. Metabolism of gibberellins in a cell-free system from immature seeds of *Phaseolus vulgaris* L. Planta 168: 190-199.
41. HEUPEL, R.C., B.O. PHINNEY, C.R. SPRAY, P. GASKIN, J. MACMILLAN, P. HEDDEN and J.E. GRAEBE. 1985. Native gibberellins and the metabolism of $[^{14}C]$gibberellin A_{53} and of $[17\text{-}^{13}C,17\text{-}^{3}H_2]$gibberellin A_{20} in tassels of *Zea mays*. Phytochemistry 24: 47-53.
42. FUJIOKA, S., H. YAMANE, C.R. SPRAY, M. KATSUMI, B.O. PHINNEY, P. GASKIN, J. MACMILLAN and N. TAKAHASHI. 1988. The dominant non-gibberellin responding dwarf mutant (*D8*) of maize accumulates native gibberellins. Proc. Natl. Acad. Sci. USA 85: 9031-9035.
43. BLANCHARD, J.S., and S. ENGLARD. 1983. γ-Butyrobetaine hydroxylase: primary and secondary tritium kinetic isotope effects. Biochemistry 22: 5922-5929.

Chapter Six

DITERPENOID PHYTOALEXINS: BIOSYNTHESIS AND REGULATION

CHARLES A. WEST, AUGUSTO F. LOIS,
KAREN A. WICKHAM AND YUE-YING REN

Department of Chemistry and Biochemistry
University of California, Los Angeles
Los Angeles, California 90024-1569, U.S.A.

INTRODUCTION

Most fungal and bacterial plant pathogens can establish systemic infections in a very limited range of host plants. Thus, higher plants in general are able to resist infection by most of the fungal and bacterial phytopathogens they may encounter. In cases of general resistance during the interaction of a non-host plant with a microbial pathogen, no obvious symptoms develop in the plant. During other cases of interaction of a non-host plant, or resistant cultivars of a host plant, with a microbial pathogen, a limited necrotic lesion develops rapidly in the host plant after penetration

Biochemistry of the Mevalonic Acid Pathway to Terpenoids
Edited by G.H.N. Towers and H. A. Stafford
Plenum Press, New York

of the microbe. This hypersensitive response is characterized by the rapid death of one or more plant cells at the site of microbial invasion, and the induction of production in still healthy plant cells immediately surrounding this site of a number of new metabolic products.[1] These products usually include low molecular weight antibiotics called phytoalexins, a number of macromolecular "barriers" such as callose (a β-1,3-glucan), hydroxyproline-rich glycoprotein and lignin-like cross-linked aromatic polymers composed of cinnamyl alcohol-derived monomer units, a family of pathogenesis-related proteins that includes chitinases and β-1,3-glucanases, ethylene, and doubtless others. Collectively, these substances are considered to be defensive agents that contribute to the hypersensitive resistance responsible for preventing the spread of the potential pathogen to other parts of the plant and a systemic infection.[2] These defensive agents are also produced in compatible interactions of a virulent pathogen with a susceptible host plant where systemic infection and eventual death of the plant occurs, but in these cases the response is delayed and more diffuse in relation to the growth of the pathogen, and thus is not effective in preventing disease.[3]

Phytoalexins have received a lot of attention in connection with research to elucidate the biochemical basis for hypersensitive resistance, although it is clear that phytoalexins are but one of the factors of importance. Phytoalexins have been defined as "low molecular weight, antimicrobial compounds that are both synthesized by and accumulated in plant cells after exposure to microorganisms."[4] A given plant species may produce several phytoalexins that are biogenetically related to one another and to a class of secondary metabolites found in that plant and closely related species. The initial studies emphasized the role of isoflavonoids as phytoalexins in members of the Leguminosae and of sesquiterpenes as phytoalexins in members of the Solanaceae. However, as the search was broadened to include plants from other Families, it became clear that many classes of secondary natural products of higher plants are represented among the phytoalexins.[5] It seems clear that independent evolutionary events were involved in establishing phytoalexins as defensive agents in different groups of higher plants. The utilization of secondary natural products as phytoalexins is consistent with the viewpoint summarized by Swain[6] that secondary metabolites in higher plants often function as regulatory agents that play specialized roles in the adaptation of a plant to its ecological niche.

For almost thirty years, starting with the gibberellin family of growth regulatory agents, investigations in our laboratory have focussed on the regulation of biosynthesis of diterpenoid substances of physiological interest for higher plants. Our attention was directed to diterpenoid phytoalexins when we discovered that the macrocyclic diterpene hydrocarbon casbene, one of several diterpenes identified as products of geranylgeranyl-PP metabolism in cell-free extracts of castor bean (*Ricinus communis* L.) seedlings,[7] possesses antimicrobial properties and is produced most readily by seedlings that have been exposed to fungal pathogens.[8] Although the specific role of casbene in disease resistance has not been investigated, we have utilized the production of this substance in castor bean seedlings exposed to fungal pathogens or elicitors derived from these pathogens as a model for studying the regulation of biosynthesis of a putative phytoalexin. Some of the results of these investigations will be summarized later in this report.

Rice (*Oryza sativa*) has also been shown to be a source of diterpenoid phytoalexins. These substances were discovered in connection with investigations of antifungal substances in rice tissues that were infected with *Pyricularia oryzae* (also known as *Magnaporthe grisea* in the imperfect stage), the causative agent of rice leaf blast disease. This disease is a major concern in rice production in most areas of the world; it has proved difficult to establish stable resistance in rice cultivars through plant breeding because of the easy mutability and genetic heterogeneity of *P. oryzae* populations.[9]

Cartwright *et al.*[10] isolated two antifungal substances from the leaves of *P. oryzae*-infected rice plants and from the cotyledons of UV-treated, dark-grown rice plants. These sustances were identified as momilactones A and B by comparing their properties with those of authentic reference compounds. The momilactones were initially isolated from rice husks and structurally characterized by Kato *et al.*[11] In addition to momilactones A and B, minor amounts of momilactone C were also recovered from the rice husks.[12] The structures and stereochemical features of the momilactones are illustrated in Fig. 1. The *trans-syn* arrangement of the substituents on ring junction carbons at positions 5, 10 and 9 for the momilactones

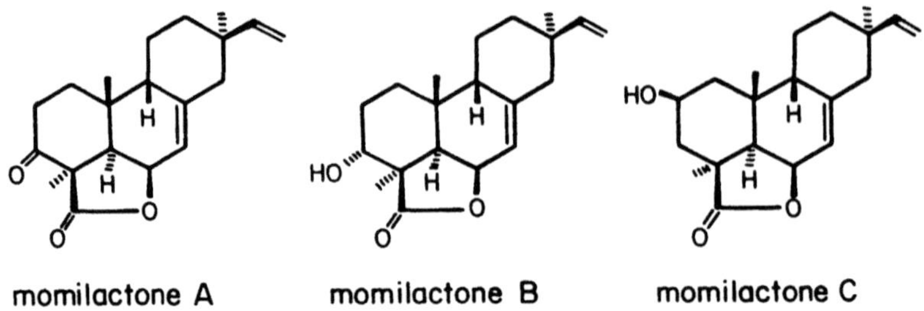

Fig. 1. Structures of momilactones A,B and C.

is unusual among natural pimaradiene derivatives, most of which possess either a normal (*nor*) or an enantio (*ent*) *trans-anti* arrangement of these substituents. Cartwright *et al.*[10] demonstrated that momilactones A and B are potent antifungal substances with ED_{50} values of about 4.8 and 0.9 μg ml^{-1}, respectively, and they could not detect these substances in healthy rice tissues. Therefore, they proposed that these substances are phytoalexins that contribute to the defense of rice plants against infection by the blast fungus.

Independent investigations of the antifungal substances of rice tissues infected with *P. oryzae* led to the identification of a different group of diterpenoid phytoalexins that have been named oryzalexins.[13-16] Four substances, oryzalexins A-D, have been characterized and shown to possess the carbon skeleton of *ent*-sandaracopimaradiene. The proposed structures of the oryzalexins are illustrated in Fig. 2. Although both the momilactones and the oryzalexins possess a pimaradiene carbon skeleton, they differ in their stereochemical features, in the position of the endocyclic double bond and in their oxygenation patterns. Since the oryzalexins inhibit *P. oryzae* spore germination and germ tube growth and they are not detected in healthy plant tissue,[13] it is suggested that they serve as phytoalexins in the defense of rice plants against blast disease.

Since each of the above investigations detected one group of diterpenoid phytoalexins, but not the other, there has been some question as to which one is the most important class of phytoalexins for rice. Although this question has not been resolved, it is interesting to note that momilactones A and B and oryzalexins A, B, C and D have recently been observed to accumulate in rice leaves after elicitation with

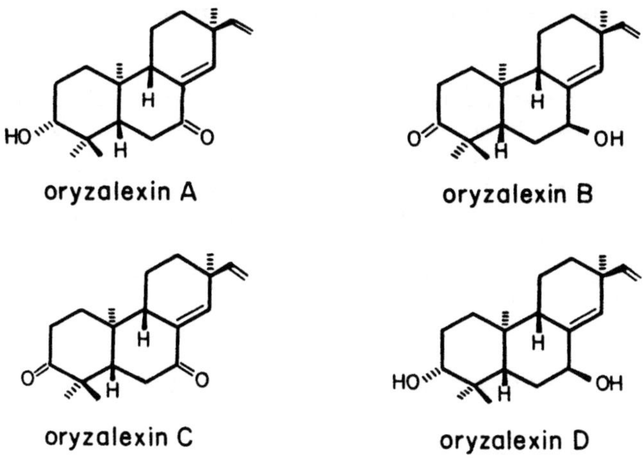

Fig. 2. Structures of oryzalexins A,B,C and D.

UV-irradiation[17] or heavy metal ions.[18] It may be that all are important in an overall sense, but some may accumulate to a greater extent than others depending on the circumstances.

Phytoalexins are either not produced, or produced only at very low levels, in plant cells until they are subjected to pathogen attack or other stresses. Also, a number of studies have indicated that phytoalexin biosynthetic capacity is stimulated during a hypersensitive response primarily, or exclusively, in those cells surrounding the site of microbial penetration into the plant, and thus phytoalexin accumulation appears to be a localized rather than systemic response. It is therefore obvious that the production of phytoalexins in a plant is closely regulated. This regulation appears to involve chemical signalling between the microbe and the plant and between cells within the plant as well. The perception of appropriate signals by a plant cell leads, by means of signal transduction mechanisms that are largely unknown, to the activation of enzymes of some metabolic pathways, including those participating in phytoalexin production, and doubtless to the down regulation of enzymes in other pathways. The net result is a metabolic differentiation of responding cells to a form that is specialized to produce substances of importance in the protection of the plant against the systemic spread of the potential pathogen. Investigations in recent years have provided some support for such a model, but there remain large gaps in our knowledge of how it might function at the molecular level.

The objective of this report is to summarize the progress that has been made in understanding the biosynthetic pathways leading to the production of diterpenoid phytoalexins in castor bean and rice cells, and in the regulation of these pathways. It is our hope that the conclusions from these studies will also have some relevance for diterpene biosynthetic pathways and for the production of secondary metabolites in higher plants more generally.

CASBENE BIOSYNTHESIS AND ITS REGULATION

The Casbene Biosynthetic Pathway

The accumulated evidence[7,19,20] supports the anticipated pathway for casbene biosynthesis from mevalonic acid illustrated in Fig. 3. The properties of the prenyl transferase activities responsible for the conversion of isopentenyl-PP and dimethylallyl-PP to geranylgeranyl-PP in castor bean seedlings have been examined. Green and West[21] were able to resolve two geranyl transferase isoenzymes with similar kinetic properties that catalyzed the specific synthesis of farnesyl-PP from either dimethylallyl-PP or geranyl-PP as the allylic donor (Step 5, Fig. 3). Subsequently, Dudley *et al.*[22] confirmed that whole tissue extracts derived from castor bean seedlings infected with the fungus, *Rhizopus stolonifer*, yielded two geranyl transferase isoenzymes, whereas whole tissue extracts from uninfected seedlings contained only one peak of geranyl transferase activity. On the basis of this information, it was proposed that the isoenzyme common to both infected and uninfected tissues was involved in sterol biosynthesis, whereas the isoenzyme found only in infected tissues participated specifically in casbene biosynthesis. A farnesyl transferase activity (geranylgeranyl-PP synthetase) was also purified from whole tissue extracts of *R. stolonifer*-infected castor bean seedlings (Step 6, Fig. 3).[22] The purified enzyme utilized farnesyl-PP as an allylic donor quite efficiently (K_m = 0.5 μM), geranyl-PP only poorly (K_m = 23 μM) and dimethylallyl-PP not at all. It was therefore concluded that this enzyme must function under *in vivo* conditions solely to elongate farnesyl-PP, which is produced by geranyl transferase, to geranylgeranyl-PP.

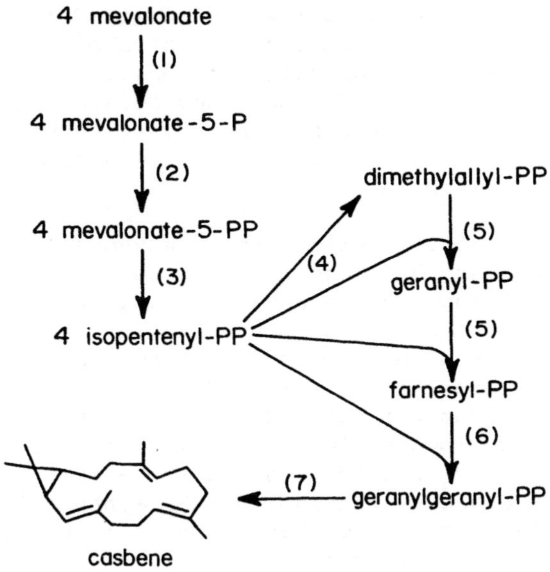

Fig. 3. Biosynthetic pathway from mevalonate to casbene. Enzymes involved are: (1) mevalonate kinase (EC 2.7.1.36); (2) mevalonate-5-P kinase (EC 2.7.4.2); (3) mevalonate-5-pyrophosphate decarboxylase (EC 4.1.1.33); (4) isopentenyl pyrophosphate Δ^3-Δ^2 isomerase (EC 5.3.3.2); (5) farnesyl pyrophosphate synthetase (geranyl transferase or geranyl pyrophosphate: isopentenyl pyrophosphate geranyl transferase) (E.C.2.5.1.1); (6) geranylgeranyl pyrophosphate synthetase (farnesyl transferase or farnesyl pyrophosphate: isopentenyl pyrophosphate farnesyl transferase); (7) casbene synthetase. From Dudley *et al.*[20]

The final step of casbene biosynthesis, the cyclization of the geranylgeranyl-PP to casbene in a single step reaction, is catalyzed by casbene synthetase (Step 7, Fig. 3). The utilization of all-*trans* geranylgeranyl-PP as the specific precursor of cyclic diterpenes appears to be general.[23] Casbene synthetase was partially purified from *R. stolonifer*-infected castor bean seedlings and some of its properties were assessed. The enzyme has an apparent M_r of approximately 53,000 as determined by molecular exclusion chromatography. It showed typical Michaelis-Menten kinetics with geranylgeranyl-PP as the substrate (K_m = 1.9 μM) and Mg^{+2} as the preferred cofactor. Casbene synthetase was subsequently purified to apparent homogeneity and utilized for the preparation of polyclonal antibodies in rabbits.[24] The similarity of the M_r of approximately 59,000 for the pure enzyme as determined on SDS-PAGE under denaturing conditions and of approximately 53,000 for the native enzyme on non-denaturing gels indicates that the active enzyme consists of a single polypeptide chain.

The subcellular localization of enzymes of the casbene biosynthetic pathway has been investigated by Dudley *et al.*[20] Techniques involving sucrose density gradient centrifugation based on both equilibrium sedimentation and velocity sedimentation were employed to resolve microsomes, mitochondria, glyoxysomes and proplastids released from *R. stolonifer*-infected castor bean tissues. The bands containing each type of organelle were located in the gradient from the positions of characteristic marker enzymes. Casbene synthetase, farnesyl transferase (geranylgeranyl-PP synthase), geranyl transferase (farnesyl-PP synthase), and isopentenyl-PP: dimethylallyl-PP isomerase activities were all associated with the proplastid fraction. Geranyl transferase activity, but none of the others cited above, was also associated with the mitochondrial fraction. However, subsequent work by Pargellis[25] suggests that this may have represented contaminating activity from broken proplastids. These results indicate that casbene biosynthesis, from at least the stage of isopentenyl-PP as an intermediate, is localized in the proplastid compartment. No effort was made to localize the enzymes responsible for the synthesis of isopentenyl-PP from mevalonate in these investigations.

A partial purification and determination of the properties of the prenyl transferases from the proplastid fraction of infected castor bean seedlings obtained by differential centrifugation was carried out by Pargellis.[25] It was not possible to resolve geranyl transferase and farnesyl transferase activities by ion exchange chromatography of proplastid extracts. This result differs from that seen with whole seedling extracts of infected tissue where two geranyl transferase activities and a farnesyl transferase were clearly resolvable, and therefore brings into question the model proposed earlier[20,22] for the involvement of substrate coupled, but separate, geranyl transferase and farnesyl transferase enzymes in the synthesis of geranylgeranyl-PP from isopentenyl-PP in these tissues.

Regulation of Casbene Biosynthesis

Transient increases in the activities of casbene synthetase and farnesyl transferase were observed in cell-free extracts of castor bean seedlings after infection with *R. stolonifer* spores, with the maximum activity occurring at about 12 hours.[20] Casbene synthetase and farnesyl transferase activities increased substantially from very low basal levels in uninfected tissues, and geranyl transferase activity approximately doubled from a higher basal level.[20] On the other hand, no significant change was detected in the overall activity for the conversion of mevalonate to isopentenyl-PP between uninfected and infected tissues.[20] Isomerase activity was not quantitated in these studies. These results suggest that the rate determining enzymes for casbene biosynthesis are those involved in the conversion of isopentenyl pyrophosphate to casbene (Steps 4-7, Fig. 3), and that these are greatly enhanced in the proplastid compartment after infection from very low levels in uninfected tissues.

These changes in the activities of enzymes involved in casbene biosynthesis in the plant after exposure to the fungus implied that there must be chemical signals (so-called elicitors) released from the fungus that are recognized by plant cells to trigger the response. A bioassay based on measuring the activity of casbene synthetase in cell-free extracts of seedlings exposed to substances to be tested for elicitor activity was developed.[26] This assay permitted the partial purification of an elicitor from culture filtrates of *R. stolonifer* grown on defined medium.[26] The elicitor activity at this stage was associated with a glycoprotein fraction and was sensitive to heating in a manner that suggested a native protein structure was required for activity.[26] An endopolygalacturonase present in the partially purified elicitor was then purified to apparent homogeneity, and the pure enzyme was shown to be responsible for the casbene synthetase elicitor activity.[27,28]

Endopolygalacturonase catalyzes the hydrolysis of homogalacturonan polymers to a mixture of oligo-1,4-D-galacturonide fragments. Evidence obtained by Bruce and West[29] indicated that the elicitor activity of *R. stolonifer* endopolygalacturonase is proportional to its catalytic activity. Therefore, it was proposed, and subsequently demonstrated, that the oligogalacturonide fragments released by the digestion of homogalacturonans from the plant cell wall were themselves elicitors of casbene synthetase activity.[29] We therefore propose that the extracellular fungal endopolygalacturonase, which is one of a group of degradative enzymes normally released by microbial pathogens to break down plant cell walls, acts on the homogalacturonan components of the castor bean cell wall pectic polysaccharides to release oligogalacturonide fragments. It is these fragments, then, that are recognized by healthy castor bean cells as endogenously-generated elicitors to trigger the activation of casbene synthetase and other enzymes of the defensive response. It is now known that such pectic fragments elicitors are active more generally in higher plants.[30]

A number of tests have been performed with the castor bean elicitor bioassay to determine the specificity of the assay for oligogalacturonides as elicitors and the structural requirements of oligogalacturonides for elicitor activity. Among the oligosaccharides known to possess elicitor activity in other assay systems, chitosan[31] is weakly active and the branched β-glucan from *Phytophthora megasperma* f.sp. *glycinea* is inactive in the castor bean bioassay. Galacturonide oligomers ranging in size from the dimer to the heptadecamer have been tested individually for activity. Only oligomers with ten or more monomer units were active, with the tridecamer showing maximum activity.[32] Methyl esterification of the carboxyl groups in galacturonide oligomers renders them inactive.[32] A group of polyanionic substances structurally unrelated to galacturonides were inactive,[32] as were oligomers containing L-guluronic acid and D-mannuronic acid derived from alginic acid (unpublished results). The conclusion from these and other tests is that the castor bean assay recognizes unmodified oligogalacturonides of appropriate size rather specifically among the substances that have been tested. These results appear to support the view that oligogalacturonides may have an *in vivo* role as elicitors during a castor bean-microbe interaction.

Three to four hours are required before elicited enzymes begin to increase in activity, and nine to twelve hours are required before they reach their maximal activity. This timing would be consistent with *de novo* enzyme synthesis as the basis for the increase in enzyme activity. A series of investigations were undertaken to test whether transcriptional activation of the casbene synthetase gene(s) was an important factor in regulating the observed increase in casbene synthetase activity after treatment of castor bean seedlings with oligogalacturonide (pectic fragment) elicitors. Moesta and West[24] utilized the antiserum directed against casbene synthetase in conjunction with *in vitro* translation in a rabbit reticulocyte lysate to measure the translatable casbene synthetase mRNA levels as a function of time after elicitation.[24] Lois[33] then prepared a cDNA library from poly A^{+}RNA of elicited castor bean tissues in the expression vector λgt11. A cDNA clone was isolated and shown to contain a partial casbene synthetase sequence by hybrid arrested and hybrid selected translation experiments. Northern analysis and slot blot hybridization with this clone were performed with RNA samples prepared from castor bean seedlings at various times after elicitation to measure total hybridizable casbene synthetase mRNA as a function of time after elicitation. Finally, nuclear run-on experiments were performed with the aid of the casbene synthetase cDNA clone to assess relative rates of transcription initiation at the casbene synthetase gene(s) in isolated nuclei as a function of time after elicitation.

The results of all of these experiments are summarized in Fig. 4. Relative values, expressed as a percentage of the maximum, for transcription initiation, total hybridizable casbene synthetase mRNA, translatable casbene synthetase mRNA and casbene synthetase enzyme activity are plotted as a function of time after treatment of castor bean seedlings with oligogalacturonide elicitors. It can be seen that transient increases for each of these occurs, with the maximum for transcription initiation of the casbene synthetase gene at approximately 5 hours, for total and translatable casbene synthetase activity at approximately 6 hours and for casbene synthetase activity at 9-10 hours. These kinetics are consistent with a model in which the activation of the casbene synthetase gene for transcription following elicitor treatment is at least partially responsible for the observed increases in casbene synthetase enzyme activity. The fact that the curves for total hybridizable casbene synthetase mRNA and translatable casbene synthetase mRNA are nearly coincident suggests that activation of an inactive pool of casbene synthetase mRNA is not a rate determining factor in this case. The basis for the relatively long lag time of two hours after elicitation before transcription of the casbene synthetase gene is not known.

These results, which indicate that transcriptional activation of plant defense genes after exposure to elicitor

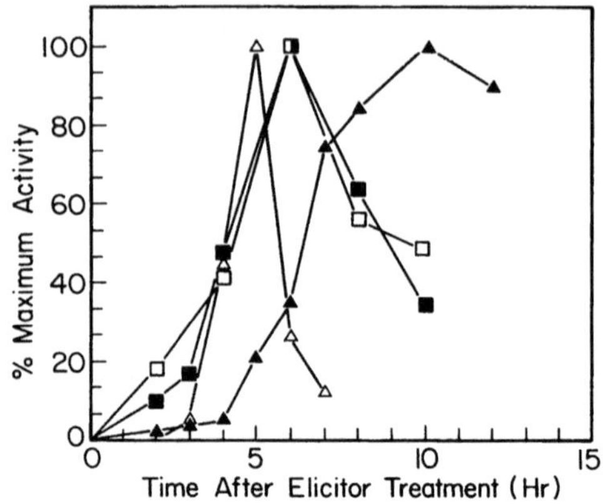

Fig. 4. Regulation of expression of the casbene synthetase gene in castor bean seedlings as a function of time after treatment with oligogalacturonide elicitors. The curves represent relative changes in transcriptional initiation (open triangles), total hybridizable mRNA levels (closed squares), translatable mRNA levels (open squares) and casbene synthetase enzyme activity (closed triangles), each expressed as a percentage of the maximum = 100%. The data for transcription initiation and total hybridizable mRNA levels are from Lois.[33] The data for translatable mRNA levels and casbene synthetase enzyme activity are from Moesta and West.[24]

is an important factor in the regulation of phytoalexin biosynthesis, are consistent with the findings of other laboratories (see for example references 34 and 35). However, this is the first demonstration that endogenous elicitors of plant cell wall origin function in this way, and it is the first case where regulation of terpenoid phytoalexin biosynthesis has been shown to involve transcriptional activation of biosynthetic genes.

Regulation of Phytoalexin Biosynthesis: Future Directions

A proposed elicitation cascade is outlined in Fig. 5 as a basis for thinking about future directions in this area of research. Fragments of fungal cell walls and fragments of plant cell walls are both known to act as elicitors to initiate the expression of plant defense products. The fungal cell wall elicitor fragments may be released through the

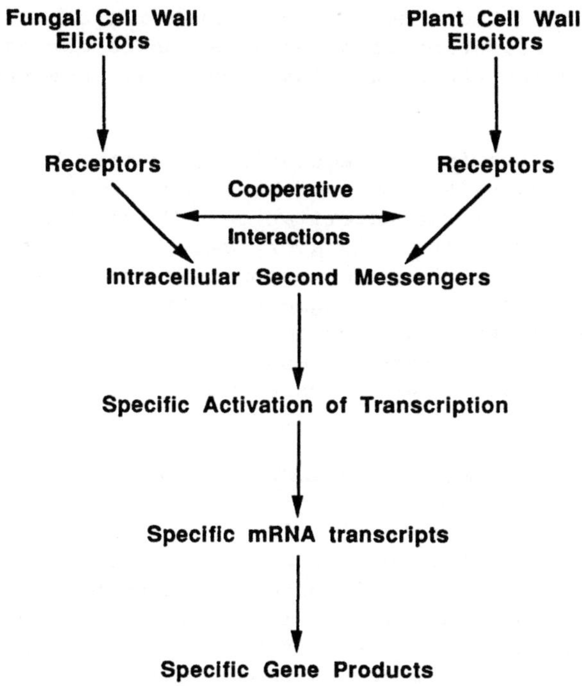

Fig. 5. Proposed elicitation cascade.

action of plant enzymes generated during a plant-microbe interaction, and plant cell wall elicitor fragments may be released either through the action of microbial enzymes generated during a plant-microbe interaction, or by autolytic breakdown stimulated in the plant under those circumstances. Both of these types of elicitors may act through specific plasma membrane receptors in responding cells to generate intracellular second messengers that participate in the signal transduction process. There is only very limited evidence at the present time regarding putative receptors for phytoalexin elicitors, and very little is known about possible signal transduction pathways that might mediate such pathways. Observations concerning the synergistic interactions of oligogalacturonides and fungal cell wall fragments as elicitors of phytoalexin production[36-38] suggest that positively cooperative interactions between the two signal transduction pathways may operate to modulate the response to these two types of signals when they are present simultaneously in the plant. Research to define the nature of these signal transduction pathways that mediate the conversion of elicitor signals into activation of defense responses during plant-pathogen interactions is currently underway in a number of laboratories.

Current work in the regulation of castor bean phytoalexin synthesis is aimed at obtaining clones of the casbene synthetase gene for characterization. We are interested in defining the *cis*-acting regulatory sequences in the flanking regions of the gene that are responsible for gene activation in the presence of pectic fragments. As a longer range goal, we would like to transfer the gene to other plant cells in order to study its expression and the role of casbene production in plant disease resistance.

BIOSYNTHESIS OF RICE PHYTOALEXINS AND ITS REGULATION

We became interested in rice phytoalexins in part because of their diterpenoid nature and therefore their relevance to our interest in diterpene biosynthesis and its regulation. Also, in spite of its obvious importance as a crop plant, relatively little is known about the mechanisms of disease resistance in rice. Although the structures of a number of phytoalexins of rice in the momilactone and oryzalexin families have been established, little is known about the biosynthetic pathways leading to these products or how they are regulated. The progress we have made in investigating this problem will be summarized in what follows in this report; it will become clear that our understanding is still in the early stages.

The Biosynthetic Pathway to Rice Phytoalexins

An outline of a proposed biosynthetic scheme for the momilactone and oryzalexin families of phytoalexins is given in Fig. 6. This scheme is based on analogy with biosynthetic pathways leading to other polycyclic diterpenes that have been characterized.[23] The skeletons of both families are those of pimaradienes, but they differ in their stereochemical features as noted earlier. Geranylgeranyl-PP is considered to be the acyclic precursor of all cyclic diterpenes. Two modes of cyclization of geranylgeranyl-PP must operate to produce the bicyclic pyrophosphorylated intermediates 9βH-labdadienyl-PP and *ent*-copalyl-PP as precursors of the momilactones and oryzalexins, respectively. Copalyl-PP is the known precursor

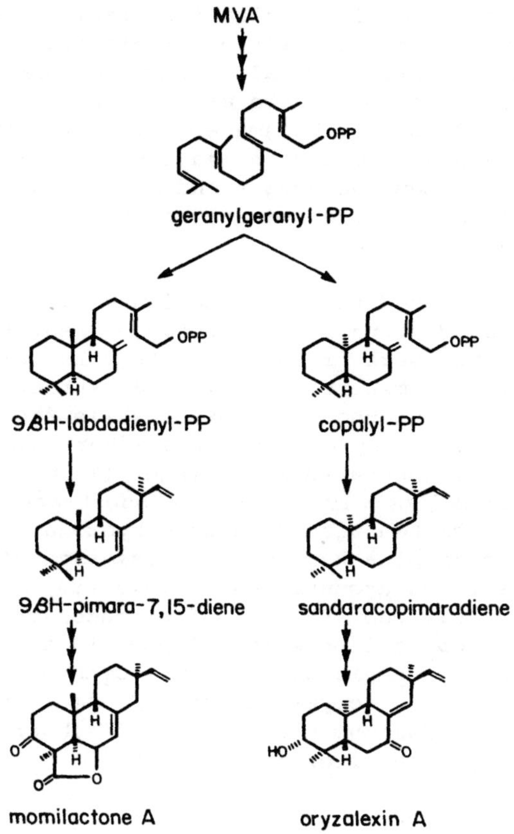

Fig. 6. Proposed pathway of diterpene phytoalexin biosynthesis in rice.

of *ent*-kaurene, *ent*-beyerene, *ent*-trachylobane and *ent*-sandaracopimaradiene in castor bean seedling extracts,[39] whereas 9βH-labdadienyl-PP has not been reported. Further cyclization coupled with pyrophosphate elimination of each bicyclic intermediate would lead to the pimaradiene isomers pictured. And finally, a series of oxygenation reactions at specific positions would logically be involved in generating the two families of phytoalexins from the hydrocarbon precursors.

Wickham[40] has tested the general validity of this scheme in our laboratory. Our approach has been to identify the cyclization products formed from geranylgeranyl-PP and copalyl-PP in cell-free extracts of rice leaves that have been elicited by exposure to ultraviolet irradiation. We chose UV-irradiation as a convenient means of elicitation on the basis of the report by Cartwright *et al.*[10] that UV-treatment stimulated the production of momilactones A and B in dark-grown rice cotyledons. Recently, Kodama *et al.*[17] have shown that UV-treatment of young rice leaves leads to the production of oryzalexins A-D and momilactones A and B along with an unknown antifungal substance. A preliminary account of some of our results has been published.[41]

It was established that cyclization activity was greatly enhanced by UV-irradiation of rice leaves. Leaves of 4- to 8-week old rice plants were irradiated with a 254 nm short wavelength UV lamp for 24 minutes. Cell-free extracts of the leaves that had been ground to a powder while frozen in liquid N_2 were prepared at an appropriate time after irradiation. These extracts were incubated with either [1-$^3H_2$]-geranylgeranyl-PP or [1-$^3H_2$]-copalyl-PP as the substrate, and the radioactivity associated with diterpene hydrocarbon products (pimaradiene isomers plus kaurene) was assessed after separation of products in an argentation TLC system. Geranylgeranyl-PP yielded 490 pmol of products in the pimaradiene fraction in comparison with 6.4 pmol in an untreated control, and copalyl-PP yielded 300 pmol of products in the pimaradiene fraction in comparison with 34 pmol in an untreated control. Maximum conversions of substrates were seen approximately 42 h after irradiation.

Large scale incubations were performed in order to accumulate enough of the diterpene products for purification and characterization. Irradiation of rice leaves and preparation of enzyme extracts were performed as described previously. Incubations of pooled enzyme extracts with low specific radioactivity geranylgeranyl-PP or copalyl-PP as substrates were scaled up 1,000-fold. The reaction products were extracted into petroleum ether and then separated into a hydrocarbon fraction and more polar products by silicic acid chromatography. The hydrocarbon fraction was further resolved by chromatography on a $AgNO_3$-impregnated silicic acid column.

In one set of experiments, irradiated rice leaves from a total of 80 flats of rice plants were utilized as the source of enzyme and a total of 9 liters of 50 mM geranylgeranyl-PP was employed as the substrate. The radioactive "pimaradiene" hydrocarbon products from these incubations were associated with three major peaks recovered from the $AgNO_3$-impregnated

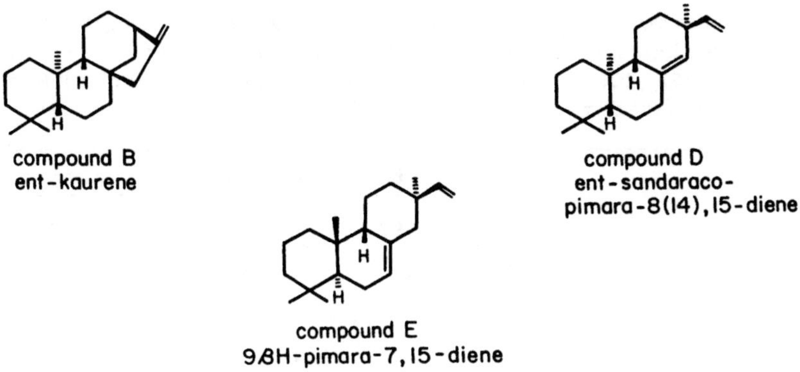

Fig. 7 Structures of compounds B, D and E recovered from incubations of rice cell-free extracts.

silicic acid column. GC analysis of these fractions on a DB-1 column revealed the presence of five putative diterpene hydrocarbons in the amounts indicated (based on radioactivity): Compound A, 25 μg; Compound B, 61 μg; Compound C, 1,200 μg; Compound D, 350 μg; and Compound E, 90 μg (Table 1, Fig. 7). A comparison of the retention times with those of standard reference compounds that were kindly made available by R. Coates of the University of Illinois, Urbana indicated tentative identifications of Compound B as kaurene, Compound D as sandaracopimaradiene and Compound E as 9βH-pimara-7,15-diene. These identifications were confirmed by GC-mass spectral analysis of the isolated substances for comparison with authentic reference compounds. The GC retention times and mass spectra of Compounds A and C also indicated that both were diterpene hydrocarbons which resembled pimaradienes in their spectral properties, but neither was identical in these properties with any of the six pimaradiene isomers available as reference compounds.

Similar larger scale incubations were performed with copalyl-PP as the substrate to provide sufficient quantities of the diterpene hydrocarbons after purification by the procedures indicated above to permit GC-MS analysis. The GC analysis indicated that Compounds B (kaurene), C (unidentified) and D (sandaracopimaradiene) were present in approximately the same relative amounts as were formed from geranylgeranyl-PP as the substrate. Compounds E and A were not present at detectable levels. If Compound E (9βH-pimara-7,15-diene)

Table 1. Diterpene products formed from geranylgeranyl-PP (GGPP) in cell-free extracts of UV-irradiated rice leaves and from copalyl-PP (CPP) in cell-free extracts of UV-irradiated and non-irradiated rice leaves.

	UV-treated tissue		Untreated tissue
	GGPP	CCP	CPP
Compound A	+[a]	nd[b]	-[c]
Compound B (Kaurene)	+	+	+
Compound C	+	+	-
Compound D (Sandaracopimaradiene)	+	+	-
Compound E (9βH-pimara-7,15-diene)	+	-	-

[a]Detectable by GC analysis.
[b]Not determined.
[c]Not detectable by GC analysis.

was formed in the same relative amounts from copalyl-PP as from geranylgeranyl-PP, it should have been detectable in this experiment.

Table 1 indicates the diterpene products that were formed in extracts of UV-irradiated rice leaves from geranylgeranyl-PP and copalyl-PP as substrates. The results of these incubations with copalyl-PP as the substrate help to define the stereochemistry of the products. Since copalyl-PP has the *enantio* stereochemical arrangement of the ring junction substituents, Compounds B and D must be *ent*-kaurene and *ent*-sandaracopimaradiene, respectively. Compound C must possess the *enantio* arrangement as well. The 9βH-pimara-7,15-diene isomer that serves as a precursor of the momilactones (Fig. 6) should not be synthesized from copalyl-PP as the substrate. The failure to detect any conversion of copalyl-PP to Compound E is therefore consistent with the conclusion that Compound E is the appropriate 9βH-pimaradiene isomer that serves as the momilactone precursor. The structures, with appropriate stereochemical designations, of the three diterpene hydrocarbon products which have been identified are shown in Fig. 7.

Table 1 also indicates that *ent*-kaurene (Compound B) was the only diterpene hydrocarbon detected by GC analysis in incubations of copalyl-PP with cell-free extracts of rice leaves that were not subjected to UV-irradiation. If *ent*-kaurene is serving as a precursor of the gibberellins, its synthesis might be expected whether or not a stress has been imposed. On the other hand, the synthesis of the phytoalexin precursors, 9βH-pimaradiene and *ent*-sandaracopimaradiene, would be expected only in plants that have received UV-treatment since these conditions are known to induce phytoalexin formation. Thus, the structural assignments made for these three diterpenes are consistent with the biosynthetic patterns seen in Table 1 and their proposed functions.

Compound C is the most abundant of these diterpene hydrocarbons whose production is dependent on UV-irradiation. The structure of C is unknown, although its GC and MS characteristics support the idea that it is a diterpene hydrocarbon in the pimaradiene group. Furthermore, the 500 MHz proton NMR spectrum revealed the presence of four vinylic protons whose chemical shifts at 4.95, 5.2, 6.35 and 5.55 ppm are similar, but not identical, to those of reference pimaradiene isomers. The first three are tentatively assigned to an A-B-X system associated with an exocyclic vinyl group, and the other to a single proton associated with a tri-substituted endocyclic double bond. To date it has not been possible to relate Compound C to a number of possible pimaradiene structures with different endocyclic double bond positions, or to some conceivable diterpenes with rearranged carbon skeletons. The function of Compound C is not known beyond the likely possibility that it is involved in stress metabolism in rice plants. It is interesting to note some recent reports of the presence of unidentified antifungal agents that appear not to be identical with known momilactones or oryzalexins in UV-irradiated[17] or blast-infected[42,43] rice leaves. The properties of some of these substances clearly indicate that they are oxygenated diterpenes.[43] We speculate that Compound C may prove to be the precursor of some of these unknown phytoalexins.

Very little has been done to date to characterize the diterpene cyclization enzymes involved in stress metabolism in rice. The sensitivity of the mixture of cyclization activities found in the crude cell-free extracts of UV-irradiated rice leaves to two plant growth retardants that are potent

Table 2. Comparison of plant growth retardant inhibitor activities for diterpene cyclization enzymes.

Inhibitor	(M)	*M. macrocarpus* kaurene syn[a,b]		Rice cyclases[a,c]	Casbene synthetase[a,d]
		A	B	AB	
Phosphon D	5×10^{-6}	100	2	10	55
	5×10^{-5}	100	8	50	91
	5×10^{-4}	100	34	98	99
AMO-1618	5×10^{-5}	100	12	<10	9
	5×10^{-4}	100	75	15	

[a]% Inhibition of activity
[b]Reference 40
[c]Reference 44
[d]Reference 45

inhibitors of *ent*-kaurene synthetase *A* activity in plants[44] has been examined by Wickham.[40] Table 2 compares the inhibitor activities of AMO-1618 and Phosphon D for rice cyclases with those for *Marah macrocarpus* kaurene synthetase *A* and *B* activities[44] and casbene synthetase.[45] It can be seen that the rice cyclases are relatively insensitive to those inhibitors in comparison with their effects on the mechanistically analogous kaurene synthetase, and are more similar in behavior to casbene synthetase. The *A* activity of the kaurene synthetase and the *A* activity from rice that gives rise to copalyl-PP catalyze the same reaction. However, it is evident that their regulatory properties are different.

The subcellular localization of the rice cyclase activities has also been examined.[40] Several methods designed to preserve organelles during cell disruption were attempted with UV-irradiated rice leaves, and the resulting particulate and cytosolic fractions were examined for cyclization activities. In no case was there evidence for significant cyclase activity that could be released from organellar fractions; all the activity present in the initial homogenates was recovered in the cytosolic fraction. Although it is difficult to be certain, our tentative conclusion is that the inducible rice diterpene cyclase activities are localized in the cytoplasm outside the organelles. The UV-inducible rice diterpene cyclases differ in this respect from casbene synthetase, which is localized in the proplastids of castor bean seedlings.[20]

Regulation of Rice Phytoalexin Biosynthesis: Elicitors

The only feature of regulation of rice phytoalexin biosynthesis that has been examined to date concerns possible biotic elicitors of the response. Y-Y. Ren has established an elicitor bioassay involving rice cell suspensions, and has surveyed the response in this bioassay to exposure to a number of cell wall materials as potential elicitors (unpublished data).

The basis for this assay is the measurement of diterpene cyclase activity in extracts of rice suspension cells that have been exposed to substances to be tested for elicitor activity. The potential elicitors are added to 50 ml of rice suspension cell culture (Japonica Sesheniski cv.) that has been grown for 6-days in B-5 medium. (Callus from which these cells were grown was kindly supplied by N.H. Chua, Rockefeller University.) After a further 40 hours, the cells from one treatment are harvested, and a cell-free extract is prepared. An aliquot of the cell extract is then incubated with [1-$^{3}H_2$]-geranylgeranyl-PP under conditions where the cyclase activity is proportional to the amount of enzyme added. The radioactive products in the incubation mixture are extracted into petroleum ether, and the radioactivity associated with the diterpene hydrocarbon fraction recovered from silicic acid chromatography is assessed.

Oligosaccharide elicitors that have been tested in this assay include: a preparation of insoluble chitin (β-1,4-N-acetylglucosaminoglycan; Sigma); nitrous acid-treated chitosan (β-1,4-D-glucosaminoglycan kindly supplied by L. Hadwiger, Washington State U.); α-1,4-D-galacturonide oligomers with an average degree of polymerization of 17[32]; and a partially purified preparation of branched β-1,3-glucan oligomers from *Phytophthora megasperma* f.sp. *glycinea*[46] (kindly supplied by M. Hahn, U. of Georgia). Fig. 8 illustrates the results of a

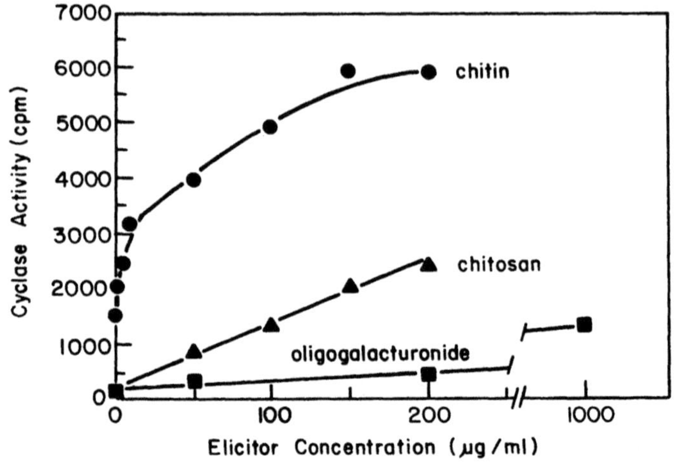

Fig 8. Bioassay of oligosaccharides as elicitors of diterpene cyclization enzymes in rice suspension cells. Chitin, chitosan and oligogalacturonides were added to rice cell suspensions and tested for elicitor activity as described in the text.

bioassay with three of these substances. Chitin was the most active; as little as 5 µg of insoluble chitin per milliliter of suspension culture gave a readily detectable response. The non-linear nature of the chitin response curve is typical. Presumably this is a consequence of the insoluble nature of the chitin supplied. Chitosan gave a much lower response and was less active in this assay than in the pea bioassay.[31] Oligogalacturonides were not significantly active in this assay. The branched β-1,4-glucan (not shown in Fig. 8) showed a low activity in this assay approximately equivalent to that of chitosan. Among these substances, chitin would appear to be the most attractive candidate as a physiological elicitor.

Chitin is a major cell wall polysaccharide in many fungi, including phytopathogens of rice. The culture medium and suspension cells utilized in the rice bioassay have been shown to contain chitinase activity, which is further induced by the addition of chitin to the medium (Ren, unpublished). We presume that chitin oligomers released by the action of chitinase on insoluble chitin are responsible for the activity of added chitin in this assay, although the activity of chitin oligomers has not been tested directly. Chitin and chitin oligomers have been shown to elicit lignin production in wheat,[47] phenolic substances in carrot[48] and chitinase activity in carrot[48] and melon.[49] However, to our knowledge, chitin has not been shown previously to be an elicitor of phytoalexins. It seems logical that chitin, along with the β-glucan and chitosan fungal cell wall components, should play a role as a source of regulatory signals during a plant-pathogen interaction.

Future Directions for Research with Rice Phytoalexins

A stronger foundation of understanding of the biosynthetic pathway to rice phytoalexins is needed in order to investigate its regulation. It is necessary to demonstrate that the predicted pimaradiene precursors do indeed serve as precursors to the momilactone and oryzalexin families of phytoalexins. The structure of compound C should be established and its possible relationship to unidentified antifungal stress metabolites in rice should be explored. The UV-inducible and elicitor-inducible enzymes, especially the prenyl transferases, cyclization enzymes, and oxygenases, should be characterized. The role of rice phytoalexins in resistance to pathogen attack needs investigation. The mode of action of chitin as an elicitor, and tests of fungal cell walls as a source of chitin oligomers should be examined. Preliminary experiments in our laboratory indicate that a heat stable endogenous elicitor is released from rice leaves by UV-irradiation. Characterization of this elicitor seems worth pursuing. The question of regulation of enzyme activities and gene expression in response to elicitation have not yet been addressed in rice. This partial list illustrates how little is known and how much remains to be done before we have some basic understanding of phytoalexin biosynthesis in rice and its physiological role.

Two problems of great current interest in connection with the molecular aspects of plant-pathogen interactions are the elucidation of mechanisms of signal transduction in plant cells responding to elicitors and the basis for microbial race and pathovar specificity in establishing resistance responses or susceptibility in host cultivars. It is our hope that rice will prove to be a useful plant in which to study these questions once more biochemical understanding of the plant responses to pathogens has been established.

CONCLUSION

(1) The complexity of phytoalexin production that can be encountered in a given plant species is well illustrated in rice. Rice plants can produce at least six different phytoalexins in two groups of oxygenated tricyclic diterpenes, and it seems likely from recent reports that additional antimicrobial substances, probably of still different structural types, are produced as well. This diversity in the array of phytoalexins may be important in increasing the range of potential pathogens to which the plant is resistant. This ability of the plant to produce more than one structural type may be particularly important in protecting against the situation where a pathogen becomes virulent by acquiring a detoxication mechanism against the host phytoalexin. In the case of rice, it might be necessary for the pathogen to acquire three or more different detoxication mechanisms to become infectious in this manner.

(2) The diterpenoid phytoalexins of castor bean and rice provide additional examples of the utilization of secondary metabolites in a defensive role that contributes to the ability of plants to adapt to and survive biological stresses. Since many diterpenes have been reported to have growth inhibitory properties, it will be surprising if other examples of diterpenes as phytoalexins are not discovered. Diterpenes take their place alongside many other classes of secondary metabolites that serve as phytoalexins and in other protective capacities for the plants that produce them.

(3) The biosynthetic pathways to casbene and the rice phytoalexins are consistent with the general pattern of biosynthesis of cyclic diterpenes that has emerged.[23] The all-*trans*-isomer of geranylgeranyl-PP appears to be the general precursor of all cyclic diterpenes by serving as the substrate for cyclization enzymes. Macrocyclic or polycyclic diterpene hydrocarbons are the products of either one or two successive cyclization reactions, respectively. Further modification of these hydrocarbons occurs in many cases by a series of oxygenase-catalyzed steps to produce the functional oxygenated diterpenoid metabolites that accumulate.

(4) The studies to localize diterpene cyclization enzymes described in this report seem to indicate that more than one subcellular compartment may house diterpene biosyn-

thetic capacity. The localization of casbene biosynthesis in the plastid is consistent with the active metabolism of geranylgeranyl-PP to pigments such as the phytyl side chains of chlorophyll and the tetraterpenes that are known to be localized there. However, the indication that rice phytoalexins are synthesized in the cytosol suggests that sources of geranylgeranyl-PP are available in that compartment as well.

(5) This research has emphasized the importance of plant cell wall fragments as phytoalexin elicitors in the castor bean plant and of fungal cell wall elicitors in rice. The latter case focuses attention on chitin fragments as an additional type of elicitor that may be of more general importance. It seems evident that different plant groups have evolved the capacity to recognize different biotic elicitors as signals to initiate their defensive responses.

(6) There are now a number of instances where two types of signalling molecules act synergistically in eliciting a defensive response, thus suggesting a positively cooperative interaction between their signal transduction pathway components. The fact that both UV-irradiation and chitin can stimulate phytoalexin biosynthetic enzyme activities in rice may be an example of "cross-talk" between the signal transduction pathways responding to two different types of stress.

(7) The demonstration that regulation of casbene biosynthesis occurs at least in part at the transcriptional level by gene activation adds to the growing body of evidence for this type of control in the phytoalexin literature. This may be the major type of regulation in the synthesis of secondary metabolites that are being produced as defensive agents in response to environmental or biological stresses, rather than effector and end product modulation of pre-formed enzyme activities. It makes sense in terms of conservation of energy and carbon for the plant cells to synthesize these defensive enzymes only when needed and only in locations where the products resulting from their activities function.

ACKNOWLEDGEMENTS

We are grateful to the following persons for making the materials indicated available to us for the research described in this report: Robert Coates, University of Illinois,

Urbana–reference standards of a number of diterpene hydrocarbons prepared synthetically in his laboratory; Nam-Hai Chua, Rockefeller University for cell cultures of rice; Lee Hadwiger, Washington State University for samples of chitosan; and Michael Hahn, University of Georgia for a sample of β–glucan elicitor derived from *Phytophthora megasperma* f.sp. *glycinea* cell walls. This research was supported in part by grants from the National Science Foundation (PCM-8302011 and DCB 88-04727) and the Competitive Grants Program of the United States Department of Agriculture (USDA 86-CRCR-12145 and FDP DA-88-37151-3616).

REFERENCES

1. DIXON, R.A. 1986. The phytoalexin response: elicitation, signalling and control of host gene expression. Biol. Rev. Cambridge Philos. Soc. 61:239-291.
2. BAILEY, J.A., J. W. MANSFIELD, eds. 1982. Phytoalexins, John Wiley & Sons, New York.
3. BELL, J.N., T.B. RYDER, V.P.M. WINGATE, J. A. BAILEY, C. J. LAMB. 1986. Differential accumulation of plant defense gene transcripts in a compatible and an incompatible plant-pathogen interaction. Mol. Cell. Biol. 6: 1615-1623.
4. PAXTON, J. D. 1981. Phytoalexins - a working redefinition. Phytopathol. Z. 101: 106-109.
5. DIXON, R.A., P.M. DEY, C. J. LAMB 1983. Phytoalexins: enzymology and molecular biology. Adv. in Enz. and Related Area of Mol. Biol. 53: 1-126.
6. SWAIN, T. 1977. Secondary compounds as protective agents. Annu. Rev. Plant Physiol. 28:479-501.
7. ROBINSON, D. R., C. A. WEST. 1970. Biosynthesis of cyclic diterpenes in extracts from seedlings of *Ricinus communis* L. II. Conversion of geranylgeranyl pyrophosphate into diterpene hydrocarbons and partial purification of cyclization enzymes. Biochemistry 9: 80-89.
8. SITTON, D., C. A. WEST. 1975. Casbene: an anti-fungal diterpene produced in cell-free extracts of *Ricinus communis* seedlings. Phytochemistry 14: 1921-1925.
9. OU, S.H. 1980. Pathogen variability and host resistance in rice blast disease. Annu. Rev. Phytopathol. 18: 167-187.
10. CARTWRIGHT, D.W., P.W. LANGCAKE, R.J. PRYCE, D.W.

LEWORTHY, J.P. RIDE. 1981. Isolation and characterization of two phytoalexins from rice as momilactones A and B. Phytochemistry 20: 535-537.

11. KATO, T., C. KABUTO, N. SASAKI, M. TSUNAGAWA, H. AIZAWA, K. FUJITA, Y. KATO, Y. KITAHARA. 1973. Momilactones, growth inhibitors from rice, *Oryza sativa* L. Tet. Lett. 3861-3864.
12. TSUNAGAWA, M., A. OHBA, N. SASAKI, C. KABUTO, T. KATO, Y. KITAHARA, N. TAKAHASHI. 1976. Momilactone C—a minor constituent of growth inhibitors in rice husk. Tet. Lett. 1157-1158.
13. AKATSUKA, T., O. KODAMA, H. SEKIDO, Y. KONO, S. TAKEUCHI. 1985. Novel phytoalexins (oryzalexins A, B and C) isolated from rice blast leaves infected with *Pyricularia oryzae*. Part I: isolation, characterization and biological activities of oryzalexins. Agric. Biol. Chem. 49: 1689-1694.
14. KONO, Y., S. TAKEUCHI, O. KODAMA, H. SEKIDO, T. AKATSUKA. 1985. Novel phytoalexins (oryzalexins A, B and C) from rice blast leaves infected with *Pyricularia oryzae*. Part II: structural studies with oryzalexins. Agric. Biol. Chem. 49: 1695-1701.
15. KONO, Y., S. TAKEUCHI, O. KODAMA, T. AKATSUKA. 1984. Absolute configuration of oryzalexin A and structures of its related phytoalexins isolated from rice blast leaves infected with *Pyricularia oryzae*. Agric. Biol. Chem. 48: 253-255.
16. SEKIDO, H., T. ENDO, R. SUGA, O. KODAMA, T. AKATSUKA, Y. KONO, S. TAKEUCHI. 1986. Oryzalexin D (3,7-dihydroxy-(+) sandaracopimaradiene), a new phytoalexin isolated from blast infected rice leaves. J. Pesticide Sci. 11: 369-372.
17. KODAMA, O., T. SUZUKI, J. MIYAKAWA, T. AKATSUKA. 1988. Ultraviolet-induced accumulation of phytoalexins in rice leaves. Agric. Biol. Chem. 52: 2469-2473.
18. KODAMA, O., A. YAMADA, A. YAMAMOTO, T. TAKEMOTO, T. AKATSUKA. 1988. Induction of phytoalexins with heavy metal ions in rice leaves. Nippon Noyaku Gakkaishi 13: 615-617.
19. DUEBER, M.T., W. ADOLF, C. A. WEST. 1978. Biosynthesis of the diterpene phytoalexin casbene. Partial purification and characterization of casbene synthetase from *Ricinus communis*. Plant Physiol. 62: 598-603.
20. DUDLEY, M. W., M. T. DUEBER, C.A. WEST. 1986. Biosynthesis of the macrocyclic diterpene in castor bean (*Ricinus communis* L) seedlings. Changes in

enzyme levels induced by fungal infection and intracellular localization of the pathway. Plant Physiol. 81: 335-342.

21. GREEN, T.R., C.A. WEST. 1974. Purification and characterization of two forms of geranyl transferase from *Ricinus communis.* Biochemistry 13: 4720-4729.
22. DUDLEY, M.W., T.R. GREEN, C.A. WEST. 1986. Biosynthesis of the macrocyclic diterpene casbene in castor bean (*Ricinus communis* L.) seedlings. The purification and properties of farnesyl transferase from elicited seedlings. Plant Physiol. 81: 343-348.
23. WEST, C.A., 1981. Biosynthesis of diterpenes. Chapter 7 In: Biosynthesis of Isoprenoid Compounds. (J. W. Porter, S. L. Spurgeon, eds.), Vol. 1, John Wiley & Sons, New York. pp. 376-411.
24. MOESTA, P., C.A. WEST. 1985. Casbene synthetase: regulation of phytoalexin biosynthesis in *Ricinus communis* L. seedlings. Purification of casbene synthetase and regulation of its biosynthesis during elicitation. Arch. Biochem. Biophys. 238: 325-333.
25. PARGELLIS, C.A.. 1983. Studies of the prenyl transferases from organelles of *Ricinus communis* L. Ph.D. Dissertation, University of California, Los Angeles.
26. STEKOLL, M., C.A. WEST. 1978. Purification and properties of an elicitor of castor bean phytoalexin from culture filtrates of the fungus *Rhizopus stolonifer.* Plant Physiol. 61: 38-45.
27. LEE, S.C., C.A. WEST. 1981. Polygalacturonase from *Rhizopus stolonifer*, an elicitor of casbene synthetase activity in castor bean (*Ricinus communis* L.) seedlings. Plant Physiol. 67: 633-639.
28. LEE, S.C., C.A. WEST. 1981. Properties of *Rhizopus stolonifer* polygalacturonase, an elicitor of casbene synthetase activity in castor bean (*Ricinus communis* L.) seedlings. Plant Physiol. 67: 640-645.
29. BRUCE, R.J., C.A. WEST. 1982. Elicitation of casbene synthetase activity in castor bean. The role of pectic fragments of the plant cell wall in elicitation by a fungal endopolygalacturonase. Plant Physiol. 69: 1181-1188.
30. WEST, C.A., P. MOESTA, D.F. JIN, A.F. LOIS, K.A. WICKHAM. 1985. The role of pectic fragments of the plant cell wall in the response to biological stresses. In: Cellular and Molecular Biology of Plant Stress, UCLA Symposia on Molecular and Cellular Biology, New Series, Volume 22, (J.L. Key, T. Kosuge, eds.), Alan

R. Liss, New York. pp. 335-349.

31. WALKER-SIMMONS, M., D. JIN, C.A. WEST, L. HADWIGER, C.A. RYAN. 1984. Comparison of proteinase inhibitor-inducing activities and phytoalexin elicitor activities of a pure fungal endopolygalacturonase, pectic fragments, and chitosans. Plant Physiol. 76: 833-836.
32. JIN, D.F., C.A. WEST. 1984. Characteristics of galacturonic acid oligomers as elicitors of casbene synthetase activity in castor bean seedlings. Plant Physiol. 74: 989-992.
33. LOIS, A.F. 1988. Regulation of the casbene synthetase gene during elicitation of castor bean seedlings with pectic fragments. Ph.D. Dissertation, University of California, Los Angeles.
34. CHAPPELL, J., K. HAHLBROCK. 1984. Transcription of plant defence genes in response to UV light or fungal elicitor. Nature 311: 76-78.
35. LAWTON, M.A., C.J. LAMB. 1987. Transcriptional activation of plant defense genes by fungal elicitor, wounding, and infection. Mol. Cell. Biol. 7: 335-341.
36. MANIARA, G., R. LAINE, J. KUC. 1984. Oligosaccharides from *Phytophthora infestans* enhance the elicitation of sesquiterpenoid stress metabolites by arachidonic acid in potato. Physiol. Plant Pathol. 24: 177-186.
37. DAVIS, K.R., A.G. DARVILL, P. ALBERSHEIM. 1986. Host-pathogen interactions. XXXI. Several biotic and abiotic elicitors act synergistically in the induction of phytoalexin accumulation in soybean. Plant Mol. Biol. 6: 23-32.
38. DAVIS, K.R., K. HAHLBROCK. 1987. Induction of defense responses in cultured parsley cells by plant cell wall fragments. Plant Physiol. 85: 1286-1290.
39. SHECHTER, I., C.A. WEST. 1969. Biosynthesis of gibberellins. IV. Biosynthesis of cyclic diterpenes from *trans*-geranylgeranyl pyrophosphate. J. Biol. Chem. 244: 3200-3209.
40. WICKHAM, K.A. 1988. The biosynthesis of diterpenes in rice. Ph.D. Dissertation, University of California, Los Angeles.
41. WICKHAM, K., C.A. WEST. 1987. Biosynthesis of diterpene phytoalexin precursors in cell-free extracts of rice. In: The Metabolism, Structure and Function of Plant Lipids. (P.K. Stumpf, J. B. Mudd, W.D. Nes, eds.), Plenum Publishing Corp., New York. pp. 123-125.
42. MATSUYAMA, N., S. WAKIMOTO. 1985. Purification and

characterization of anti-blast substance, S-1, formed mainly in blast-resistant lower rice leaves. Ann. Phytopath. Soc. Japan 51: 498-500.
43. MATSUYAMA, N., S. WAKIMOTO. 1988. Isolation and identification of diterpenoid anti-blast substances produced in the blast-infected rice leaves. Nippon Shokubutsu Byori Gakkaiho 54: 183-188.
44. FROST, R.G., C.A. WEST. 1977. Properties of kaurene synthetase from *Marah macrocarpus*. Plant Physiol. 59: 22-29.
45. DUEBER, M.T. 1979. Studies of phytoalexin biosynthesis in castor beans: purification and characterization of casbene synthetase. Ph.D. Dissertation, University of California, Los Angeles.
46. DARVILL, A.G., P. ALBERSHEIM. 1984. Phytoalexins and their elicitors—a defense against microbial infection in plants. Annu. Rev. Plant Physiol. 35: 243-275.
47. BARBER, M.S., J.P. RIDE. 1988. A quantitative test for induced lignification in wounded wheat leaves and its use to survey potential elicitors of the response. Physiol. Mol. Plant Pathol. 32: 185-197.
48. KUROSAKI, F., M. AMIN, A. NISHI. 1986. Induction of phytoalexin production and accumulation of phenolic compounds in cultured carrot cells. Physiol. Mol. Plant Pathol. 28: 359-370.
49. ROBY D., A. GADELLE, A. TOPPAN. 1987. Chitin oligosaccharides as elicitors of chitinase activity in melon. Biochem. Biophys. Res. Commun. 143: 885-892.

Chapter Seven

TERPENOID ANTI-HERBIVORE CHEMISTRY OF *ENCELIA* SPECIES (ASTERACEAE)

MURRAY B. ISMAN*, PETER PROKSCH† AND CURTIS CLARK§

Department of Plant Science*
University of British Columbia
Vancouver, B.C., Canada V6T 2A2

Institut fur Pharmazeutische Biologie†
Technische Universität
D-3300 Braunschweig, F.R.G.

Biological Sciences Department§
California State Polytechnic University
Pomona, California 91768, U.S.A.

INTRODUCTION

The role of terpenoid natural products in plant defense against herbivory by insects has been well established.[1-2] This phenomenon extends to arid and semi-arid plants, even though taxa from these latter habitats have been less frequently examined.[3] In this chapter, we compare the terpenoid chemistry of desert sunflowers of the genus *Encelia*, with the performance of a generalist insect feeding on extracts of these plants, in an attempt to define the role of the terpenoid allelochemicals as protective agents against herbivory.

The genus *Encelia* Adans. includes approximately twenty taxa of shrubby perennials that inhabit the Sonoran and Mojave deserts of the southwestern United States and adjacent areas of Mexico, particularly Baja California; however, there are two disjunct species that occur in South America. Several of the species are widespread and are frequently dominant elements of the desert floras. There is scant information on natural enemies of *Encelia*, but the common brittle bush, *E. farinosa* A. Gray, is known to be attacked by a specialist leaf beetle, *Trirhabda geminata* (Chrysomelidae),[4] and a generalist grasshopper, *Cibolacris parviceps* (Acrididae) (R.F. Chapman, personal commumication).

The terpenoid chemistry of *Encelia* has been the subject of considerable investigation, with respect to both chemotaxonomy[5,6] and quantitative variation.[6-8] The insecticidal and antifeedant actions of one class of *Encelia* terpenoids, the benzopyrans/benzofurans, have been well documented against several types of insects.[9-11] However, these previous reports have been based solely on laboratory bioassays utilizing isolated compounds, without establishing the exact role of these natural products as defensive agents in the plants themselves.

TERPENOID CHEMISTRY OF ENCELIA

Most species of *Encelia* are characterized by two distinct classes of terpenoids, sesquiterpene lactones and benzopyrans/benzofurans. The latter group are biogenically related, both arising from prenylation of a phenolic ring by a C_5 unit from the mevalonic acid pathway. The phenolic ring including the C-acetyl substituent has recently been shown to be derived from the shikimic acid pathway, via cinnamic acid.[12] Formation of the heterocyclic ring can take place in two ways, resulting in 2,2-dimethylchromenes (benzopyrans, **2-4**) (Fig. 1), or 2-isopropylbenzofurans (**5-6**) (Fig. 1).

The quantities of benzopyrans and benzofurans accumulated vary widely both between and within species of *Encelia*.[6-7] Qualitative variation occurs interspecifically, but the overall pattern of these compounds intraspecifically is relatively constant, which makes them useful as chemotaxonomic markers.[6] This approach has allowed a subgeneric division of the genus into three groups (Table 1); these groupings based on benzopyran/furan patterns are well supported by morphological characters.

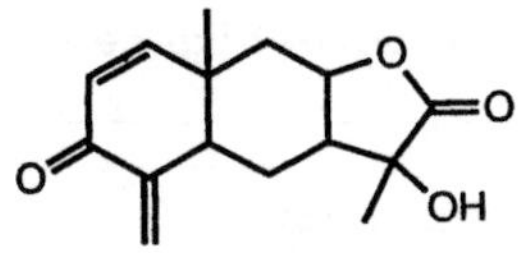

Farinosin (1)

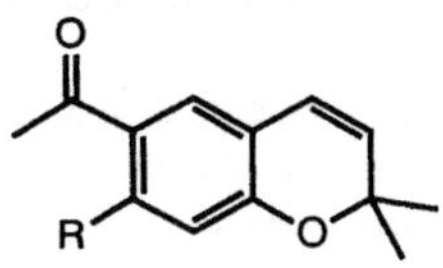

R		
H	Demethoxyencecalin	(2)
OH	Demethylencecalin	(3)
OCH_3	Encecalin	(4)

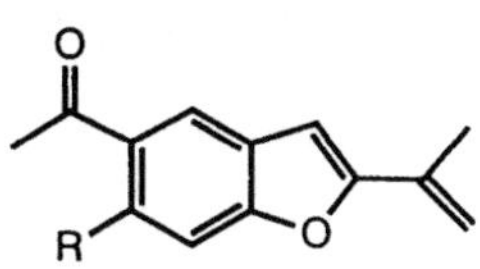

R		
OH	Euparin	(5)
OCH_3	Methyleuparin	(6)

Fig. 1. Structures of the major terpenoid (benzopyrans/ benzofurans) constituents of *Encelia* species.

As part of our investigation, we quantified the major benzopyrans/furans from the foliage of fourteen taxa of *Encelia* (Table 1) by reverse-phase HPLC. Our system included a 3.9 mm x 15 cm NovaPak™ C18 column (4 μ), a linear gradient from 10-60% aqueous acetonitrile in 20 minutes at a flow rate of 1.5 ml/min, and UV detection at 254 nm. Crude methanolic extracts of air-dried foliage were prepared for bioassay (see section on bioactivity of foliar extracts); a small aliquot of each sample was saved for analysis. With the exception of samples of *E. palmeri* and *E. farinosa*, plant material was harvested from plants growing in an experimental garden at Pomona, California.

Table 1. List of *Encelia* taxa investigated.

Group I	CAL*	*E. californica* Nutt.
	CON	*E. conserpse* Benth
	DEN	E. densifolia Clark & Kyhos
	FAR	*E. farinosa* A. Gray
	PHE	*E. farniosa* var. *phenicodonta* Blake
	HAL	*E. halimfolia* Cav.
	LAC	*E. X laciniata* Vasey & Rose
	PA1	*E. palmeri* Vasey & Rose
	RAD	*E. radians* Brandegee
	VEN	*E. ventorum* Brandegee
Group II	ACT	*E. actoni* Elmer
	RES	*E. frutescens* A. Gray var. *resinosa* M. E. Jones
Group III	ASP	*E. asperifolia* Clark & Kyhos
	FRU	*E. frutescens* var. *frutescens* A. Gray

Taxa grouped according to Proksch and Clark.[6]

*species codes used in Figures 2 and 3.

Material of *E. palmeri* was collected from plants in Baja California. Material of *E. farinosa* was a composite of several samples collected in Arizona.

Five of the fourteen taxa analyzed contained relatively large quantities of the dominant benzopyran, encecalin (**4**) (Fig. 2). For *E. palmeri*, *E. densifolia* (also see Fig. 4) and *E. conspersa*, encecalin constitutes greater than 90% of the total benzopyrans/furans in foliar extracts.

E. californica has almost equally large amounts of encecalin and methyleuparin (**6**) (also see Fig. 4). *E. halimifolia* has little encecalin, but considerable quantities of demethylencecalin (**3**) and methyleuparin (**6**). *E. frutescens* var. *resinosa* was notable in that no benzopyrans/furans were detected in the sample analyzed (Fig. 2). All of the other taxa had minor amounts of these compounds. It is noteworthy that the sample of *E. farinosa* analyzed contained little encecalin, but an appreciable quantity of demethylencecalin. This

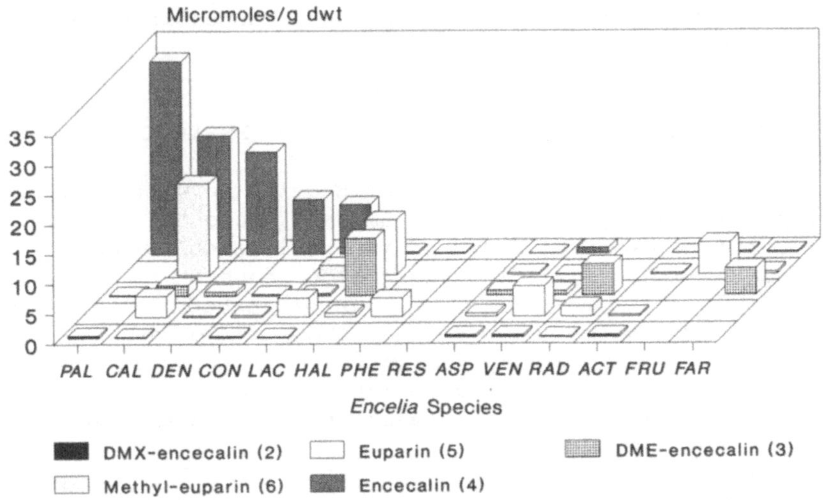

Fig. 2. Concentration of benzopyrans and benzofurans in foliar extracts of *Encelia* species, based on HPLC analysis of methanol extracts. DMX = demethoxy; DME = demethyl. Numbers in parentheses refer to structures in Figure 1, codes for species refer to Table 1 (dwt = dry weight).

species has been documented to be extremely variable with respect to benzopyran/furan content,[6,7] existing as two races or chemotypes based on encecalin content. Populations of *E. farinosa* from the Mojave desert of California generally have concentrations of encecalin ten-fold greater than those of populations from the Sonoran desert of Arizona.[6] An earlier analysis of a sample from the Mojave desert indicated that it contained approximately 20 µmol/g dwt (unpublished data).

Four structurally related sesquiterpene lactones of the eudesmanolide skeletal class have been isolated from species of *Encelia*.[13] These arise from an eremophilane C15 precursor. The predominant compound of this group, farinosin (**1**) (Fig. 1) has been quantified, along with encecalin (**4**) and euparin (**5**) in several populations of *E. farinosa*.[7] Values given for the latter two compounds in the report appear erroneously high (foliar encecalin concentrations averaging 20% dwt!), but the salient observations are that, on average, there is about one-eighth as much farinosin as encecalin, but slightly more farinosin than euparin.[8] These authors report that for Sonoran populations, farinosin is the predominant terpenoid compound in foliage of *E. farinosa*.

BIOACTIVITY OF ENCELIA TERPENOIDS TO INSECTS

Pure benzopyrans isolated from *Encelia* have contact insecticidal action against several types of insects. Encecalin (**4**) is toxic to neonate larvae of the variegated cutworm (*Peridroma saucia*, Noctuidae) via residue contact on glass surfaces (LD_{50} = 1.4 µg/cm^2) or when incorporated into an artificial diet (LD_{50} = 2.4 µmol/g fwt). This latter value is equivalent to 12 µmol/g dwt, suggesting that at least five species of *Encelia* should be more than adequately defended chemically against this insect based on their encecalin content alone (see Fig. 2).

In both tyes of bioassay (residue-contact and dietary incorporation), demethylencecalin (**3**) is significantly less toxic than encecalin.[9] This differential toxicity is also apparent for both the milkweed bug (*Oncopeltus fasciatus*, Lygaeidae)[14] and the mosquito (*Culex pipiens*, Culicidae).[15] In the case of the migratory grasshopper (*Melanoplus sanguinipes*, Acrididae), demethylencecalin is somewhat more toxic than either encecalin or demethoxyencecalin when the compounds are applied topically to neonate nymphs.[11] In contrast, the benzofurans euparin and methyleuparin lack toxicity in all bioassays used for the above mentioned insects. However, these compounds have been shown to have antimicrobial properties[14] and thus they may serve in a defensive capacity against plant pathogens. It is noteworthy that the benzofurans *antagonize* the toxicity of encecalin in bioassays using the cutworm, grasshopper and milkweed bug.[11]

The antifeedant action of the *Encelia* benzopyrans and euparin was recently assessed, using a leaf disc choice test[16] with 5th instar variegated cutworms (Table 2). All three of the benzopyrans tested have considerable antifeedant activity; the EC_{50} (concentration reducing feeding by 50%) for encecalin corresponds to an approximate value of 12 µmol/g dwt (cf. Fig. 2). As the sensitivity of this insect to allelochemicals is known to diminish with age,[16] it would be expected that smaller larvae would be deterred by lesser foliar concentrations. In contrast, euparin has only weak antifeedant activity, being about seven times less active than encecalin in this bioassay (Table 2).

Table 2. Antifeedant action of *Encelia* benzopyrans/furans to 5th instar larvae of *Peridroma saucia*.

Compound*	EC_{50}	
	µg/cm²	nmol/cm²
encecalin (**4**)	17.1	73.9
demethylencecalin (**3**)	17.0	78.1
demethoxyencecalin (**2**)	13.1	68.8
euparin (**5**)	114.8	531.3

Based on a cabbage leaf disc choice test.

*numbers in parentheses refer to structures in Figure 1.

In the only report of bioactivity of farinosin against insects, this sesquiterpene lactone did not significantly deter feeding or growth of the corn earworm (*Heliothis zea*, Noctuidae) when added to an artificial diet at 5.3 µmol/g fwt.[17] We have tested a related eudesmanolide sesquiterpene lactone, helenin (a mixture of alantolactone and its isomer isoalantolactone) against the variegated cutworm, and found that this mixture inhibited growth and reduced survivorship of neonate larvae in a dose-dependent manner with an EC_{50} and LD_{50} of approximately 2.5 and 3.0 µmol/g fwt respectively (unpublished data). Helenin is therefore similar in bioactivity to a series of pseudoguaianolide sesquiterpene lactones previously tested.[18-20]

BIOACTIVITY OF FOLIAR EXTRACTS OF *ENCELIA* SPECIES

To assess the role of benzopyrans in the chemistry of defense against insect herbivory in *Encelia* species, HPLC analysis of crude foliar extracts as described previously was combined with parallel bioassay of the extracts against neonate larvae of the variegated cutworm, a generalist herbivore. This insect, which feeds on an extremely wide range of plants including many econoomically important vegetable and fruit crops, ornamental plants and even seedling conifers, has been used as a test insect for many other plant extracts and purified natural products.[19-22] Dry, powdered foliar material of *Encelia* species was extracted once with hot methanol, and

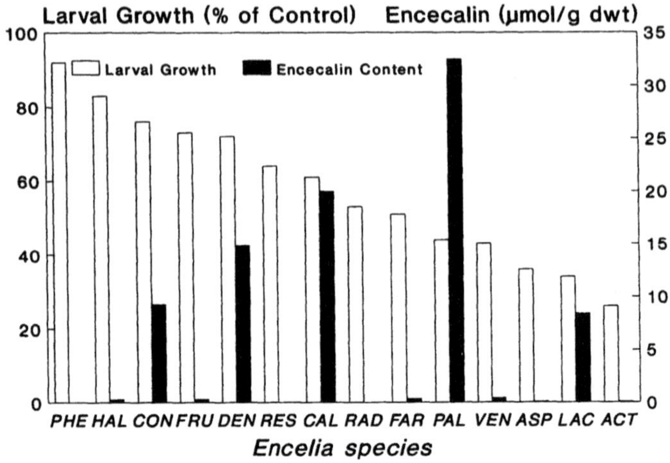

Fig. 3. Larval weight of *Peridroma saucia* larvae feeding on artificial diets spiked with foliar extracts of *Encelia* species at 25% of natural concentration. Encecalin con-centrations in respective species are shown for comparison. Codes for species refer to Table 1.

following concentration of the extracts, they were added to an artificial medium in the standard manner[21] such that the final concentration in the diet represented 25% of the natural concentration (i.e. extract from 1 g dwt foliage added to 4 g dwt of diet ingredients). Preliminary bioassays indicated that this concentration would permit some larval growth but might indicate differences between species with respect to their ability to support larval growth. Phytochemical analyses (results shown in Fig. 2) were conducted on an aliquot of the same extracts which were used for bioassay, allowing for direct comparison. For each plant species, forty neonate larvae were reared (two per cup) on diet containing extract for 7 days at 27^0, after which time all larvae were weighed. Our rationale for using crude foliar extracts (as opposed to adding leaf powder to the diet) is that this method permits us to evaluate the effect of the major extractable substances without interference from physical characteristics of the plants and without altering the nutritional value of the diet.

Results of the experiment are shown in Figure 3. Diets spiked with foliar extracts of the different *Encelia* taxa differ markedly in their ability to support larval growth, with mean larval growth for the treatments ranging from 92% of controls (*E. farinosa* var. *phenicodonta*) to 26% of controls (*E. actoni*). At this extract concentration, there was no appreciable mortality in any of the treatments. However, when an extract of *E. palmeri* was bioassayed against neonate larvae at 20, 40, 60 and 80% of natural concentration, all larvae died at the highest concentration and only 65% survived at 60% nat.conc. These observations collectively suggest that foliage of at least half of the species investigated would not be expected to support larval growth, based on extractable substances.

What is the relationship between encecalin (and/or benzopyrans/furans) and larval growth on diets containing crude foliar extracts? The data plotted in Figure 3 indicate unambiguously that encecalin does not play a major role in the relative suitability of *Encelia* extracts to the variegated cutworm. This observation is paradoxical in light of the considerable toxicity and deterrency of pure encecalin to this insect in several laboratory bioassays.[11] The best current explanation of our data is that other substances, such as the benzofurans (but probably other compounds) anatagonize or 'mask' the toxicity/deterrency of encecalin in this type of bioassay. One exception may be *E. palmeri*, which is relatively inhibitory to the cutworm (though not the most) and possesses the highest concentration of encecalin amongst the taxa examined.

SOURCES OF BIOACTIVITY IN INHIBITORY SPECIES

The most inhibitory foliar extracts in our bioassay were those of *E. actoni*, *E. X laciniata*, *E. asperifolia* and *E. ventorum* (Fig. 3) Of these, only *E. X laciniata* has appreciable quantites of benzopyrans/furans (Figs. 2,3). That this species possesses considerable concentrations of encecalin is not surprising as it consists of hybrids between *E. palmeri* (which has the highest concentration of encecalin in the collection) and *E. ventorum*.[23] Does chromatographic analysis provide any clues as to the substances in these inhibitory species that may be responsible for their bioactivity?

Chromatographic profiles of *E. actoni* and *E. asperifolia*, two of the most inhibitory species, have in common a dominant peak occurring at just under 5 minutes in our HPLC system (Fig. 5) We have isolated this peak using preparative HPLC, and have tentatively identified this substance as the sesquiterpene

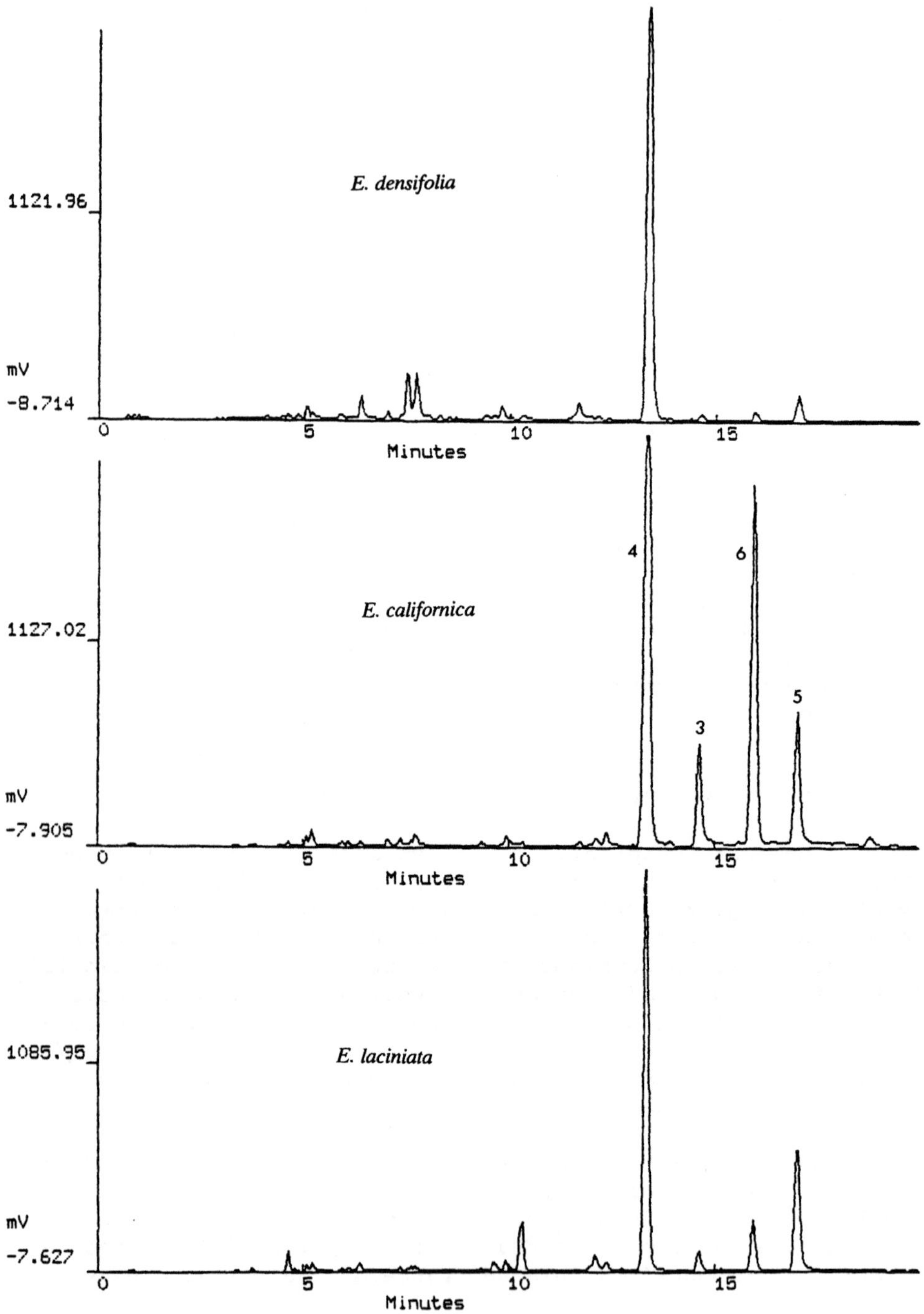

E. densifolia
1121.96
mV
-8.714
0
5
10
15
Minutes
E. californica
4
6
3
5
1127.02
mV
-7.905
0
5
10
15
Minutes
E. laciniata
1085.95
mV
-7.627
0
5
10
15
Minutes

lactone farinosin, based on ^{1}H-NMR and GC-MS analyses (unpublished data). To determine if this compound is the source of bioactivity in the foliar extracts of these two plants, we fractionated each extract using preparative HPLC, and bioassayed the fractions as before.

In the case of the *E. asperifolia* extract, the fraction corresponding to 2.5-5.0 minutes (Fig. 5), which contains the peak tentatively identifed as farinosin, has considerable larval growth inhibiting activity, however, the following fraction (5.0-7.5 minutes), containing the only other UV-visible substances in the extract, is even more biologically active, though not extremely so. Additionally, we analyzed and bioassayed a second collection of *E. asperifolia* (from a different locale). The extract of this latter sample contained approximately 60% as much 'farinosin' as the first sample, did not have the second major peak, but most important, did not significantly inhibit larval growth. In the case of *E. actoni*, the fraction containing the 'farinosin' peak had only modest activity (though statistically signifiant, $p < 0.05$), whereas again the fraction representing 5.0-7.5 minutes was the most active. However, in this species, the following two fractions (7.5-10 and 10.0-12.5 minutes) were also quite active. However, in this species, the following two fractions (7.5-10 and 10.0-12.5 minutes) were also quite active, though less so than the 5.0-7.5 minute fraction. Collectively these observations suggest that although the peak corresponding to farinosin contributes to the bioactivity of the most inhibitory extracts, it cannot account for most of the activity observed.

The extract of *E. ventorum* contains a series of moderate sized peaks in the 5.0-7.5 minute range (Fig. 5), which provides further circumstantial evidence that the important bioactive substances of *Encelia* species reside in this area ofthe chromatogram. *E. X laciniata* is a complete anomaly. This species has no apparent UV-visible (232/254 nm) substances which elute in less than 10 minutes (Fig. 4), yet this is one of the most inhibitory taxa (Fig. 3). The only feasible explanation for the pronounced bioactivity of this species (and, to a lesser extent, the aforementioned species) is that

---

Fig. 4. HPLC chromatograms of foliar extracts of *E. densifolia* (top), *E. californica* (middle), and *E. laciniata* (bottom), monitored at 254 nm. Numbered peaks refer to structures in Figure 1.

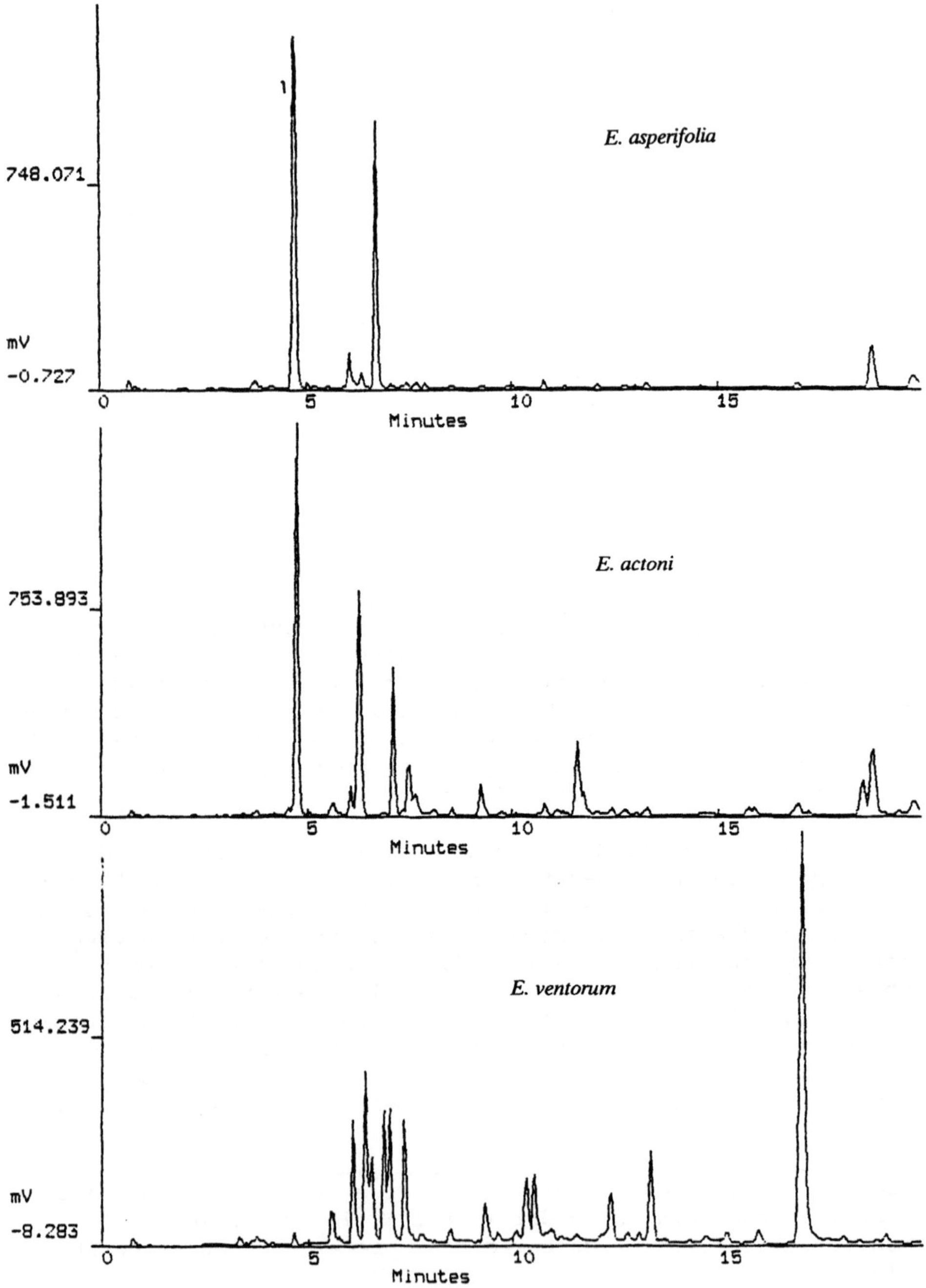
1
E. asperifolia
748.071
mV
-0.727
0
5
10
15
Minutes
E. actoni
753.893
mV
-1.511
0
5
10
15
Minutes
E. ventorum
514.239
mV
-8.283
0
5
10
15
Minutes

the substance(s) accounting for the bioactivity are UV-invisible or irreversibly bound to the chromatographic column.

CONCLUSION

The present investigation serves to underline the undesirable, but well documented, discrepancy between the bioactivity of isolated natural products in laboratory bioassays and the bioactivity of these substances in planta.[24]

In *Encelia*, the lack of a relationship between encecalin content and bioactivity in foliar extracts is especially surprising given the degree of biological activity of this compound against different types of insects. It should be noted, however, that amongst noctuid species, *P. saucia* is known to be relatively less sensitive to many allelochemicals than other frequently used bioassay species,[20] and therefore it is probably a very conservative model for this type of study. In spite of the results presented here, it may well be that the benzopyrans/furans serve in a defensive capacity for *Encelia* species against other insect herbivores. For example, a mixture of terpenoids (farinosin, encecalin and euparin) representative of the compounds in *E. farinosa* (of which encecalin made up 85%) significantly reduced larval growth and survival of the beetle *Trirhabda geminanta*, even though this insect feeds exclusively on this host plant.[4] Nonetheless, the key unanswered questions from our study (i.e. the nature of the bioactive constituents) indicate that there remains considerable scope for phytochemical investigation in the genus *Encelia*.

ACKNOWLEDGMENTS

We wish to thank Anna Luczynski and Nancy Brard for their expert technical assistance and unflagging enthusiasm. Opender Koul conducted the antifeedant bioassay, the results of which are presented in Table 2. We also thank Don Champagne for

---

Fig. 5. HPLC chromatograms of foliar extracts of *E. asperifolia* (top), *E. actoni* (middle), and *E. ventorum* (bottom), monitored at 254 nm. The numbered peak (**1**) has been tentatively identified as farinosin (see Fig. 1 for structure).

assistance with spectroscopic analyses, and for preparation of Figure 1, and D.W. Kyhos for field collection of seeds used to establish the experimental garden. In addition to those mentioned above, we thank Tom Lowery and Brian Ellis for critical comments on the manuscript. Supported by grants from NSERC (GP-2729, E-6850) to M.B.I. and the DFG to P.P.

REFERENCES

1. HARBORNE, J.B. 1986. Recent advances in chemical ecology. Natural Product Reports 1986: 323-344.
2. VAN BEEK, T.A., A.E. DE GROOT. 1986. Terpenoid antifeedants, part I. An overview of terpenoid antifeedants of natural origin. Recl. Trav. Chim. Pays-Bas 105: 513-527.
3. RODRIGUEZ, E. 1985. Insect feeding deterrents from semi-arid and arid land plants. In: Bioregualtors For Pest Control. (P.A. Hedin, ed.), A.C.S. Symposium Series 276, American Chemical Society, Washington, D.C., pp. 447-453.
4. WISDOM, C.S. 1982. Doctoral Dissertation, University of California, Irvine.
5. STEELINK, C., G.P. MARSHALL. 1979. Structures, syntheses and chemotaxonomic significance of some new acetophenone derivatives from *Encelia farinosa* Gray. J. Org. Chem. 44: 1429-1433.
6. PROKSCH, P., C. CLARK. 1987. Systematic implications of chromenes and benzofurans from *Encelia* (Asteraceae). Phytochemistry 26: 171-174.
7. WISDOM, C.S., E. RODRIGUEZ. 1982. Quantitative variation of the sesquiterpene lactones and chromenes of *Encelia farinosa*. Biochem. Syst. Ecol. 10: 43-48.
8. WISDOM, C.S., E. RODRIGUEZ. 1983. Seasonal age-specific measurements of the sesquiterpene lactones and chromenes of *Encelia farinosa*. Biochem. Syst. Ecol. 11: 345-352.
9. ISMAN, M.B., P. PROKSCH. 1985. Deterrent and insecticidal chromenes and benzofurans from *Encelia* (Asteraceae). Phytochemistry 24: 1949-1951.
10. ISMAN, M.B., J-Y. YAN, P. PROKSCH. 1986. Toxicity of natural chromene derivatives to a grasshopper. Naturwissenschaften 73: 500-501.
11. ISMAN, M.B. 1989. Toxicity and fate of acetylchromenes in pest insects. In: Insecticides of Plant Origin. (J.T. Arnason, B.J.R. Philogene and P. Morand, eds.),

A.C.S. Symposium Series 387, American Chemical Society, Washington, D.C., pp. 44-58.

12. SIEBERTZ, R., P. PROKSCH, V. WRAY, L. WITTE. 1989. Accumulation and biosynthesis of benzofurans in root cultures of *Eupatorium cannabinum*. Phytochemistry 28: 789-793.
13. SEAMAN, F.C. 1982. Sesquiterpene lactones as taxonomic characters in the Asteraceae. Bot. Rev. 48: 121-595.
14. PROKSCH, P., M. PROKSCH, G.H.N. TOWERS, E. RODRIGUEZ. 1983. Phototoxic and insecticidal activities of chromenes and benzofurans from *Encelia*. J. Nat. Prod. 46: 331-334.
15. KLOCKE, J.A., M.F. BALANDRIN, R.P. ADAMS, E. KINGSFORD. 1985. Insecticidal chromenes from the volatie oil of *Hemizonia fitchii*. J. Chem. Ecol. 11: 701-712.
16. KOUL, O., M.B. ISMAN. 1990. Antifeedant and growth inhibitory effects of *Acorus calamus* L. oil on *Peridroma saucia* (Lepidoptera: Noctuidae). Insect Sci. Appl. (in press).
17. WISDOM, C.S., J.T. SMILEY, E. RODRIGUEZ. 1983. Toxicity and deterrency of sesquiterpene lactones and chromenes to the corn earworm (Lepdioptera: Noctuidae). J. Econ. Entomol. 76: 993-998.
18. ISMAN, M.B., E. RODRIGUEZ. 1983. Larval growth inhibitors from species of *Parthenium* (Asteraceae). Phytochemistry 22: 2709-2713.
19. ARNASON, J.T., M.B. ISMAN, B.J.R. PHILOGENE, T.G. WADDELL. 1987. Mode of action of the sesquiterpene lactone, tenulin, from *Helenium amarum* (Asteraceae) against herbivorous insects. J. Nat. Prod. 50: 690-695.
20. ISMAN, M.B., N.L. BRARD, J. NAWROT, J. HARMATHA. 1989. Antifeedant and growth inhibitory effects of bakkenolide-A and other sesquiterpene lactones on the variegated cutworm, *Peridroma saucia* (Lep., Noctuidae). J. Appl. Entomol. 107: 524-529.
21. SALLOUM, G.S., M.B. ISMAN. 1989. Crude extracts of asteraceous weeds. Growth inhibitors for the variegated cutworm. J. Chem. Ecol. 15: 1379-1389.
22. CHAMPAGNE, D.E., M.B. ISMAN, G.H.N. TOWERS. 1989. Insecticidal activity of phytochemicals and extracts of the Meliaceae. In: Insecticides of Plant Origin. (J.T. Arnason, B.J.R. Philogene and P. Morand, eds.), A.C.S. Symposium Series 387, American Chemical Society, Washington, D.C., pp. 95-109.
23. KYHOS, D.W., C. CLARK, W.C. THOMPSON. 1981. The hybrid

nature of *Encelia laciniata* (Compositae: Helianthiae) and control of population composition in post-dispersal selection. Syst. Bot. 6: 399-411.

24. ISMAN, M.B., S.S. DUFFEY. 1982. Phenolic compounds in foliage of commercial tomato cultivars as growth inhibitors to the fruitworm, *Heliothis zea*. J. Amer. Soc. Hortic. Sci. 107: 167-170.

Chapter Eight

TERPENOIDS FROM SELECTED MARINE INVERTEBRATES

RAYMOND J. ANDERSEN*, E. DILIP DE SILVA,
ERIC J. DUMDEI, PETER T. NORTHCOTE,
CHARLES PATHIRANA AND MARK TISCHLER

Departments of Chemistry and Oceanography*
University of British Columbia
Vancouver, B.C.
Canada V6T 1W5

INTRODUCTION

Oceans cover over 70% of the surface area of the earth. They are populated by an extremely large number of plant and animal species that are uniquely marine. These marine organisms represent a vast, largely untapped, biochemical resource. During the last three decades, natural products chemists have been actively exploring this resource. They have been motivated by a practical desire to find new leads to drugs and an intellectual desire to understand the richness of nature's biosynthetic capabilities and the role that secondary metabolites play in the biology of marine organisms.

An analysis of the recent marine natural products literature showed that roughly two-fifths of the new compounds reported (1700 over an 8 year period) were isolated from

Biochemistry of the Mevalonic Acid Pathway to Terpenoids
Edited by G.H.N. Towers and H. A. Stafford
Plenum Press, New York

marine plants (algae) and the remaining three-fifths from marine invertebrates.[1] Two invertebrate phyla, the Porifera (sponges) and the Cnidaria (Coelenterata), accounted for about one half of all the reported metabolites. Further analysis showed that 58% of the algal metabolites, 85% of the Cnidarian metabolites and 37% of the Poriferan metabolites were terpenoids.

These statistics demonstrate that in terms of sheer numbers, marine organisms are an extremely rich source of new terpenoids. What they don't reveal is the degree of novelty that one encounters. Some features of marine terpenoid chemistry that make it of great interest to natural products chemists are: i) the large number of new terpenoid carbon skeletons, ii) the extensive bromination and chlorination in algal terpenoids, iii) the isonitrile and isothiocyanate substituents on sponge terpenoids, and iv) the emergence of sesterterpenoids as a major class of sponge metabolites.[2-5]

A number of years ago, we initiated structural studies on secondary metabolites extracted from marine invertebrates found in the cold temperate waters of the northeastern Pacific ocean. Recently we have broadened the scope of our investigations to include tropical species. The main objective of our research program is simply to discover new metabolites which are of interest because of the novelty of their biogenesis and/or their biological activities. An extensive screening program employing *in vitro* bioassays for antimicrobial, cytotoxic, herbicidal, and insecticidal activities has been used to provide leads to new compounds of interest to the pharmaceutical and agricultural industries. We have also focussed our attention on metabolites that are apparently employed by the source organisms as defensive allomones to thwart predation. Our efforts to date have been concentrated on invertebrates belonging to the phyla Mollusca, Porifera and Cnidaria. During the course of our studies, we have discovered some very interesting new terpenoids. Some of the highlights of this terpenoid chemistry are presented below.

METABOLITES FROM NUDIBRANCHS AND THE INVERTEBRATES IN THEIR DIETS

Dorid nudibranchs are delicate, slow-moving, shell-less, and often strikingly colored molluscs that appear to be ill-equipped to ward off predators. In spite of this apparent vulnerability, nudibranchs have few recorded predators. It is

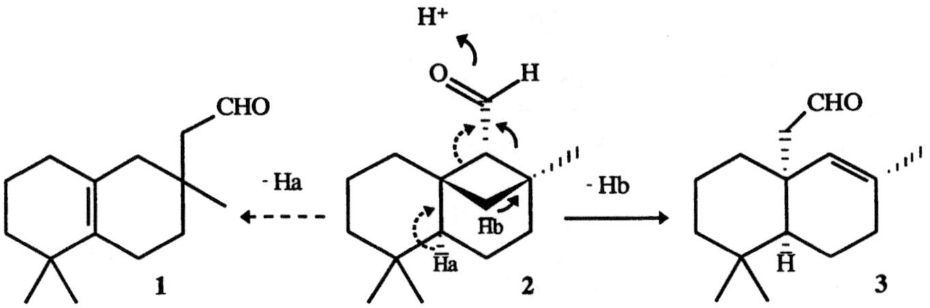

Scheme 1

now well documented that nudibranchs use organic metabolites stored in glands on their dorsums as one means to thwart predation. The defensive substances are usually sequestered from the sponges, bryozoans, hydroids or soft corals that constitute the nudibranch's diet. A few species are known to have the ability to synthesize their defensive allomones *de novo*.

Approximately eighty species of nudibranchs have been reported from the coastal waters of British Columbia. Many of these species are represented by sizable populations in the shallow intertidal zone. This easily accessible and abundant source of organisms prompted us to examine the local species.

One of the first nudibranchs that we studied was *Acanthodoris nanaimoensis*. We were attracted to this animal by the extremely fruity odor given off by freshly collected specimens because we felt that the odor was a likely indication of the presence of terpenoids. Three sesquiterpenoid aldehydes, nanaimoal (**1**),[3] acanthodoral (**2**) and isoacanthodoral (**3**),[7] were isolated from the skin extracts of *A. nanaimoensis*. Each of these compounds represented a new sesquiterpene carbon skeleton for which we proposed the names nanaimoane, acanthodorane and isoacanthodorane, respectively. Scheme 1 shows that nanaimoal (**1**) and isoacanthodoral (**3**) can be readily formed by opening the cyclobutane ring in acanthodoral (**2**).

Two common nudibranchs in the genus *Archidoris*, namely *A. odhneri* and *A. montereyensis*, were found to contain the family of terpenoic acid glycerides **4** to **13**.[5,6,7] The farnesic acid glycerides **4** to **6** were the major components in extracts of *A. odhneri*,[6] while the diterpenoic acid glycerides **7** to **9** were the major components in extracts of *A. montereyensis*.[8,10] Specimens of *A. montereyensis* collected in La Jolla,

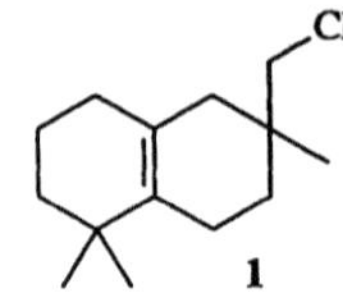

1

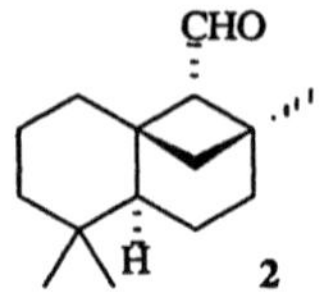

2

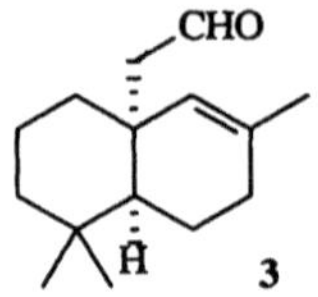

3

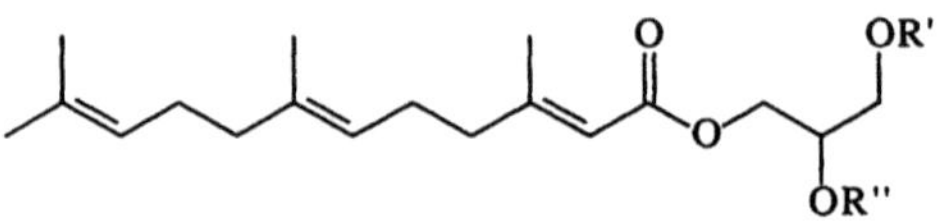

4 R' = R'' = H

5 R' = Ac R'' = H

6 R' = H R'' = Ac

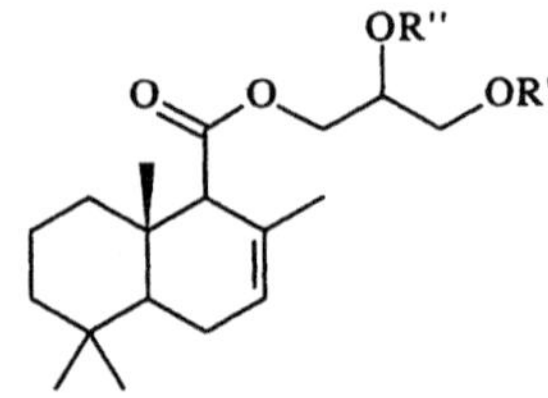

10 R' = R'' = H

11 R' = Ac R'' = H

12 R' = H R'' = Ac

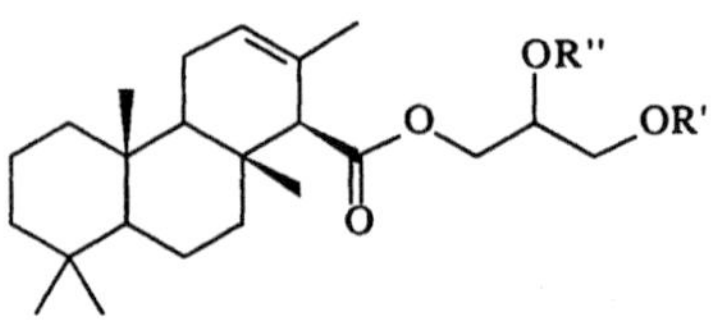

7 R' = R'' = H

8 R' = Ac R'' = H

9 R' = H R'' = Ac

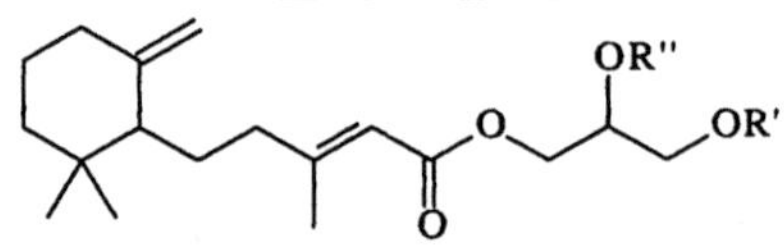

13 R'' = R' = H

California contained the same mixture of glycerides found in the B.C. specimens.[5] We were able to demonstrate that the nudibranchs were capable of incorporating ^{14}C labelled mevalonic acid into the terpenoid portion of the glycerides **4**, **7** and **10**.[7] The geographic invariability in metabolite content and the incorporation studies together support the notion that nudibranchs whose skin chemistry is identical at widely differing geographic locations are likely synthesizing their metabolites *de novo*. Bioassays showed that the drimane derivative **10** had fish antifeedant activity, while compounds **4** and **7** were inactive.[10]

Tochuina tetraquetra, which commonly grows to a length of 30 cm, is one of the largest nudibranchs in the world. It occurs from the Kuril Islands, USSR (where the natives are reported to eat them) to the Santa Cruz Islands, California. We have made collections of *T. tetraquetra* from both Port Hardy and Barkley Sound, British Columbia. Skin extracts of

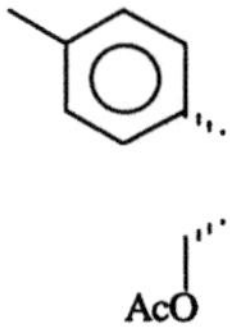
14

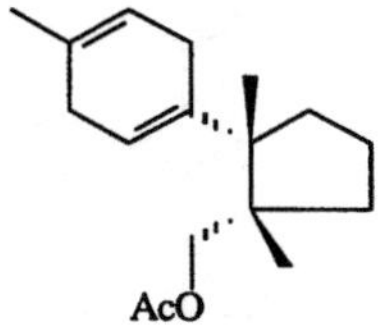
15

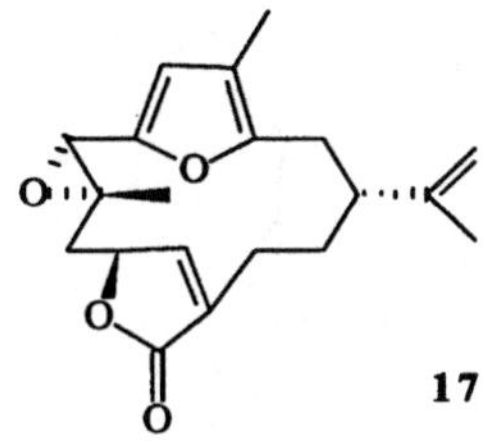
17

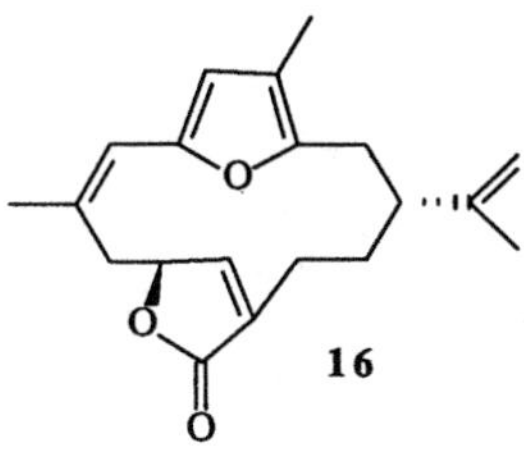
16

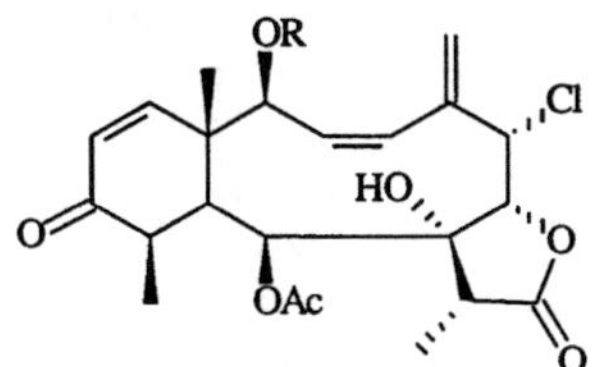
18 R=Ac
19 R=Butanoate

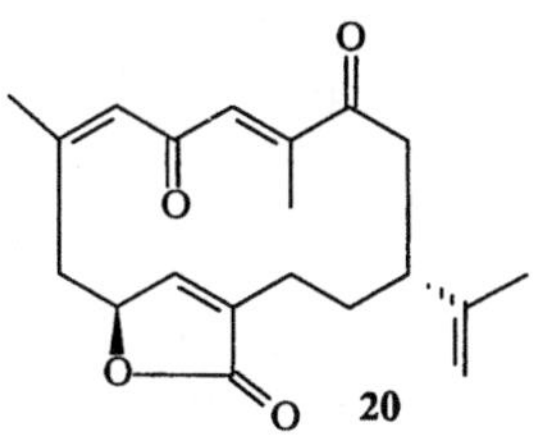
20

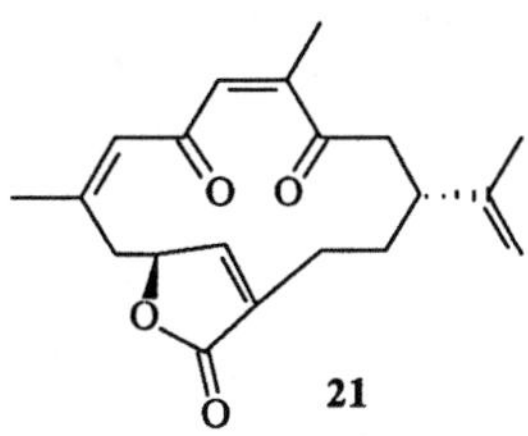
21

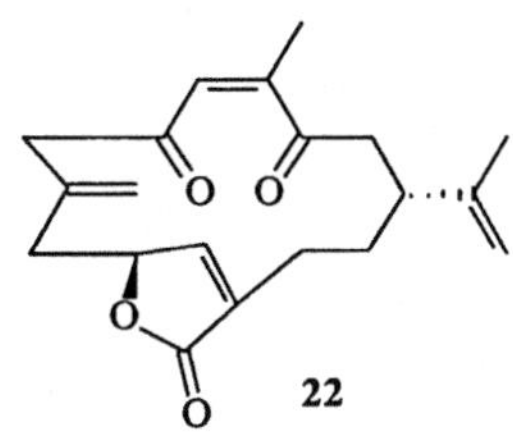
22

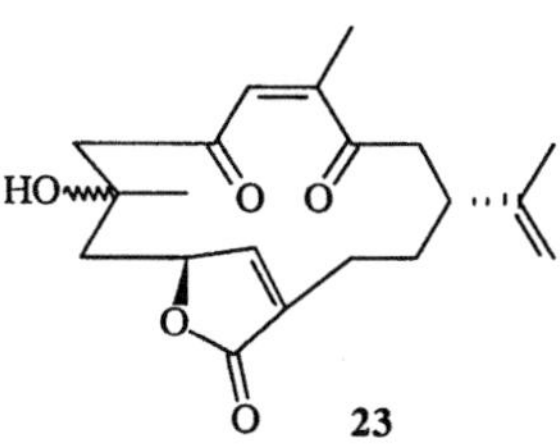
23

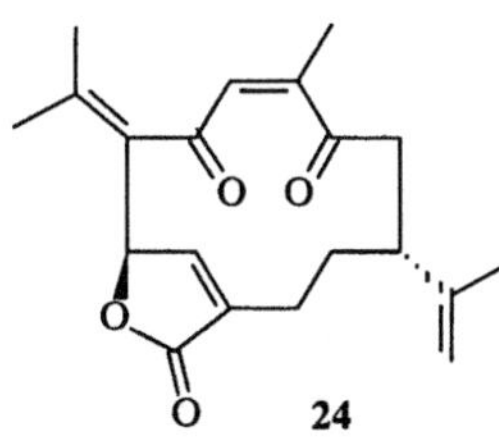
24

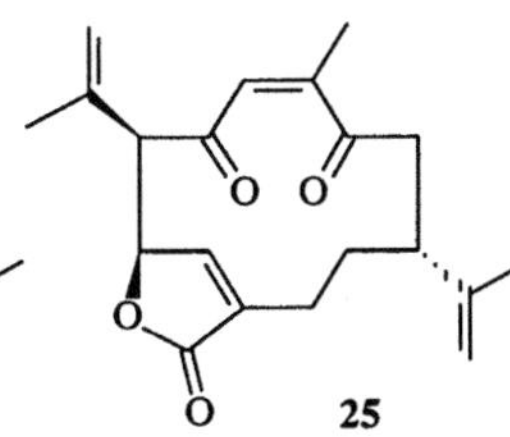
25

the Port Hardy specimens contained the two cuparane sesquiterpenoids, tochuinyl acetate (**14**) and dihyrotochuinyl acetate (**15**), as well as the cembrane diterpenoids, rubifolide (**16**) and pukalide (**17**).[8] The Barkley Sound specimens yielded only the briarein diterpenoids, ptilosarcenone (**18**) and its butanoate analogue (**19**). *T. tetraquetra* feeds exclusively on Cnidarians. Its preferred food is the soft coral *Gersemia rubiformis,* however when that is not available the nudibranch will eat the sea pen *Ptilosarcus gurneyi;* ptilosarcenone (**18**) had been previously isolated from *P. gurneyi,* establishing that the Barkley Sound specimens were feeding exclusively on

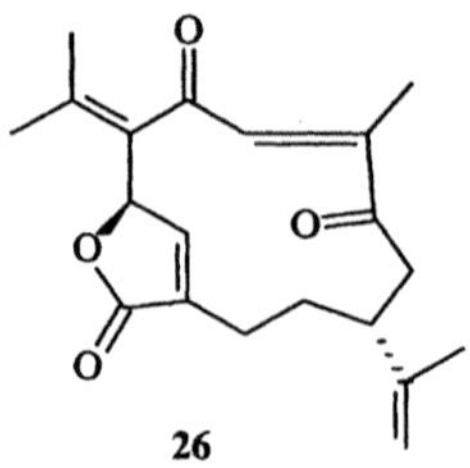
26

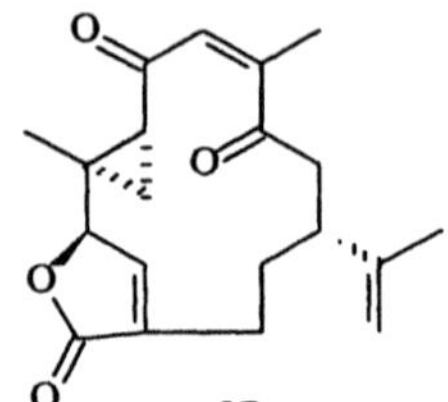
27

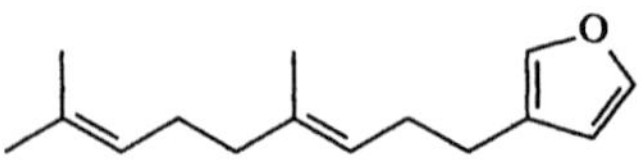
28

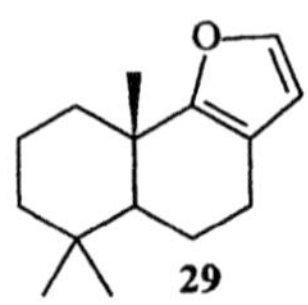
29

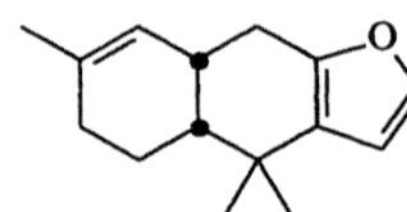
30

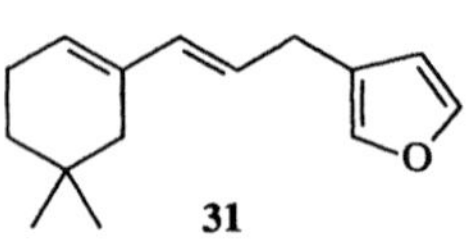
31

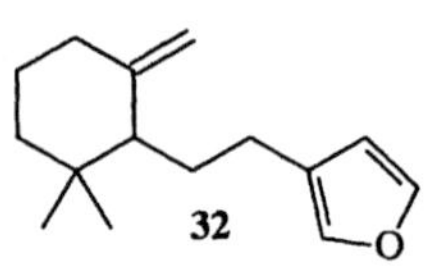
32

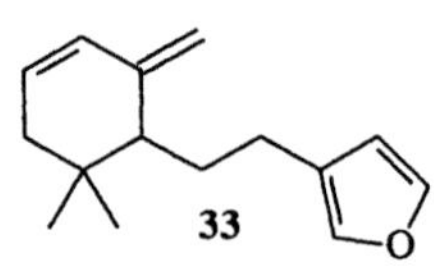
33

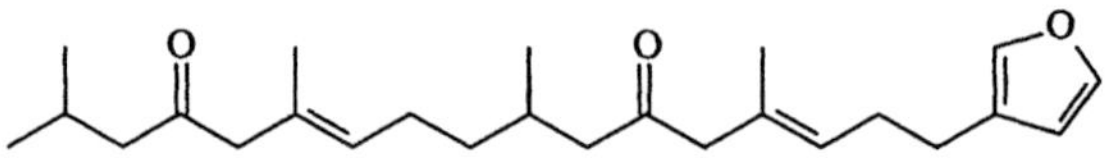
34

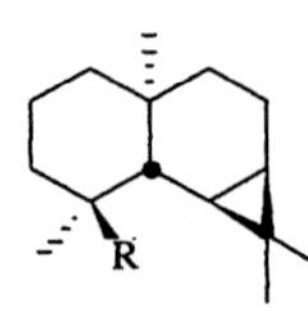
35 R= -NC
36 R= -NCS
37 R= -NHCHO

the sea pen. This was expected since *G. rubiformis* is not found in Barkley Sound. We undertook an investigation of the chemistry of *G. rubiformis*, collected at Port Hardy, and showed that metabolites **14**, **15** and **16** were being produced by the soft coral.[12,13] Pukalide (**17**) was not present in any of the samples of *G. rubiformis* that we have examined and its origin remains unclear.

Further examination of *G. rubiformis* extracts uncovered a fascinating array of diterpenoids in addition to rubifolide (**16**), which had been found in the skin extracts of *T. tetraquetra*. This included the four cembranoids **20** to **22** and the three pseudopteranoids **24** to **26**.[12,13] The pseudopterane skeleton, which had previously been encountered in a gorgonian metabolite,[11] is unique to the marine environment. Perhaps the most interesting metabolite found in *G. rubiformis* was gersolide (**27**), which had the new gersolane carbon skeleton.[12] The co-occurrence of cembrane, pseudopterane and gersolane diterpenoids in *G. rubiformis* suggested that all three carbon skeletons share a common biogenetic origin. Bandurraga et al. have proposed that the pseudopterane skeleton is formed via a ring contraction of a cembrane precursor[14] and it seems likely that the pathway to the gersolane skeleton also passes through a cembrane intermediate. It is interesting to note that *T. tetraquetra* selectively sequesters metabolites from its diet, concentrating some compounds and ignoring others. *T. tetraquetra* is an excellent example of how the skin chemistry of a nudibranch that sequesters metabolites from dietary organisms can vary from one collecting site to another.

Cadlina luteomarginata is one of the most frequently encountered nudibranchs all along the west coast of North America from the Queen Charlotte Islands, B.C. in the north to Punta Eugenia, Baja California in the south. Freshly collected specimens of *C. luteomarginata* have an intense fruity odor caused by the terpenoid metabolites on its dorsum. The skin chemistry of *C. luteomarginata* also varies quite dramatically from one collecting site to another indicating a dietary source for the compounds. Specimens collected in La Jolla California by Faulkner and co-workers yielded the seven furanoterpenoids **28** to **34** as well as the isonitrile **35**, the isothiocyanate **36** and the formamide **37**.[16] Sponges that constitute the diet of *C. luteomarginata* were shown to be the source of several of these metabolites.

Our early work on *Cadlina* collected in B.C. waters showed that both Howe Sound and Barkley Sound specimens contained albicanol (**38**), the corresponding acetate (**39**)[17] and luteone (**40**),[18] a degraded sesterterpenoid that is responsible for the odor. The luteane carbon skeleton of luteone has not been found in any other metabolite isolated from a living organism, however hydrocarbons having this skeleton have been found in petroleum.[19] We also found that some, but not all, of the specimens of *C. luteomarginata* collected in Barkley Sound contained small quantities of furodysin (**39**), furodysinin (**40**) and microcionin-2 (**43**).[17] A few animals that we collected in the Queen Charlotte Islands yielded marginatafuran (**44**), which had the new marginatane diterpenoid carbon skeleton.[17] Compounds **39, 41** and **42** were shown to have fish antifeedant activity.[17]

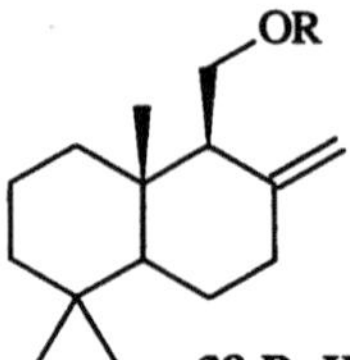

38 R=H
39 R=Ac

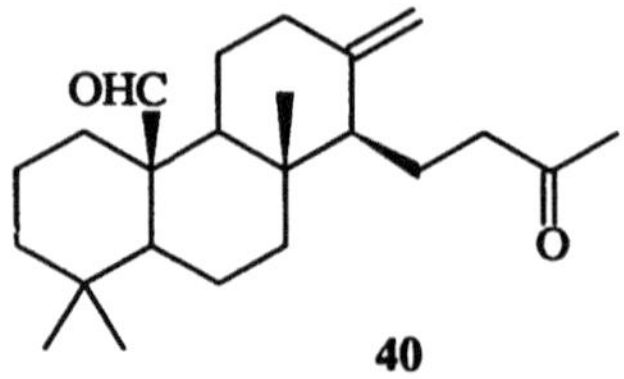

40

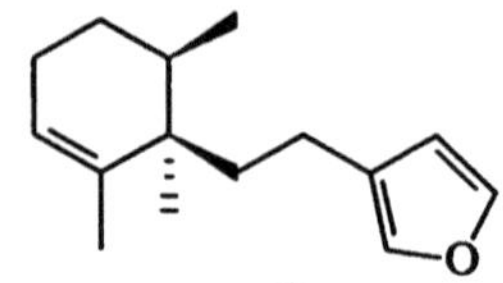

43

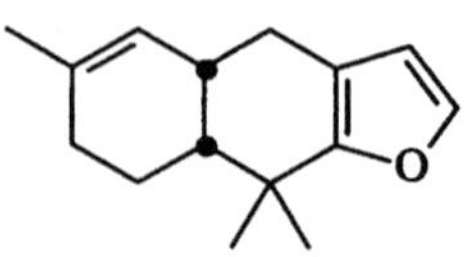

41

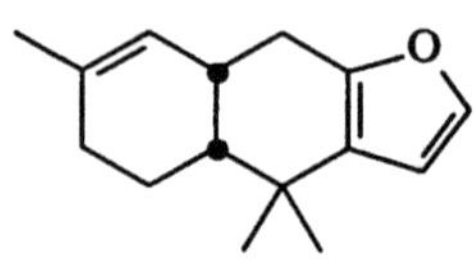

42

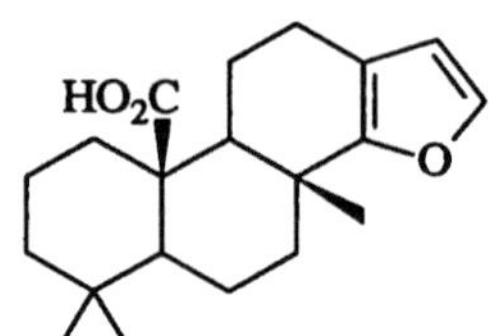

44

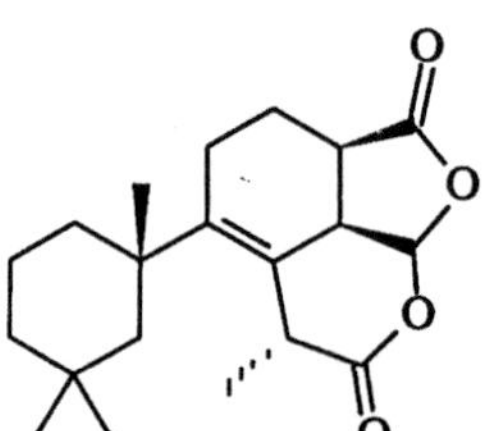

45

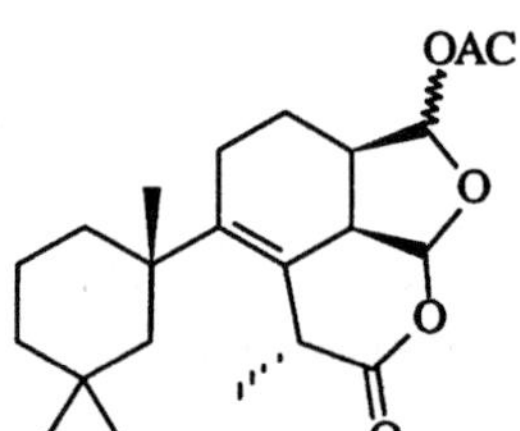

46

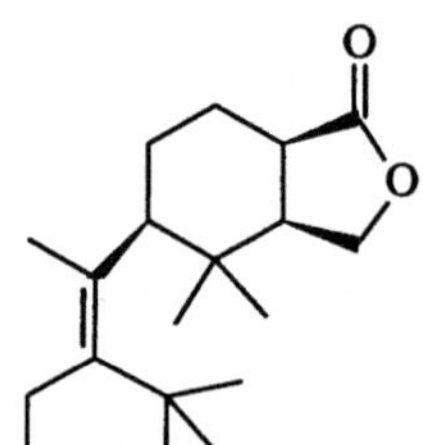

47

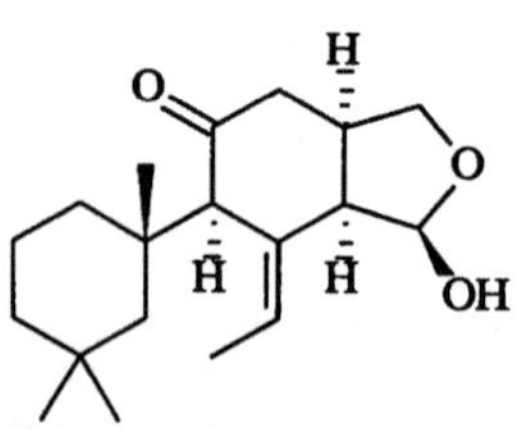

48

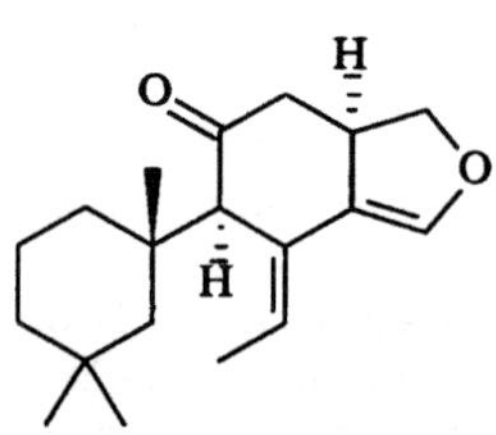

49

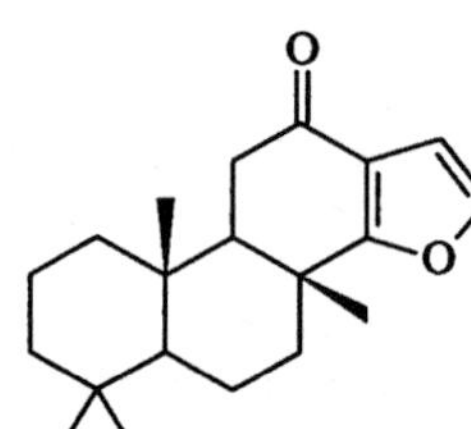

50

We have continued to make collections of *C. luteomarginata* in an attempt to learn more about the variation of its skin chemistry with locale. A small collection of individuals was recently made in an exposed surge channel on Sanford Island, Barkley Sound, where the nudibranchs were found feeding on the sponge *Aplysilla glacialis*. The skin extracts of the Sanford Island specimens contained the two rearranged spongian diterpenoids, cadlinolide A (**45**) and the corresponding acetate **46**, as well as the rearranged and degraded spongian diterpenoid, glaciolide (**47**).[21] The carbon

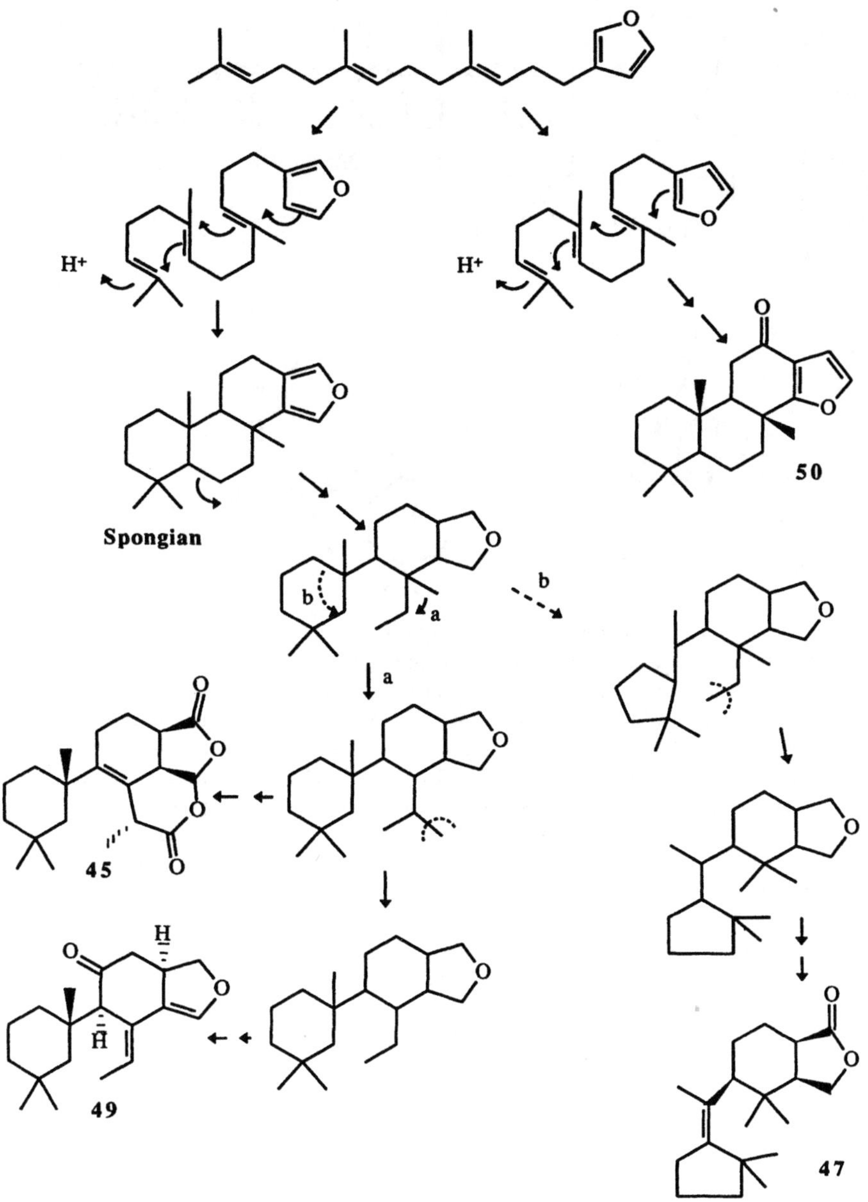

Scheme 2

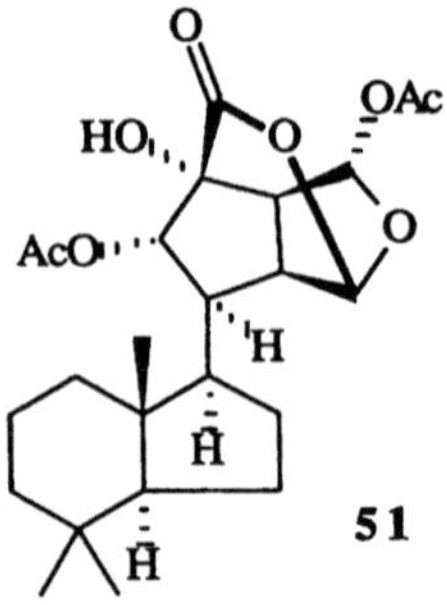

51

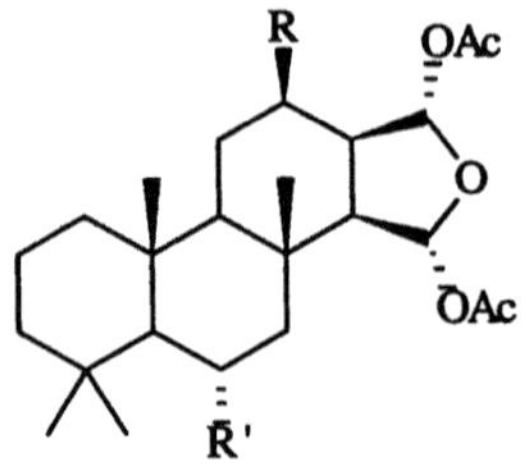

52 R=OAc R'=H
53 R=H R'=OAc
54 R=R'=OAc

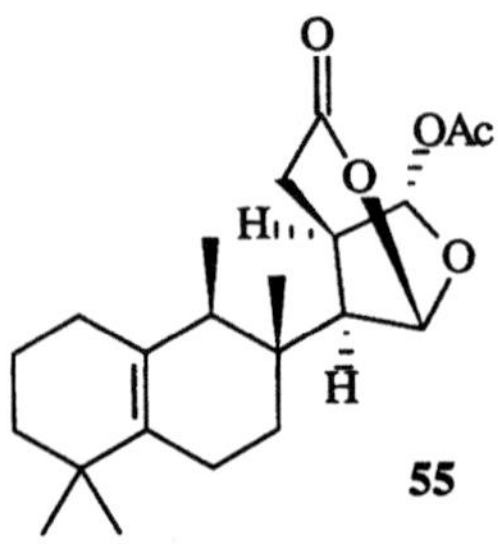

55

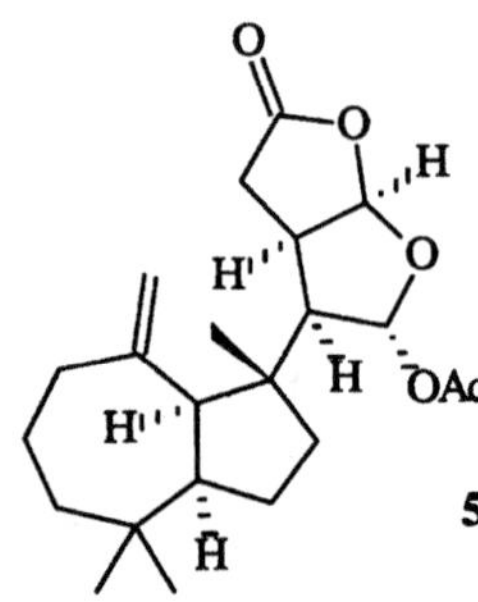

56

skeletons of **45** and **46** had been previously encountered in several sponge metabolites, however glaciolide (**47**) had the new glaciane carbon skeleton. Examination of extracts of the sponge *A. glacialis* revealed the presence of cadlinolide A (**45**) and glaciolide (**47**) as well as the new diterpenoids aplysillolide A (**48**), aplysillolide B (**49**) and marginatone (**50**). Marginatone (**50**) is only the second example of a marginatane diterpenoid and its discovery in *A. glacialis* suggests that a sponge in the genera *Aplysilla* is the source of marginatafuran (**44**) originally found in the Queen Charlotte specimens of *C. luteomarginata*. It is interesting to note that we found no trace of the acetate derivative **46** in the sponge, even though it was relatively abundant in the skin extracts of the nudibranchs feeding on the sponge. We plan to investigate the possibility that the nudibranchs are capable of chemically altering sponge metabolites such as **45** to give **46**. Scheme 2 shows a proposed biogenesis for the *A. glacialis* metabolites, starting from an acyclic furanoditerpenoid precursor.

We have recently had an opportunity to examine the skin extracts of a number of tropical nudibranchs in the genus *Chromodoris* that were collected in the coastal waters of Sri

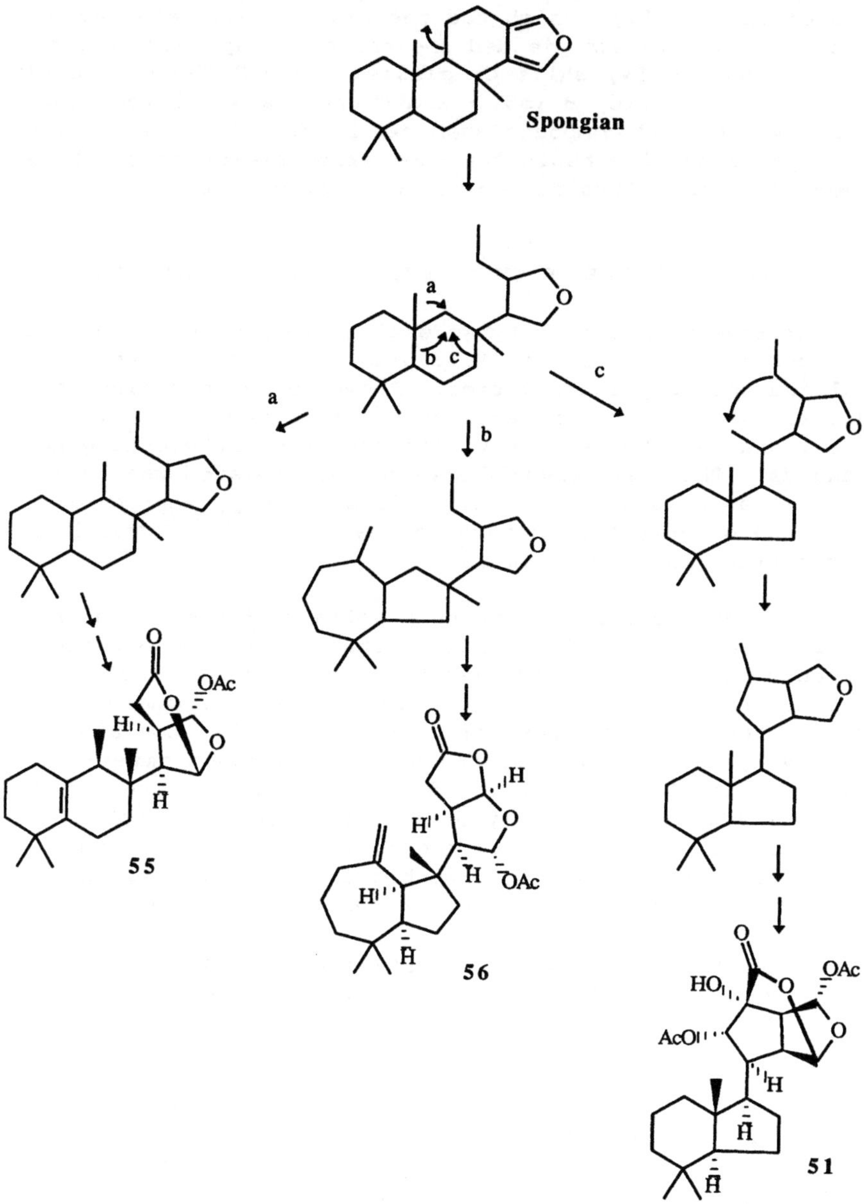

Scheme 3

Lanka. The major metabolite isolated from *C. cavae* was chromodorolide A (**51**), which had the new chromodorane carbon skeleton.[22] *C. geminus* yielded the three new spongian derivatives, **52**, **53** and **54**, while *C. gleniei* yielded the previously reported dendrillolide B (**56**)[23] and *C. annulata* yielded the previously reported compound shahamin F (**55**).[24] Scheme 3 outlines a proposed biogenesis for the carbon skeletons of these *Chromodoris* metabolites from a spongian precursor.

METABOLITES FROM THE SPONGE *XESTOSPONGIA VANILLA*

Our routine bioactivity screening program revealed that crude extracts of the sponge *X. vanilla* had moderate antibacterial and antifungal activities. Prompted by this finding, we undertook a chemical investigation of *X. vanilla* extracts and we found them to be a very rich source of interesting new terpenoids. The first metabolite that we isolated was xestodiol (**57**), a new apocarotenoid.[24] Surprisingly, xestodiol (**57**) is one of only two apocarotenoids isolated to date from either marine plants or animals.

The second group of compounds which we found in the nonpolar fractions of the *X. vanilla* extracts were the nineteen and twenty carbon containing metabolites xestenone (**58**),[22] secoxestenone (**59**) and xestolide (**60**).[27] Each of these three compounds represented a new terpenoid carbon skeleton and initially we assumed that **58** and **59** were degraded diterpenoids, coming from a rearranged precursor such as xestolide (**60**). However, the discovery of a third group of metabolites in the very polar fractions of the *X. vanilla* extracts made our initial assumptions about the biogenesis of xestenone (**58**), secoxestenone (**59**) and xestolide (**60**) untenable.

The major metabolite in the polar fractions was xestovanin A (**61**), an antifungal triterpenoid glycoside.[28] A closely related compound secoxestovanin A (**62**) was also isolated from the polar fractions. Xestovanin A (**61**) and secoxestovanin A (**62**) both had new triterpenoid carbon skeletons for which we proposed the names xestovanane and secoxestovanane. Once the structures of the triterpenoid glycosides were known, it became apparent that xestenone (**58**), secoxestenone (**59**) and xestolide (**60**) were most likely degradation products of the triterpenoids. Scheme 4 shows a proposed biogenesis for the degraded *X. vanilla* metabolites. Triterpenoid glycosides are extremely common in marine

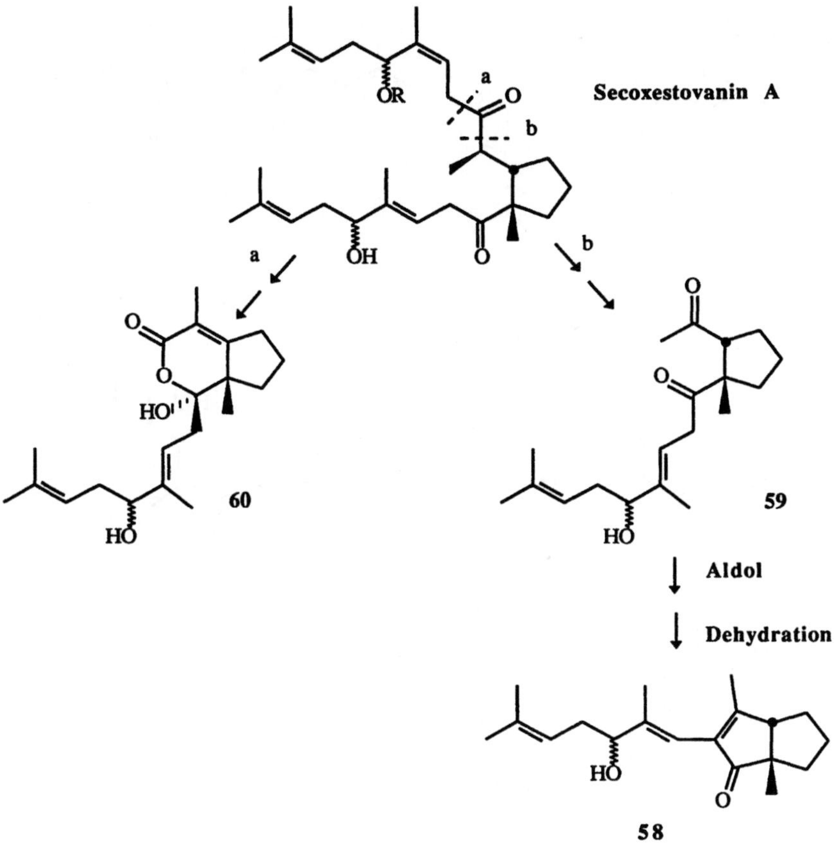

Scheme 4

invertebrates belonging to the phylum Echinodermata (starfishes and sea cucumbers), however *X. vanilla* is only the second sponge known to contain this type of metabolite.

A METABOLITE OF MIXED BIOGENESIS FROM THE HYDROID (CNIDARIA) *HYDRALLMANIA FALCATA*

Hydroids are small colonial marine invertebrates that belong to the phylum Cnidaria. Many species produce delicate upright colonies that resemble tiny bushes, trees or feathers. Most hydroids are extremely difficult to collect in any

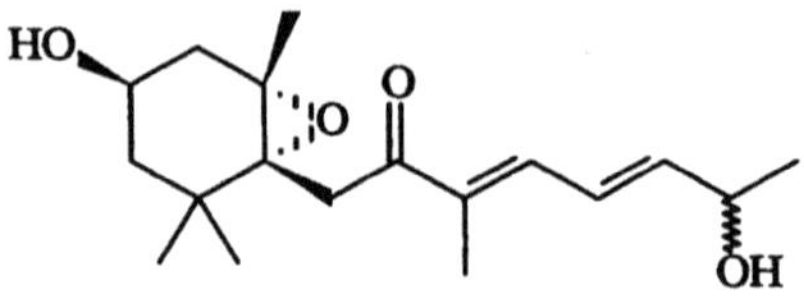

57

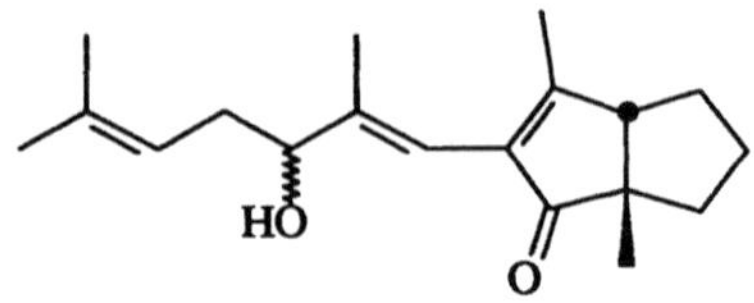

58

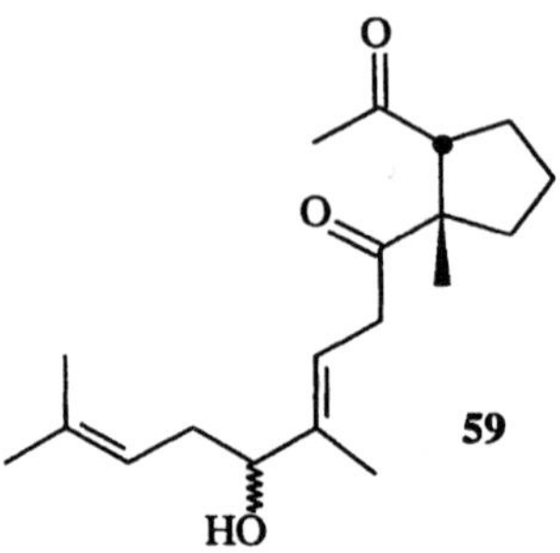

59

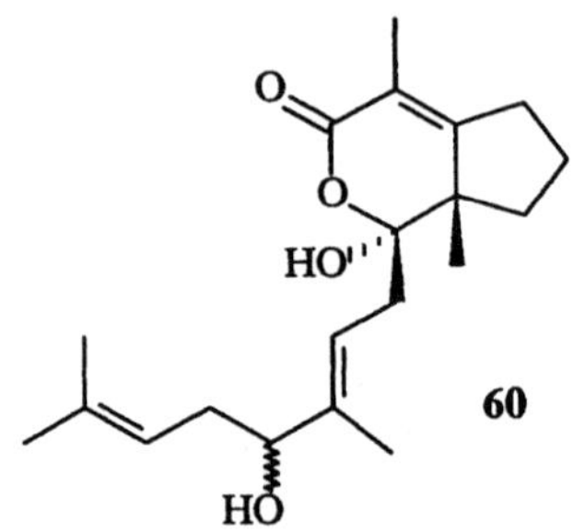

60

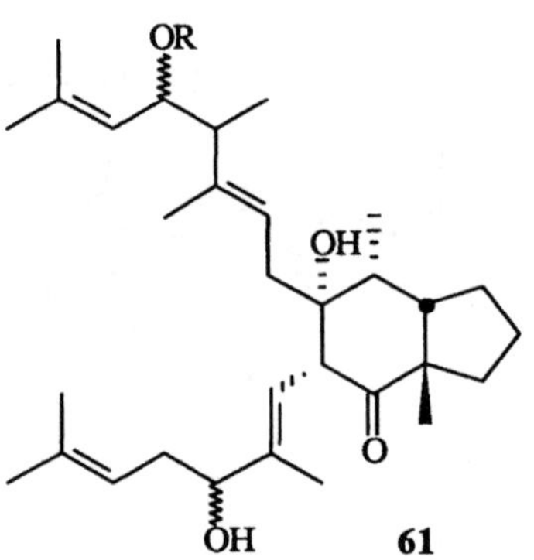

61

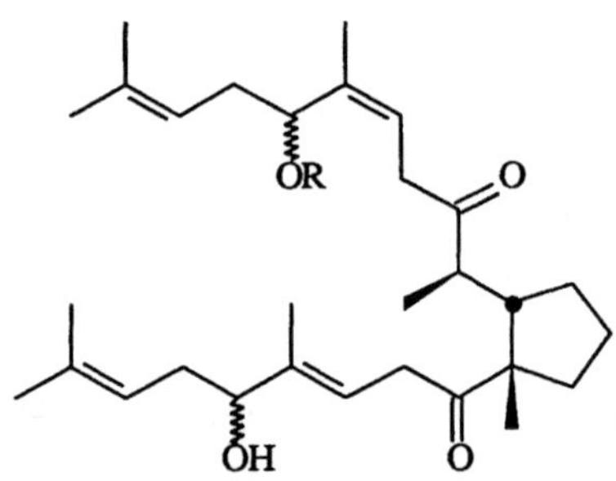

62

R=

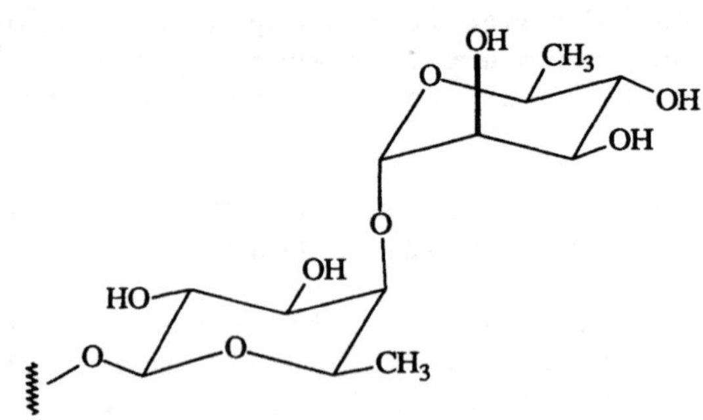

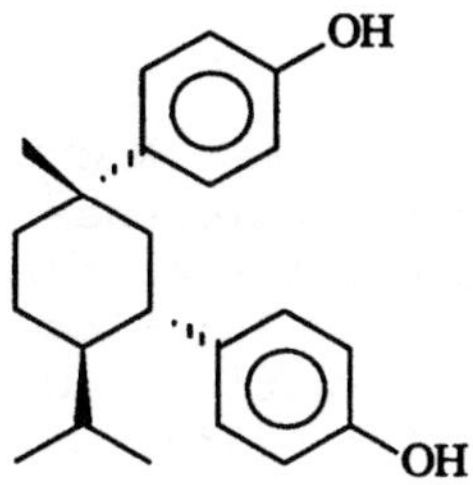

63

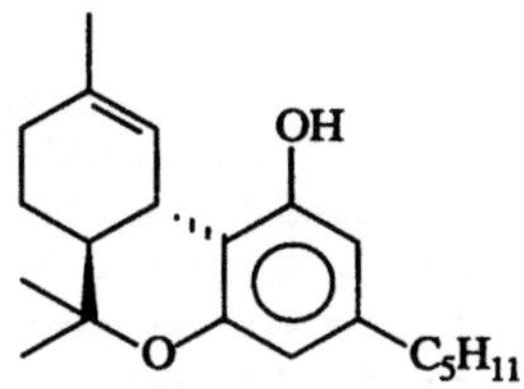

64

abundance. As a consequence, very few chemical studies have been conducted on hydroids. We became interested in the extracts of the hydroid *Hydrallmania falcata* collected in the North Atlantic Ocean because of their cytotoxicity. A family of mildly cytotoxic metabolites was found in the extracts, however the small quantities available permitted the identification of only hydrallmanol A (**63**), the major component.[29] Hydrallmanol A (**63**) is a diphenyl-p-menthane derivative that is structurally related to Δ^1-3,4-*trans* tetrahydrocannabinol (**64**) a well known metabolite from *Cannabis sativa*. Hydrallmanol A is apparently the first metabolite of this class to be isolated from an animal source.

CONCLUSION

The study of terpenoids from marine invertebrates is a very rewarding enterprise. Discovering new carbon skeletons is almost the norm rather than the exception and a wide variety of interesting functional groups adorn these new carbon skeletons. Many of the terpenoids also play a central role in the biology of the source organisms. The challenge for the future is to elaborate the specific details of the biosynthesis of these interesting new metabolites.

ACKNOWLEDGEMENT

Financial support was provided by grants to RJA from NSERC and the National Cancer Institute of Canada. The authors wish to thank Mike Le Blanc for assisting the collection of the marine invertebrates.

REFERENCES

1. IRELAND, C.M., ROLL, D.M., MOLINSKI, T.F., MCKEE, T.C., ZABRISKE, T.M., SWERSY, J.C. 1988. Uniqueness of the marine chemical environment: categories of marine natural products from invertebrates. In: Biomedical Importance of Marine Organisms. (Fautin, D.G. ed.), California Academy of Sciences, San Francisco, pp. 41-57.
2. FAULKNER, D.J. 1984. Marine natural products: metabolites of marine algae and herbivorous marine molluscs. Natural Product Reports 1: 251-280.
3. FAULKNER, D.J. 1984. Marine natural products: metabolites of marine invertebrates. Natural Product Reports 1: 552-598.
4. FAULKNER, D.J. 1986. Marine natural products. Natural Product Reports 3: 1-33.
5. FAULKNER, D.J. 1988. Marine natural products. Natural Product Reports 5: 613-663.
6. AYER, S.W., HELLOU, J., TISCHLER, M., ANDERSEN, R.J. 1984. Nanaimoal, A sesquiterpenoid aldehyde from the dorid nudibranch *Acanthodoris nanaimoensis*. Tetrahedron Letters 25: 141-144.
7. AYER, S.W., ANDERSEN, R.J., CUN-HENG, H., CLARDY, J. 1984. Acanthodoral and isoacanthodoral, Two Sesquiterpenoids with new carbon skeletons from the dorid nudibranch *Acanthodoris nanaimoensis*. J. Org. Chem. 49: 2653-4.
8. GUSTAFSON, K., ANDERSEN, R.J., CHEN, M.H.M., CLARDY, J., HOCHLOWSKI, J.E. 1984. Terpenoic acid glycerides from the dorid nudibranch *Archidoris montereyensis*. Tetrahedron Letters 25: 11-14.
9. ANDERSEN, R.J., SUM, F.W. 1980. Farnesic acid glycerides from the nudibranch *Archidoris odhneri*. Tetrahedron Letters 21: 797-800.
10. GUSTAFSON, K., ANDERSEN, R.J. 1985. Chemical studies of British Columbia nudibranchs. Tetrahedron 41: 1101-8.
11. WILLIAMS, D.E., ANDERSEN, R.J. 1987. Terpenoid metabolites from skin extracts of the dendronotid nudibranch *Tochuina tetraquetra*. Canadian J. Chem. 65: 2244-47.
12. WILLIAMS, D.E., ANDERSEN, R.J., VAN DUYNE, G.D., CLARDY, J. 1987. Cembrane and pseudopterane diterpenes from the soft coral *Gersemia rubiformis*. J. Org. Chem. 52: 332-335.
13. WILLIAMS, D.E., ANDERSEN, R.J., KINGSTON, J.F., FALLIS,

A.G. 1988. Minor metabolites of the cold water soft coral *Gersemia rubiformis*. Canadian J. Chem. 66: 2928-34.

14. BANDURRAGA, M.M., FENICAL, W., DONOVAN, S.F., CLARDY, J. 1982. Pseudopterolide, an irregular diterpenoid with unusual cytotoxic properties from the Caribbean sea whip *Pseudopterogorgia acerosa* (Pallas) (Gorgonacea). J. Amer. Chem. Soc. 104: 6463-5.
15. WILLIAMS, D.E., ANDERSEN, R.J., PARKYANI, L., CLARDY, J. 1987. Gersolide, a diterpenoid with a new rearranged carbon skeleton from the soft coral *Gersemia rubiformis*. Tetrahedron Letters 28: 5079-80.
16. THOMPSON, J.E., WALKER, R.P., WRATTEN, S.J., FAULKNER, D.J. 1982. A chemical defense mechanism for the nudibranch *Cadlina luteomarginata*. Tetrahedron 38: 1865-1873.
17. HELLOU, J., ANDERSEN, R.J., THOMPSON, J.E. 1982. Terpenoids from the dorid nudibranch *Cadlina luteomarginata*. Tetrahedron 38: 1875-1879.
18. HELLOU, J., ANDERSEN, R.J., RAFII, S., ARNOLD, E., CLARDY, J. 1981. Luteone, a twenty three carbon terpenoid from the nudibranch *Cadlina luteomarginata*. Tetrahedron Letters 22: 4173-76.
19. CYR, T.D., STRAUSZ, O.P. 1983. The structures of tricyclic terpenoid carboxylic acids and their parent alkanes in the Alberta oil sands. J. Chem. Soc. Chem. Comm.: 1028-30.
20. GUSTAFSON, K., ANDERSEN, R.J., CUN-HENG, H., CLARDY, J. 1985. Marginatafuran, a furanoditerpene with a new carbon skeleton from the dorid nudibranch *Cadlina luteomarginata*. Tetrahedron Letters 26: 2521-24.
21. TISCHLER, M., and ANDERSEN, R.J. 1989. Glaciolide, a degraded diterpenoid with a new carbon skeleton from the nudibranch *Cadlina luteomarginata* and the sponge *Aplysilla glacialis*. Tetrahedron Letters 30: 5717-20.
22. DUMDEI, E.J., DE SILVA, E.D., ANDERSEN, R.J., CHOUDHARY, M.I., CLARDY, J. 1989. Chromodorolide A, a rearranged diterpene with a new carbon skeleton from the Indian ocean nudibranch *Chromodoris cavae*. J. Amer. Chem. Soc. 111: 2712-3.
23. SULLIVAN, B., FAULKNER, D.J. 1984. Metabolites of the marine sponge *Dendrilla* sp.. J. Org. Chem. 49: 3204-3206. (The original structure assignment was incorrect. Faulkner and ourselves have both revised the structure as shown.)
24. CARMELY, S., COJOCARU, M., LOYA, Y., KASHMAN, Y. 1988.

Ten new rearranged spongian diterpenes from two *Dysidea* species. J. Org. Chem. 53: 4801-07.
25. NORTHCOTE, P.T., ANDERSEN, R.J. 1987. Xestodiol, a new apocarotenoid from the sponge *Xestospongia vanilla*. J. Nat. Prod. 50: 1174-77.
26. NORTHCOTE, P.T., ANDERSEN, R.J. 1988. Xestenone, a new bicyclic C19 terpenoid from the marine sponge *Xestospongia vanilla*. Tetrahedron Letters 29: 4357-60.
27. NORTHCOTE, P.T., ANDERSEN, R.J. 1989. Xestolide and secoxestenone, degraded triterpenoids from the sponge *Xestospongia vanilla*. Canadian J. Chem. 67: 1359-62.
28. NORTHCOTE, P.T., ANDERSEN, R.J. 1989. Xestovanin A and secoxestovanin A, triterpenoid glycosides with new carbon skeletons from the sponge *Xestospongia vanilla*. J. Amer. Chem. Soc. 111: 6276-80.
29. PATHIRANA, C., ANDERSEN, R.J., WRIGHT, J.C.L. 1989. Hydrallmanol A, an interesting diphenyl-*p*-menthane derivative of mixed biogenetic origin from the hydroid *Hydrallmania falcata*. Tetrahedron Letters 30: 1487-90.

Chapter Nine

CONTROL OF STEROL BIOSYNTHESIS AND ITS IMPORTANCE TO DEVELOPMENTAL REGULATION AND EVOLUTION

W. DAVID NES

Plant and Fungal Lipid Research
Plant Physiology Research Unit
Russell Research Center
950 College Station Rd.
Athens, GA 30613, U.S.A.

INTRODUCTION

Sterols are virtually ubiquitous constituents of all living systems at some point in their life history.[1,2] In a few organisms they are replaced by pentacyclic triterpenoids.[3-5] Although sterols may differ structurally from each other they share similar amphiphathic properties which make them suitable as membrane components. The principal function for sterols and sterol-like molecules is thought to be a non-metabolic one as

Biochemistry of the Mevalonic Acid Pathway to Terpenoids
Edited by G.H.N. Towers and H. A. Stafford
Plenum Press, New York

an architectural component of membranes; while at the same time at sites not yet clear, there appears to be additional developmentally regulated functions for sterols to control the cell cycle.[6-10] The multiple sterol functions appear to be species specific, but which sterol is essential in sterol-controlled ontogenetic events[6] is not always predictable with the current data base. This chapter will be concerned with approaches developed in this laboratory to interrupt sterol biosynthesis and to assess the functional importance of sterols in the regulation of growth and reproduction of plants and fungi. It is not my intent to examine the literature in a comprehensive manner, therefore the interested reader may elect to consult several recent reviews which examine control of sterol production with a focus different from the one considered here.[11-14] Since biosynthesis and function (biological activity) are parameters intimately associated with structural features of the molecule, the structure and stereochemistry of sterols and sterol-like molecules is examined first.

STRUCTURE AND STEREOCHEMISTRY

Orienting Remarks

Cholesterol is a "one of a kind" biomolecule. While its constitution allows for 2^8 possible stereoisomers only one is known to occur in nature.[15] Whenever there exists such a dramatic example of structural regularity amongst so many choices, the question necessarily arises as to what evolutionary mechanisms operate to govern the chiral outcome and hence 3-dimensional shape of the molecule. The sterol molecule is composed of three very different units: they are a 3β-OH group, a tetracyclic nucleus and a side chain. The representation of cholesterol shown in Figure 1 is seemingly tailored to fit the lipid bilayer. However, this view[16,17] underscores certain assumptions about conformation, biosynthesis and function.

Conformational Analysis

In a conformational view of a sterol (Fig. 1), the molecule appears flat, the OH-group lying essentially in the plane of the ring system tilted somewhat away from the angular C-19 group and the side chain orienting to the right with the C-22 group through C-26[27] in a staggered conformation. This conformation is thought to be the functionally significant one

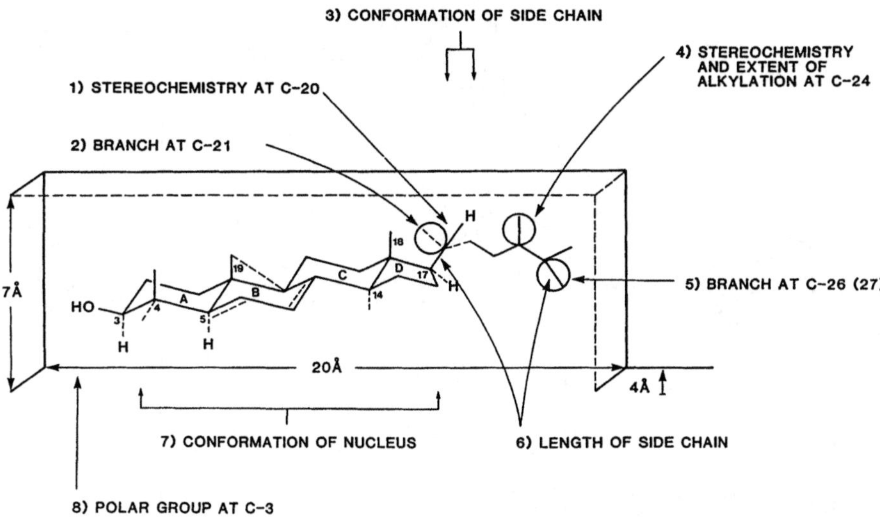

Figure 1. Conmformational view of sterols. Dotted lines are used to indicate stereochemical relationships and optical structural features.

when the molecule occurs in membranes, although, other nuclear and side chain variants can be envisaged. Sterol-like molecules are assumed to orient into pseudoplanar conformations in the membrane, analogous to sterols. We assess the conformation in membranes through examination of several chemical and biological parameters: 1)chromatographic behavior; 2) acid-induced rearrangements; 3) molecular modeling and mechanics; 4) X-ray crystallography (solid-state properties); 5) NMR and related spectroscopic determinations of compounds (solution properties); 6) NMR and related spectroscopic studies of sterols mixed with other lipids to form monolayers, bilayers, etc; and 7) structure-activity behavior using sterol auxotrophs.

Origin of Sterol Asymmetry

The evolutionary mechanisms by which sterols obtained their 3-dimensional shape and particular chirality are not clear; it has been suggested that steroids were produced through abiotic processes.[18] In order to produce a steroid in the complete absence of a living system, the appropriate precursors in an abiotic process must have been furnished in the anaerobic environment of the prelife earth ca., four

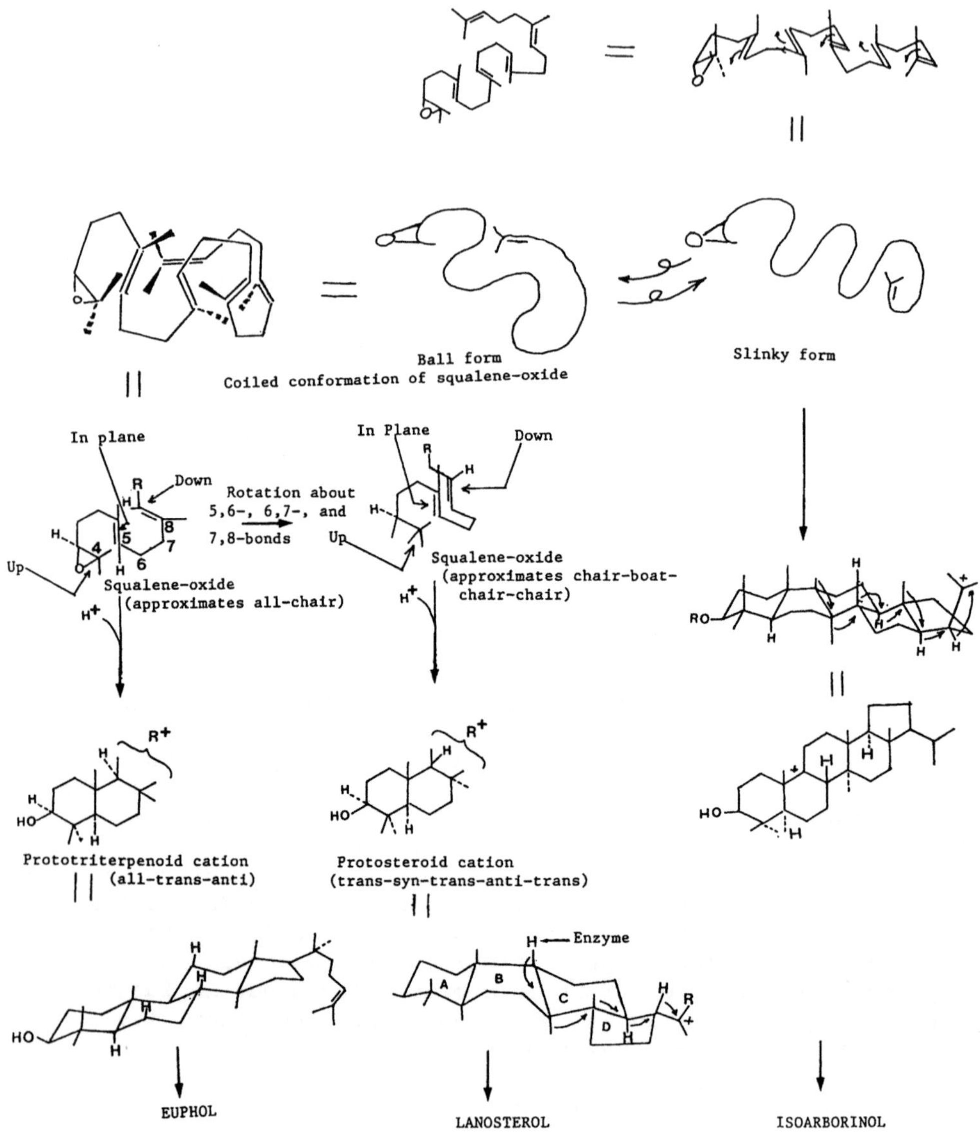

Fig. 2. Conformational sequences involved in the cyclization of squalene-oxide to tetracyclic and pentcyclic products. Rotation about carbon-carbon bonds is shown to illustrate the changes which occur during cyclization that influence the conformational outcome of the ring systems.

billion years ago. According to Calvin, through volcanic action methane and propane under intense heat in the primordial reducing atmosphere may have polymerized into a series of isoprene homologues, including squalene.[19] As squalene accumulated in the oceans it is likely that it underwent oxidation, aggregated into lipid droplets which formed part of the protocell (coacervate) membranes. In highly polar aqueous-organic media the hydrophobic effect would cause the cyclic olefin to spontaneously spiral into one or another clumps (ball or slinky forms), thereby setting the stage for pi-lobe overlap and acid-induced cyclization to tetracyclic or pentacyclic products (Fig. 2). Experiments on the cyclization of oxygenated derivatives of squalene have been carried out by van Tamelen[20] and Johnson.[21]

In Nature, six tetracyclic products are formed (Fig. 3) from squalene, whereas a score more pentacyclic products can be generated.[22] These reactions could have arisen at a very early stage of anaerobic evolution. If so, then chiral mixtures of C-4 methylated compounds would have been among the first components in the coacervate. In the tetracyclic series of compounds the racemic outcome of cyclization would likely be a mixture of C-20*R* and C-20*S* compounds. Euphol and tirucallol may be vestigial examples of C-20 epimers found in nature. Neither of these compounds can be converted to Δ^5-sterols. Nevertheless, based on their structural similarities with other known sterol-like membrane inserts, these compounds or their rearranged dammarane products may have acted as sterol surrogates in membranes of the primitive cells. The chirality introduced into euphol and tirucallol relative to the stereoisomer-lanosterol, is enantiomeric at C-13, C-14 and C-17. This stereochemical condition coupled with the boat conformation in the C-ring, allows for acid-induced rearrangements to new end products, eg., isoeuphenol, which as Dreiding models show, are flatter than the parent tetracycle. The process of transformation, however, fails to create functionalization that optimizes the H-bond capabilities of the OH-group or allows for subsequent metabolism to Δ^5. This sort of metabolism occurs in conversion of the steroidal stereoisomer, cycloartenol and lanosterol, to their corresponding rearranged products, eg., lanost-7-enol. The centers of asymmetry in tetracyclic isopentenoids can be divided into two categories: 1) those produced through cyclization that remain invariant after metabolism and 2) those which are transformable. The invariant features may have originated from an anaerobic event, but some chiral

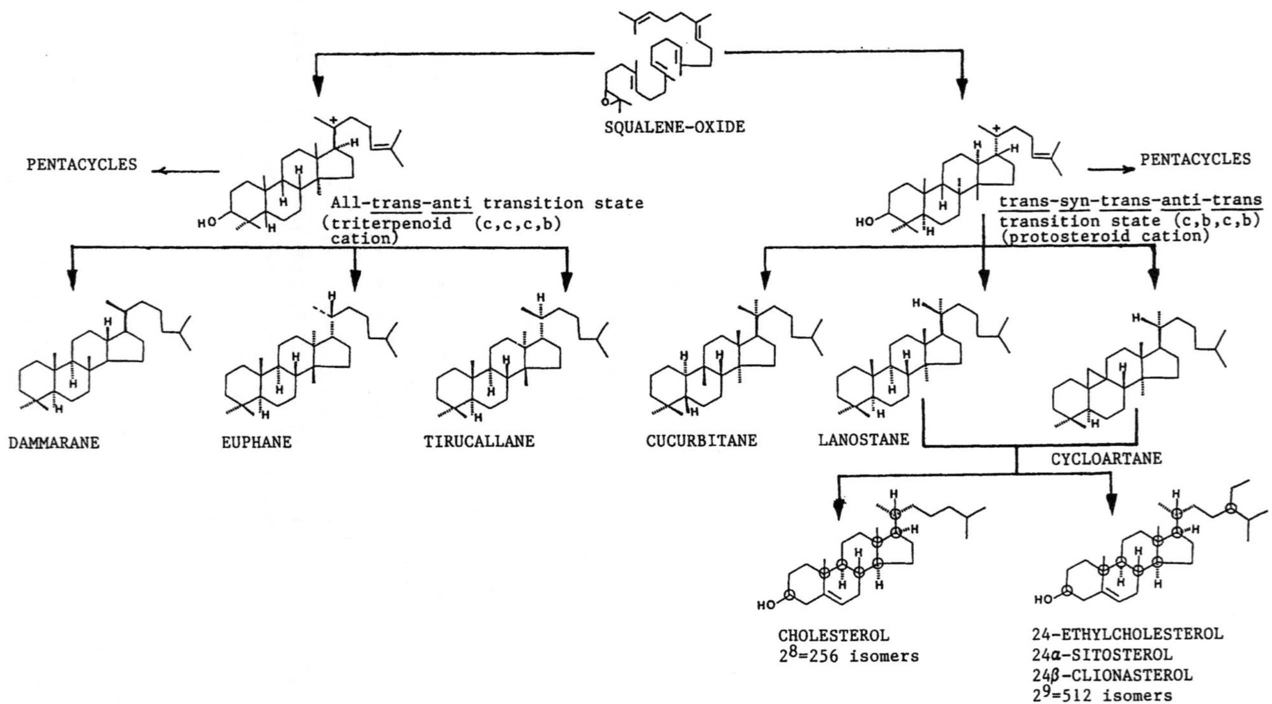

Fig. 3. Chirality in sterols produced through biosynthesis. Of the six tetracyclic products of squalene-oxide cyclization only the stereoisomers with the cycoartenol and lanosterol skeletons are known to be converted to sterols.

features in Δ^5-sterols can only have resulted from aerobic metabolism, eg., nuclear demethylation. Consequently, the origin of sterol asymmetry is rooted in anaerobic and aerobic stereoselections which progressed through a series of stepwise transitional states of increasing complexity as life on the planet developed.

PHYLOGENESIS AND BIOSYNTHESIS

It is becoming increasingly apparent from observations about phylogenesis and biosynthesis[4,5,11,15-19] that one unsolved mystery is the importance of sterol biochemistry in the evolutionary process. Life, as we know it today, probably could not have come about without sterols and their metabolites (steroid hormones). Our first clue that sterols may be involved in development is that in completely removing their production from the yeast and oomycetous fungi leads to a debilitating change in cell size, number, survival capability and/or diminished fecundity, a compromised sexual cycle or a change in phenotype,[6,23-28] even though biosynthesis may continue to form replacement molecules eg., cholesta-5,7,24-trienol rather than ergosterol.[29] Sterols have in fact been detected in each of these systems but the level varies between 0.1 fg/cell (bacteria) to 3,000 fg/cell (animals and plants). The inability of earlier investigators to detect or to have incorporated radiolabeled precursors into sterols in, for instance, cyanobacteria (blue-green algae)[30,31] is now known[32-34] to be a problem with their failure to have cultured sufficient cell mass on synthetic media for sterol analysis. Where sterols are not found at ≤ 0.1 fg/cell, sterol-like molecules (hopanoids) often occur at higher levels.[3]

In contrast to earlier speculations (cf. 35), no single sterol or sterol-like compound characterizes a particular taxonomic group. For instance, the typical "zoosterol" - cholesterol and its precursors eg., cholest-8-enol- have recently been isolated from non-photosynthetic bacteria,[36-38] as well as to less -and more- advanced fungi eg., *Saprolegnia ferax*[26,39] and *Saccharomyces cerevisiae*[29,40] respectively. They have also been isolated from vascular plants, eg., sorghum[41] and mammals.[42] Except in animals where cholesterol is the major sterol, the amount of cellular cholesterol produced by other organisms is usually low i.e., at hormonal levels ca., 0.1-1.0 fg/cell. From these new findings we interpret that the genes for cholesterol and hence the sterol pathway originated in the prokaryotes.

Opinions differ on the phylogeny of the sterol pathway.[3,5,11,16,17,43] Some of the criteria by which primitive pathways have been surmised are the following: 1) elucidation of molecular fossils; 2) distribution studies involving chirality of the C-24 alkyl group; 3) whether the pathway is aerobic or anaerobic; 4) whether the pathway is complete; 5) the number of overall steps involved in the synthesis of end products and what sorts of intermediates are formed after the cyclization of squalene-oxide; 6) whether a low- or high- energy route is used; 7) the number of functions for a particular compound during the life cycle and the extent to which the function can be duplicated by a sterol mimic; 8) substrate specificity in enzymatic catalysis; 9) similarities between laboratory type abiotic synthesis and biochemical formation; and 10) "archeological mapping" of the amino acid sequences in enzymes and DNA contained in genetic materials regulating sterol production.

The entropy-gradient involved in cyclization may be the primary bioorganic determinant of primitiveness. If energetically less expensive processes occurred before the more expensive ones, we might expect, in contrast to previous speculations on this matter[17] that the cyclization to tetracycles preceded the formation of pentacycles.

There are two major ways in which cyclization of squalene (and its 2,3-oxide) could occur, either through the protosteroid or the prototriterpeneoid cations (Fig. 3). The rearrangement pathways do not proceed to a single set of products having the greatest thermodynamic stability but to several stereoisomers of differing stabilities. The act of cyclization is driven by the free-energy of coiled conformation that is gained by delocalizing the electrons in the bonding orbitals and the hydrophobic effect resulting from the polar-aqueous milieu in which the reaction occurs. The free-energy of the packing arrangement is dependent on the energy barrier between transition states activated through the sequential formation of rings (kinetic control) and the overall evergy differences involved in proceeding into either the ball (leading to tetracycles) or slinky (leading to pentacycles) forms (thermodynamic control). There are two opposing views on whether squalene-oxide cyclization leading to tetracycles and pentacycles is similar; one is a nonstop concerted process (cf. 22), another a step-wise[44,45] process

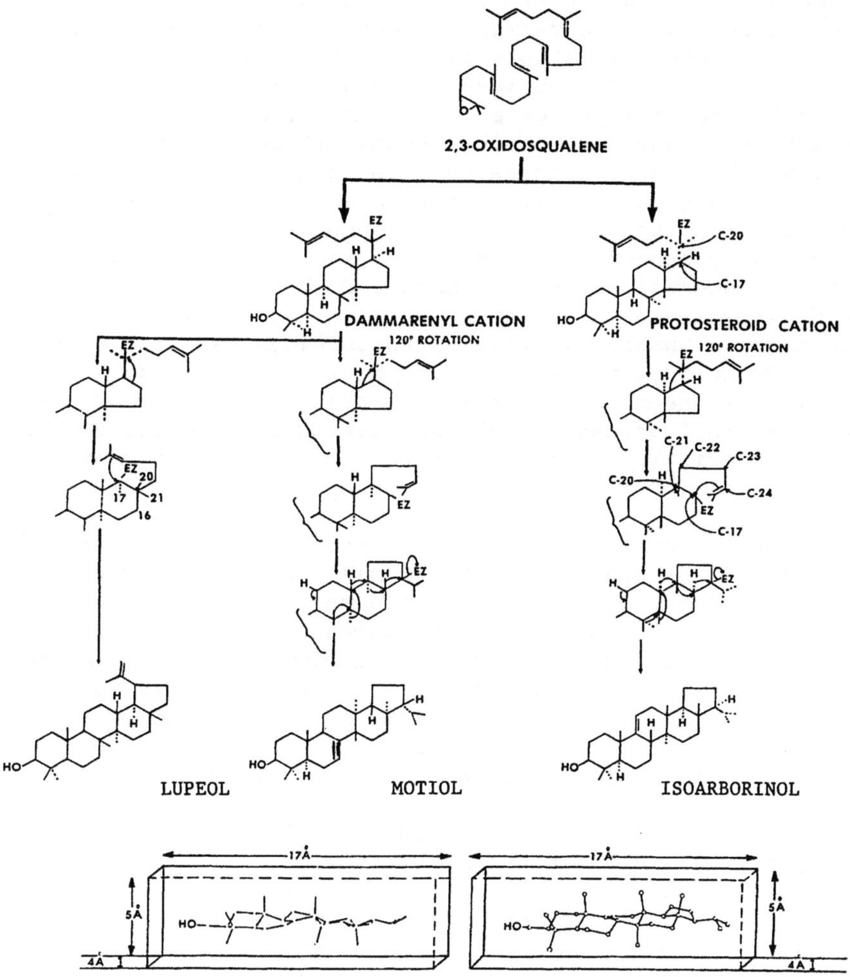

Fig. 4. Hypothetical mechanisms of squalene-oxide cyclization to isoarborinol and motiol. Note that the ring enlargement is considered to proceed from the (13)17 bond to C-20 (steroid numbering system) for motiol and isoarborinol, whereas lupeol proceeds through the 16(17) bond to C-20.

involving discrete carbocationic intermediates (Fig. 4). We believe that all pentacycles are derived by modifications of the mechanisms shown in Figure 2 and do not proceed even in certain cases eg., arboranes, through a nonstop concerted process. In the most stable packing the coils will spiral into chair-chair conformational sequences resulting in a tetracyclic dammarane-type product. Alternatively, cyclization through trans-syn systems produce energetically less favorable intermediates because of diaxial and other steric interactions. While rotation about carbon-carbon bonds close to the point of acid-induced annulation can be energetically expensive in localized regions, it allows for greater change which become apparent downstream from the epoxide bridge, notably with respect to cyclization of the fifth ring or whether it remains an unfolded isooctyl side chain. Because of the high-energy barriers in slinky formation, it seems likely that the boat intermediate formed in the terminal stages of ball annulation to a tetracycle acts as the conformational driving force for ring enlargement producing the fifth ring.

Johnson and coworkers have recently speculated that cyclases, rather than having to impose a profound and highly restrictive conformational influence on the substrate, may simply provide an electronic environment which guides and sustains the cyclization process.[46] Therefore, the differences between the tetracyclic and pentacyclic cyclases may be minimal.[17] From entropic considerations based on molecular modeling of several cyclized products of squalene-oxide cyclization (Fig. 5), it seems possible that the first tetracycles (dammaranes) were probably derived in an anaerobic atmosphere from the prototriterpenoid cyclase through changes in cofactor availability and membrane structure (affecting the hydrophobic effect) would give rise to pentacyclic products such as diplopterol and tetrahymanol. The advent of oxygen in the atmosphere may then have induced the operation of the protosteroid cation pathway. The following sequence is hypothesized: early conditions or abiotic processes gave rise to (→) dammaroid skeletons, the latter acted as templates in living systems which gave rise to the following: hopanoid skeletons (tetrahymanol) → steroids → protosteroid cation derived triterpenoids (isoarborinol). The sequential formation of the compounds was codified perhaps after their functions were determined. Codification of the pathways may be through one of two evolutionary mechanisms - punctuated equilibria[47] or gradual transmutation.[48] Unfortunately, there

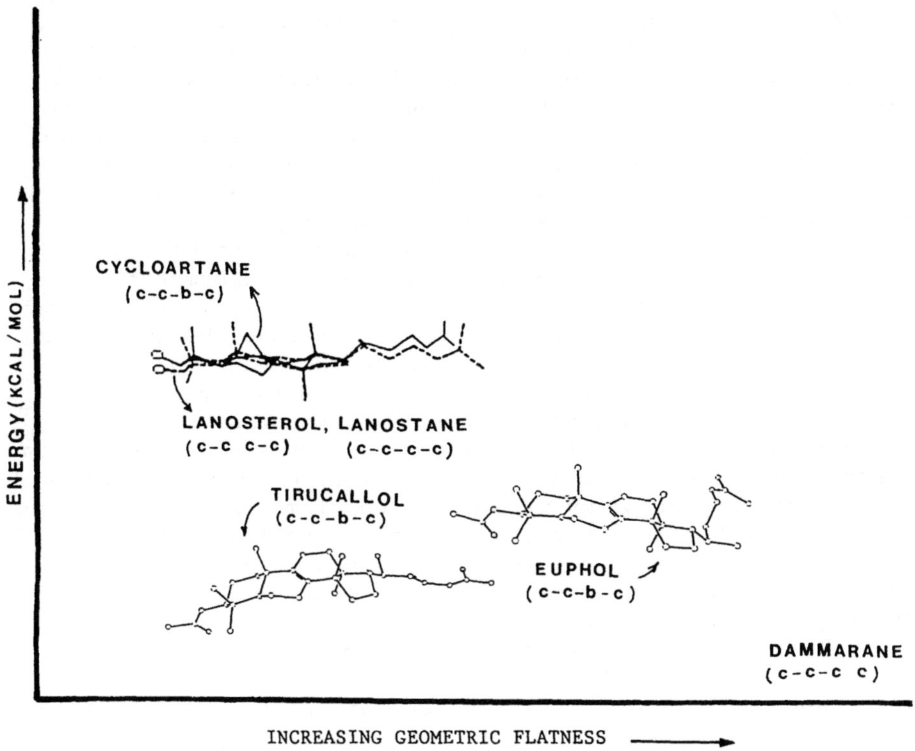

Fig. 5. Schematic representation of energy relationships of tetracyclic stereoisomers. Molecular modeling is based on solid state properties of the corresponding 3β-OH compound (unpublished results).

are no data on the extent to which the codification process is affected by the energetics of cyclization. Because steroids are functionally more versatile than hopanoids eg., through their effects on reproduction,[25] their formation may have been prerequisite to the origin and further evolution of eukaryotes.[6]

Let us return for a moment to the evolution of bacteria where steroids have been isolated in cyanobacteria and methylotroph forms. The use of either a cycloartenol or a lanosterol-based steroid pathway links groups of organisms with an evolutionary history of oxygenic photosynthesis, eg., as detected through cycloartenol biogenesis and metabolism of the 9β,19-cyclopropyl group, are distinguished from those which lack such an ancestry.[3,9,15] This bifurcation in the steroid pathway is one of the more interesting phylogenetic markers in biochemistry because the biosyntheses of the two stereoisomers have no apparent influence on the structure of the functional steroid at the end of the pathway. It so happens that blue-green algae (cyanobacteria) are oxygenic photosynthetic organisms and the methylotrophs are non-photosynthetic organisms. Which came first, the blue-green bacteria or the methylotrophs? Based on a conformational analysis of the cyclization reactions giving cycloartenol and lanosterol shown in Figure 6, it is clear that the inclusion of an extra conformation leading to cycloartenol is more costly than the entropy gradient leading directly to lanosterol. Note that the cyclization mechanism to cycloartenol and cucurbitacin given in Figure 6, differs from the mechanisms proposed by others[49] which is shown in Figure 7. A detailed account of my proposed mechanisms will be discussed elsewhere. Thus for thermodynamic and kinetic reasons lanosterol probably preceded cycloartenol in evolution, so that non-photosynthetic organisms preceded oxygenic photosynthetic ones which, as it turns out, is the view held by evolutionists[48,50] (and ref. cited therein).

Another bifurcation in the steroid pathway exists which has assumed phylogenetic significance viz., alkylation and reduction of the 24,25-bond in the sterol side chain (Fig. 8). Reduction is energetically much less expensive than C-24 alkylation which costs 15 ATP equivalents as shown in yeast (cited in 51). Therefore, reduction of the Δ^{24}-bond leading to the cholesterol side chain presumably preceded the formation of C-24 alkylation of the Δ^{24}-bond in bacteria. The chirality of the C-24 substituents has also been used as a chemotaxonomic marker for judging whether an organism is primitive or advanced.[15] Primitive organisms usually produce 24β-alkyl sterols while more advanced organisms synthesize 24α-alkyl sterols. While the frequency of occurrence seems to be inversely related to the phylogenetic hierarchy, there are cases where C-24 diastereoisomers are found in algae, as well as vascular plants.[15,52] Furthermore, the data on the occurrence of α- versus β- substituents are no longer straight-forward even with respect to the distribution of marine sterols. These unusual compounds, eg., the triple alkylated ones at C-24, C-25 and those containing a triple bond at C-22, are found in algae and tracheophytes in trace amounts[53,54] (Nes et al. unpublished data). Thus, the

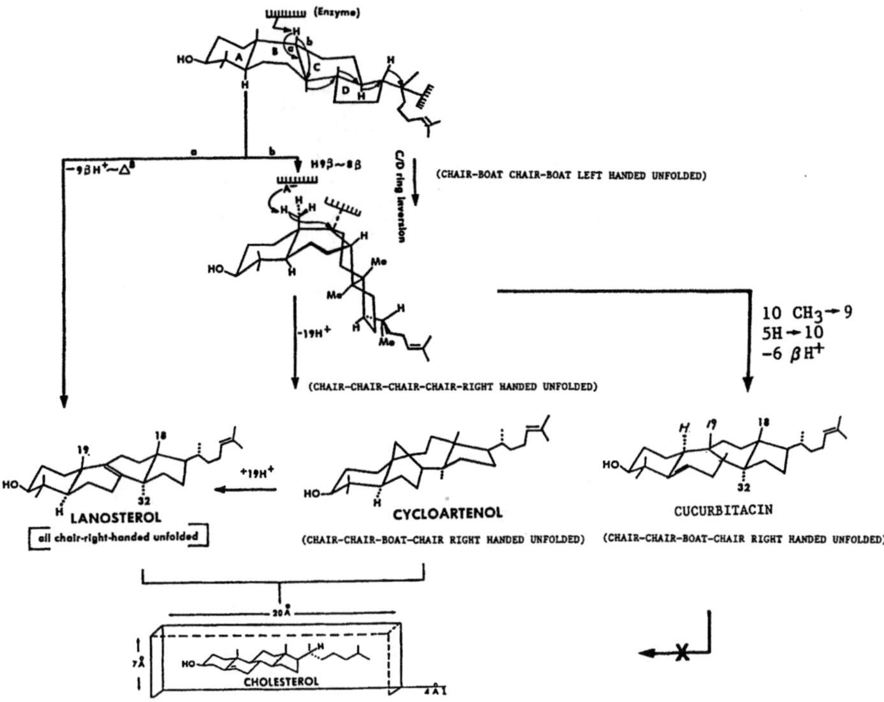

Fig. 6. Hypothetical cyclization mechanisms of squalene-oxide to sterols. In photosynthetic plants bothj cycloartenol and lanosterol (an isomerization product of cycloartenol) are converted to cholesterol. In animal and fungal systems, only lanosterol is converted to cholesterol; cycloartenol (obtained through the diet) metabolism ceases with the 9β,19-cyclopropyl to Δ^8-isomerization step.[39,43]

mechanistic details of the alkylation route rather than the extent of C-24 alkylation or the nature of the introduced chirality may be a better guide to primitive character.

As shown in Figure 8, there are three mechanisms for C-24 alkylation.[15,52,53] For steric reasons, formation of the $\Delta^{24(28)}$ sterol would be energetically less favorable than the

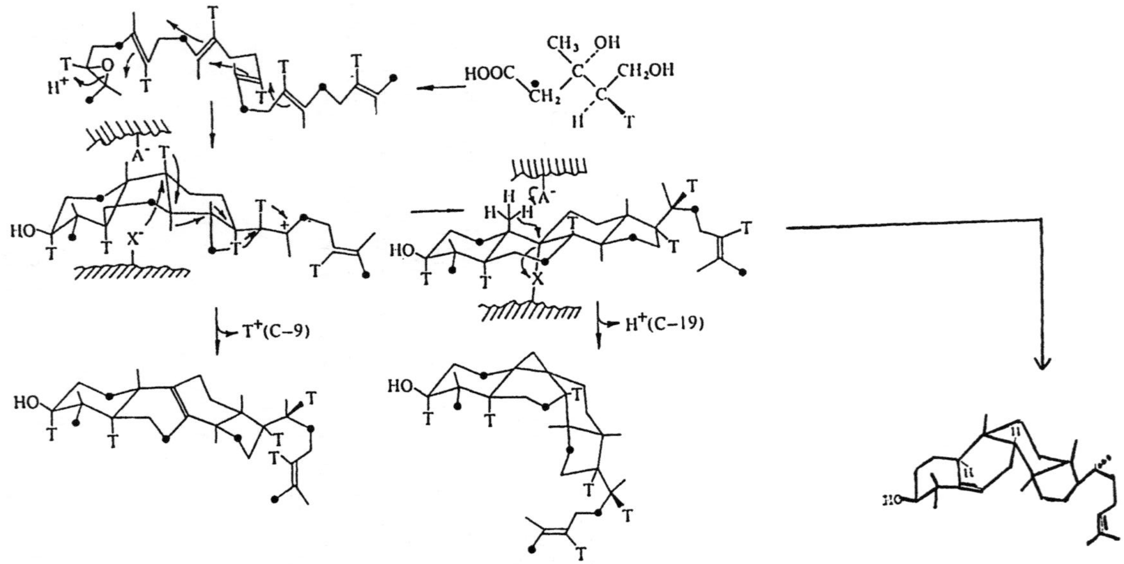

LANOSTEROL (C,C,C,C)

CYCLOARTENOL (C,B,C,C)

CUCURBITACIN (C,B,C,C)

Fig. 7. Classical view of cyclization sequence to lanosterol, cycloartenol and cucurbitacin with their assumed conformational views, cf. Goad and Goodwin, 1973.[49]

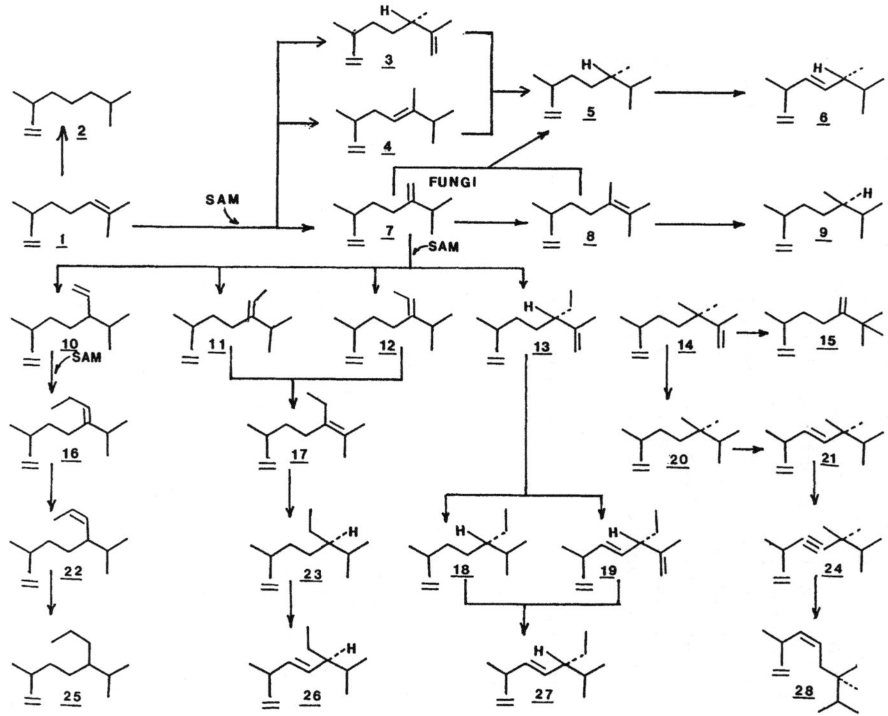

Fig. 8. Suggested pathways for sterol side chain alkylation and further modification in plants. No consideration is given to the metabolism of the nucleus relative to side chain Δ^{24}-metabolism. Sterols with side chain structures **3**, **4**, and **7** are the known products of C-24 monoalkylation. Side chain structures **3** and **4** are metabolized to 24β-methyl sterols. Side chain structure **7** serves as an intermediate to 24α-methyl sterols, 24α/24β-ethyl sterols and triply alkylated side chains.

$\Delta^{25(27)}$ sterol. However, the advantage of forming the $\Delta^{24(28)}$ ethylenic bond, compared with initially introducing the chirality at the time of C-24 alkylation viz., 24 β-methyl $\Delta^{25(27)}$ sterol, is that many additional choices become available for metabolism i.e., rearrangement, additional akylation, etc. to modified side chains of greater complexity. Interestingly, fungi which synthesize exclusively 24β-alkyl sterols, use the same C-24 alkylation mechanism and produce $\Delta^{24(28)}$-sterols. The latter intermediate may undergo isomerization to a $\Delta^{24(25)}$-methyl sterol which then can be reduced to form 24β-methyl cholesterol but not ergosterol which is formed by reduction of the $\Delta^{24(28)}$ bond.[55] The stereospecific reduction of the $\Delta^{24(25)}$-methyl sterol in fungi is opposite to the reduction in plants which form 24α-methylsterols. As we have suggested the isomerization-reduction pathway may be used as a new genealogical marker to separate fungi from photosynthetic plants.[55]

REGULATION OF STEROID AND TRITERPENOID BIOSYNTHESIS

Orienting Remarks

A well-known type of feedback regulation of a biosynthetic pathway is where the end product of the pathway inhibits the first enzyme unique to the pathway. This phenomenon, however, requires knowledge of whether a specific compound is an intermediate or whether it is an end product of the pathway. An intermediate has a high turnover rate, low steady-state concentration and no apparent biological function, save its ability to be converted onto a physiologically essential compound with regulatory activity. Alternatively, an end product exhibits a low turnover rate, a high steady-state concentration and some temporally expressed role eg., architectural. There are situations, however, where a compound may appear to have a high turn-over rate and low steady-state concentration and still play a non-metabolic but regulatory (hormonal) role eg., as in the case of one of the multiple sterol functions in yeast[56] or *Gibberella fujikuroi*.[24,57] Feedback regulation phenomena associated with the branched isopentenoid pathway may be divided into four types: 1) specific feedback inhibition of multiple enzymes; 2) sequential feedback inhibition; 3) concerted (multivalent) feedback inhibition; and 4) cumulative feedback inhibition. The extent of coarse control of the pathway (regulation of the amounts of enzymes present in the cell) or fine control of the enzyme activities (modulated by the metabolite concentrations and possible reversible covalent modification of the enzymes[13,58]), is measured by assessing the level of intermediates/end products in the pathway as they appear during development.

Ontogenetic Control

Regulation of the flux of metabolites through a multibranched pathway may be genetically determined. However, when this occurs, the number of molecules that pass through a given branch per unit time may differ from cell to cell and from plant to plant. Ontogenetic control of sterol and triterpenoid biosynthesis in tracheophytes can be examined by following the changes in composition and biosynthesis of steroids and triterpenoids in 1) whole plant parts, 2) tissues and 3) tissue cultures of plants harvested at different stages of maturity.

Using approach (1), we have studied sorghum leaf blades from germination to flowering (Fig. 9). The data indicated that pentacyclic triterpenoid production was delayed relative to steroid production. The reverse situation operates in germinating pea seeds.[59] As the sorghum plants matured and the leaves became fully expanded, the synthesis of pentacyclic triterpenoids was greatly amplified and sterol production decreased[41,61] as in the developing pea.[60] This indicates either that a switching mechanism operates at the cyclase level or that new cyclases are synthesized. The triterpenoids were preferentially localized in the surface wax of epidermal cells together with fatty alcohols[62] and cholesterol (Fig. 10), while the 24-alkylated sterols were contained in membranes of mesophyll cells of the blade.[41,61] Apparently, the switching mechanism operates to produce and transport end products for different functions in membranes *versus* surface wax. In Cucurbitaceae a different sort of developmental change occurs involving sterol compositional changes with respect to the configuration of the C-24 alkyl group.[63,64] In *Kalanchoe* different tissues and organs have also been found to possess sterols differing in their stereochemistry (Table 1). In contrast to *Kalanchoe*, sorghum does not show any change in the level of C-24 epimers with plant maturity although the level of Δ^{22}-sterols increased with plant age, accompanied with a change in the ratio of sitosterol to stigmasterol.[41]

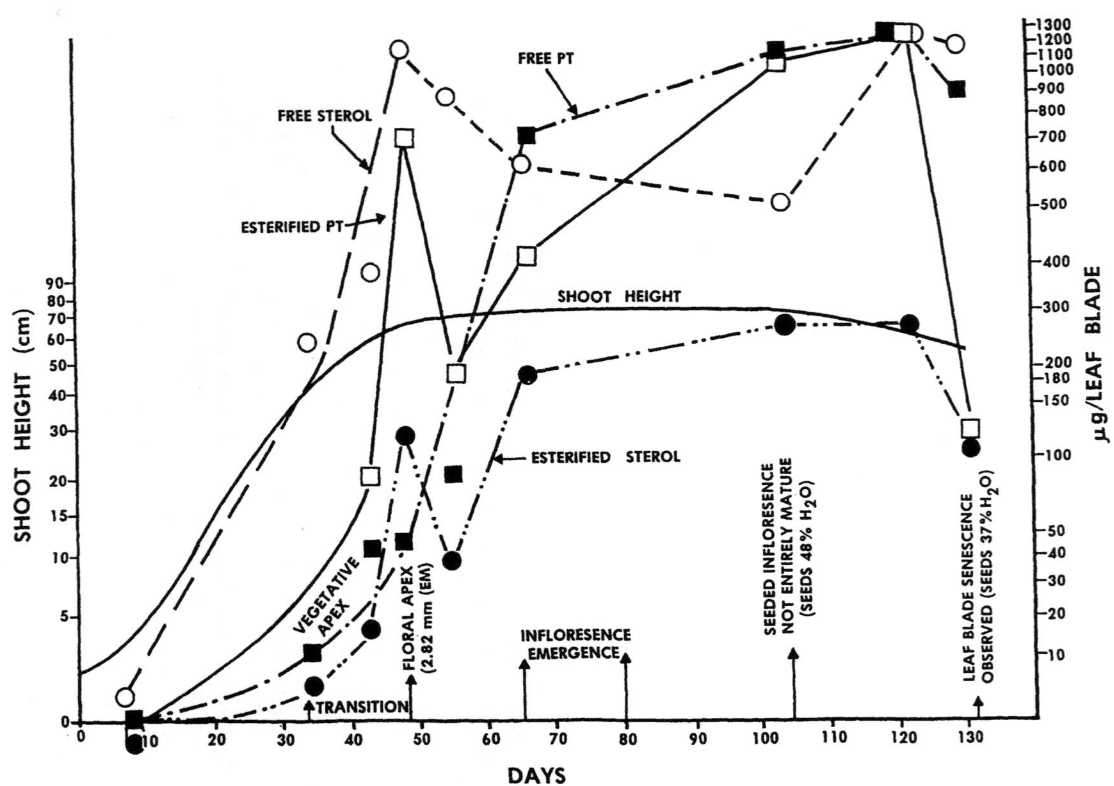

Fig. 9. Developmental changes in sterols and triterpenoids synthesized by sorghum (Heupel, Sauvaire and Nes, unpublished data). Amplification of triterpenoids synthesis beginning at 40 days is associated with the transition from vegatative growth to floral induction.

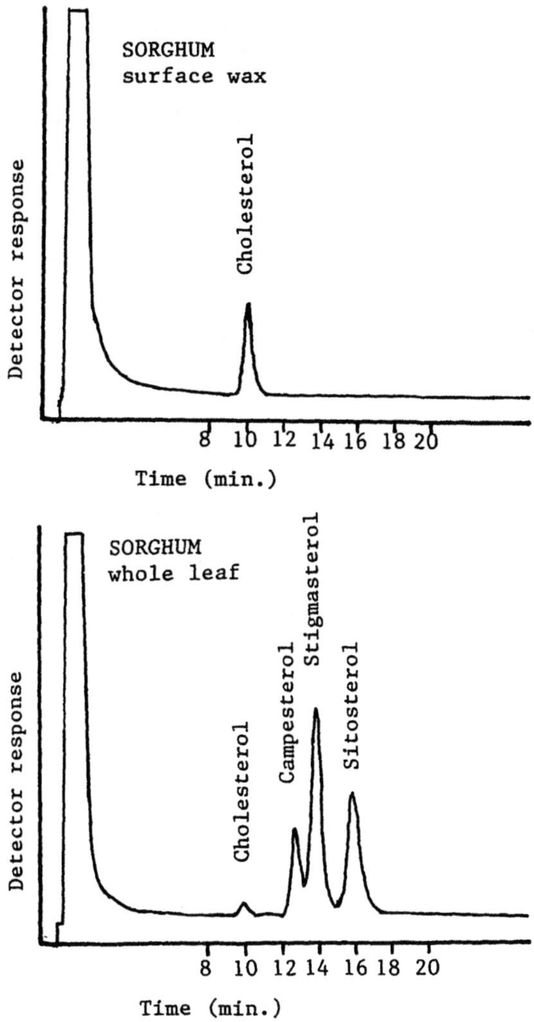

Fig. 10. GLC profile of sterol obtained from whole leaf blade and surface wax of sorghum (Nes, et al., unpublished data). Sterols were isolated from a 30 sec hexane dip of leaves (top) or after a 24 hour acetone extraction of the macerated leaves (bottom). GLC was performed on a 3% SE-30 packed column operated at 245°C.

Table 1. Sterol composition of vegetative parts of vascular plants and algae.[a]

	Sunflower[b]			*Sorghum*[b]		*Kalanchoe*[b]			*Codium*[c]	*Prototheca*[c]
Sterol[d]	Leaves	Roots	Stems	Leaves	Roots	Leaves	Roots	Stems	whole organism	cells
Ergosterol **(6)**	-	-	-	-	-	-	-	-	-	67
Cholesterol **(1)**	2	5	2	tr	2	tr	tr	1	-	2
24α-methylcholesterol (campesterol) **(9)**	4			10				5	-	-
24β-methylcholesterol **(6)**	3	16}	16}	7	19}	1}	18}	3	-	-
24(28)methylene-Cholesterol **(7)**	-	-	-	-	-	tr	-	tr	-	-
24-methyl cholesta-5,24-Dienol **(8)**	-	-	-	-	-	tr	-	tr	-	-
24-methyl cholesta-5,23-Dienol **(4)**	-	-	-	-	-	tr	-	tr	-	-
Sitosterol **(23)**	53	50	49	40	27	-	62	5	-	-
Stigmasterol **(26)**	32	29	33	43	42	-	-	-	-	-
24α-ethylcholest-7-enol **(23)**	6	-	-	-	-	-	-	-	-	-
Isofucosterol **(11)**	-	-	-	-	-	-	-	12	-	-
24-ethyldesmosterol **(17)**	-	-	-	-	-	-	2	11	-	-
Poriferasterol **(27)**	-	-	-	-	-	3	10	11	-	tr

Clionasterol **(18)**	-	-	-	-	-	4	1	38	-	1
Clerosterol **(13)**	-	-	-	-	-	37	5	21	99	-
22-dehydroclerosterol **(19)**	-	-	-	-	-	53	1	5	-	-
Codisterol **(3)**	-	-	-	-	-	tr	-	tr	-	-
24-dimethyl,25(27)-Dehydrocholesterol **(14)**	-	-	-	-	-	tr	-	tr	-	-
24-methylene,25-methyl Cholesterol **(15)**	-	-	-	-	-	tr	-	tr	-	-
Others	-	-	-	-	tr[e]	2	1	-	1[f]	30

[a]As percent total sterol unpublished data obtained by the Plant and Fungal Lipid Group. Confirmation of structure was by GS-MS, UV, and ^{1}HNMR; ()-configuration at C-24 not determined

[b]Angiosperm

[c]Alga

[d]The number following name of compound corresponds to the side chain illustrated in Figure 8.

[e]In roots of sorghum saturated 24-alkylated sterols and ^{4}-3-one compounds were present in low amounts (ca. 10% of mixture).

[f]A minor compound was present in the sterol TLC band corresponding to cholesterol. The RF (0.33), RRT_c (2.03) and MS(M+,426) of the unknown are consistent with the sterol having no methyl groups in the nucleus and triple C-24 alkylation in the side chain.

As to approach (2), we have determined the sterol and triterpenoid composition of isolated cells obtained from different sunflower tissues (Table 2). Once again we found a developmentally delayed production of pentacyclic triterpenoids. These compounds appeared in copius amounts at a later stage of development in the flower mesophyll cells. Similarly, Karunen et al., have recently found esterified triterpenes in lipid globules in mesophyll cells of aging leaves of a bryophyte.[65] Cholesterol, present only in trace amounts, was accompanied by significant levels of 24-alkyl sterols in whole leaf preparations in the leaf wax of both sorghum and sunflower (Fig. 10). The cholesterol levels fluctuated with plant maturation and leaf ontogeny from trace amounts to 50 micrograms per leaf. In other recent work cholesterol has been shown to be the major sterol of flowering *Poinsettia*[66] where presumably it occurred in the leaf wax. A reexamination of rape (*Brassica napus*) leaves also indicated that cholesterol was sequestered in the surface leaf lipids.[67] The movement of cholesterol from the site of synthesis (epidermal or mesophyll cells) to the site of depostion in surface wax is under investigation at the Russell Center. The novelty of our finding that cholesterol predominates in surface wax can be attributed to the chromatographic methods employed here which separate fatty alcohols from cholesterol in TLC and GLC.[68]

In approach (3), we have studied tissue cultures of sunflower plants. Cultured suspension cells which are derived from calli of wounded tissues have, through dedifferentiation, lost the chemical? morphological? identity of the parent explant. Consequently, regulatory strictures imposed on individual cell-types by the integrated system of the intact tissue may not exist in single cells. Certain details of the switching and homeostasis mechanisms operative during the ontogeny of the cell can be examined using suspension and callus cultures. Essentially the same sterols are found in sunflower cells analyzed at the stationary phase as occur in whole leaves (Table 2). However, no pentacyclic triterpenoids or increased production of cholesterol was detected in the cultures. The genes for these compounds were originally present in the cells but may have been down-regulated. If so, a secondary role for cholesterol and triterpenoids in mature cells may not be essential in growing undifferentiated cells and tissues.

Table 2. Sterols of individual cell types isolated from sunflower plant.[a]

	Leaf Suspension Cultures[b]			Leaf		Stem		Ligule flower	
	(7-day)		(21-day)	(1-month)		(1-month)		(3 month)	
Sterol	C	IL	C	Mesophyll cells[c]	Wax[d]	Parenchyma cells	Epidermal cells	Mesophyll cells[c]	Wax[d]
Cholesterol	1	tr	tr	17	99	2	3	1	1
Campesterol	4	tr	13	3	nd	15	12	2	tr
Stigmasterol	67	33	45	38	tr	56	46	22	5
Sitosterol	20	8	42	38	tr	26	38	12	1
Stigmast-7-enol	6	tr	tr	3	nd	tr	tr	nd	nd
Cycloartenol	tr	47	tr	1	tr	1	tr	tr	3
24(28)-Methylene-cycloartenol	1	nd	nd	tr	nd	1	tr	tr	tr
Pentacyclic triterpenoids	nd	nd	nd	tr	tr	nd	nd	62	89

[a] As percent of total sterol and triterpenoids; unpublished data obtained by the Plant and Fungal Lipid Group.
[b] C = control, IL = epiminolanosterol treatment.
[c] Cells obtained after macerase digest.
[d] Wax obtained after 30 sec. HPLC-grade hexane dip.
tr = trace amount; nd = not detected

Functional Control

Intermediates and other structural mimics are known to replace Δ^5-sterols for some vegetative tissues, eg., in a bulk membrane role,[16,67,68] suggesting that cells can utilize a range of structural polymorphs in some sterol-controlled life cycle events. If so, the question then arises as to what governs feedback mechanisms.

There are four approaches to assess functional feedback on biosynthesis which we are currently in the process of developing at the Russell Center: 1) feeding various dietary supplements to genetically altered or leaky sterol auxotrophs and measuring endogenous end product levels; 2) disrupting sterol biosynthesis with inhibitors and monitoring the resultant effects on growth and reproductive responses; 3) through mutagenesis; and 4) through molecular cloning techniques.

Structure-Activity Relationships

Although we now know that sterols play multiple roles,[69-72] the detailed biological properties of sterols and sterol-like molecules are not necessarily predictable. For this reason, the difference between *relative* and *absolute* functions must be defined. A relative sterol function is one in which broad structural features of the molecule determine efficacy. Some of these features are modifiable within certain limits as in the bulk membrane role. An absolute function is one in which specific structural features cannot be altered without impairment to growth and reproduction. An example of an absolute function is the recent demonstration of sterol-controlled fungal growth[6,25] disassociated from sterol-mediated reproduction.[24,26] For instance, *Saprolegnia ferax*, an oomycetous fungus, synthesized four sterols - cholesterol, desmosterol, 24-methylene cholesterol and 24Z-ethylidene cholesterol (fucosterol); however, fucosterol is not essential for growth, but only as a precursor to the steroid hormones - antheridiol and oogoniol, which are involved in oosporogenesis.

In the various test systems used to evaluate sterol functions and structural efficacy, the rate of change in cell number or density of cells of stationary phase cultures, or simply changes in dry weight until growth arrest, is often measured. These structure-activity data are then compared with sterol-controlled physiochemical properties of artificial membranes. It is not clear, however, which characteristic of growth or reproduction, rate, yield, or morphology, is the true measure of *in situ* sterol mediated change in membrane fluidity. This view is important to recognize when multiple roles of sterols are under investigation. We have noted a direct relationship between organismic size and sterol content and an inverse relationship between organismic size and sterol-controlled population densities (Fig. 11). Perhaps as the cells increase in size they synthesize more sterol;or possibly cells that produce more sterols increase in size. However, studies performed here and at Drexel University (W.R. Nes, unpublished data), in which yeast mutants were fed increasing levels of dietary cholesterol and wild-type yeast were incubated with sterol biosynthesis inhibitors, indicated that the size of the cell was not significantly affected by changes in sterol levels. Instead, the high accumulation of sterol inside the cell resulted in the sequestering of these compounds into lipid droplets as the C-3 ester (unpublished data). Some other eukaryotes naturally produce high concentrations of sterols (pollen and oospores), but the additional materials are deposited into the surface wax. High concentrations of pentacyclic triterpenoids in bacteria can also show up in the extra cellular matrix surrounding the cells.[73] A direct relationship between increasing cell size and sterol content would indicate a membrane associated feedback on sterol biosynthesis in which limitations are imposed on the kind and amount of end products produced.

Stereochemistry and Length of Side Chain

At the outset let me pose a question which others have posed[70] - what is the significance of sterol side chains? From work in physical biochemistry and mammalian and bacterial structure-growth studies, it is generally believed that no sterol is superior to cholesterol as a membrane insert.[16,74] If this is true why do plants utilize an energy-expensive process viz. alkylation on the Δ^{24}-bond in a stereospecific manner, to produce for instance, 24α-ethylcholesterol (sitosterol) for their membranes? Although it is well known that the genes for cholesterol synthesis exist in every cell, its synthesis is down-regulated in plant cells that are actively proliferating. This implies that C-24 alkylation compared with Δ^{24}-reduction of the 24,25-double bond may be important to sterol-mediated plant physiology, whereas apparently it is not essential in animal physiology. Another concern is whether the number and geometry of C-5 units in

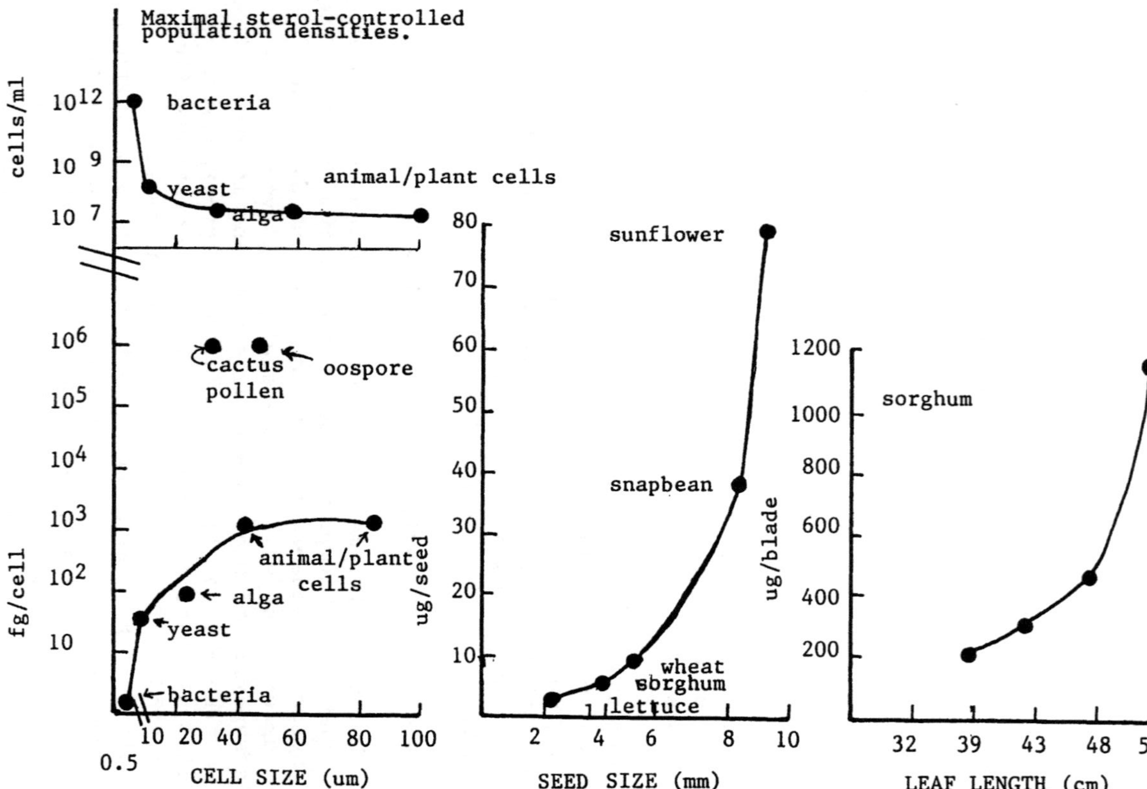

Fig. 11. Relationship between organismic/tissue size and sterol content.

triterpene assembly vary, leading to a polymerized series of isoprene variants? These compounds conceivably could compete with squalene in epoxidation and cyclization.

Studies have recently been performed using cell-free systems, with a series of synthetically prepared squalene variants (75,76 and ref. cited therein). The results demonstrated that the epoxidase and cyclase are fairly non-specific and may produce sterols with stereochemically and otherwise (length) modified side chains (Fig. 12). However, these types of compounds do not occur naturally; the precursors are invariant farnesyl PP dimers. Perhaps the synthetic aberrant sterols may not fit properly in the lipid bilayer. To test this idea, in collaboration with the late W.R. Nes who provided us with several sterols with modified side chains, we fed 19 Δ^5-sterols to the fungus *Phytophthora cactorum*,[77-82] a pathogenic sterol auxotroph which attacks crop plants. This species synthesizes squalene but fails to epoxidize the olefin.[25,81] Sterols are not required for mycelial growth but induce sexual reproduction and are required in the developing oosphere membrane.[81] *P. cactorum* is derived from a sterol biosynthesizing lineage[6] that lost part but not all of the terminal steps in the pathway. Thus $\Delta^{8(9)}$-, Δ^7-sterols are converted to Δ^5-sterols.[82]

Each of the 19 Δ^5-sterol supplements shown in Figure 13 were accumulated to similar extents by the mycelia and induced the formation of sexual structures. Aberrant effects on development were apparent[80-82] when the side chain length was shorter or longer than the one which approximated the length of the cholesterol side chain in the staggered conformation. Viable oospore maturation and hence oosphere membrane biogenesis was possible only within a narrow range of side chain variants. The results from the sterol feedings to *P. cactorum* also demonstrated that sitosterol (compound 14, Fig. 13) was more effective than cholesterol in inducing oospore formation. The superior quality of sitosterol in effecting oosporogenesis is not unexpected since *P. cactorum* obtains its sterol from plants.[79] Mycoplasmas are animal pathogens which are auxotrophic for sterol. The host provides sterol and it has been found that cholesterol is somewhat better than sitosterol in growth-support of the bacteria.[74] In sterol-auxotrophs of cultured animal cells, plant sterols are inferior to cholesterol in supporting growth.[83] Thus in systems adapted to cholesterol, it is the best suited sterol

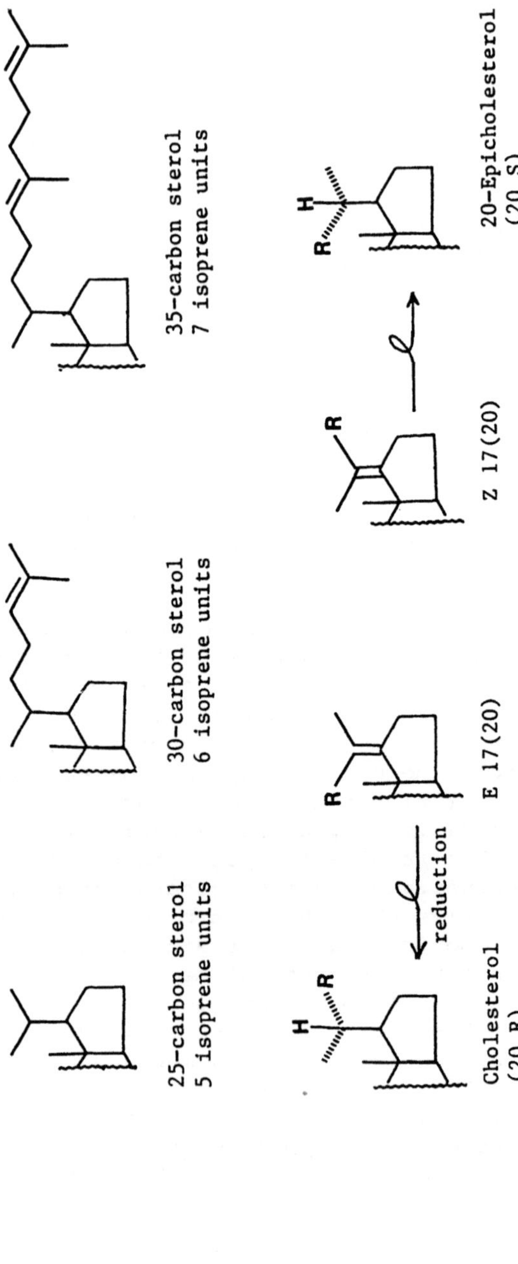

Fig. 12. Relationship between squalene-oxide structure and length of side chain.

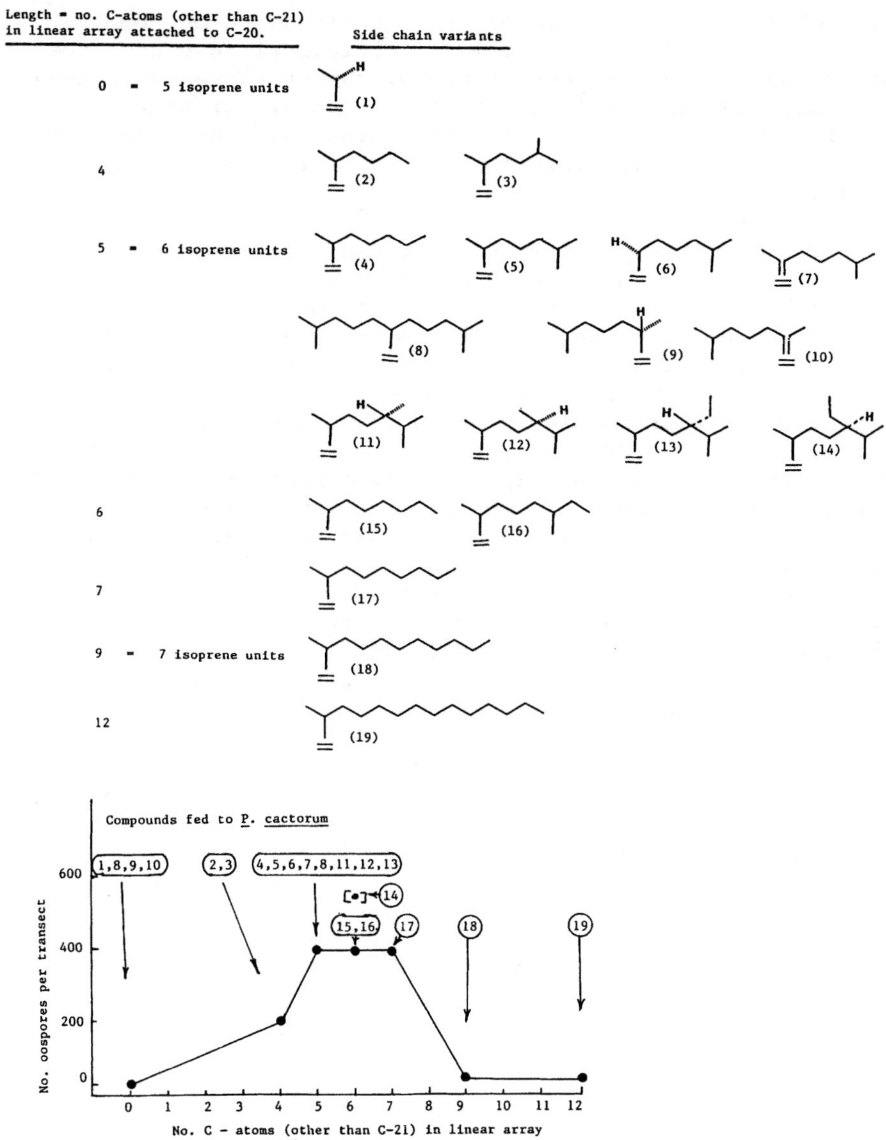

Fig. 13. Relationship between sterol structure and viable oospore production in *Phythophthora cactorum*. Data obtained from ref.[25,77-82]. All sterols were found to be accumulated by the mycelia to similar extents and to induce oospore production. Low or no activity indicates that most of the spores formed during the early phase of oosporogenesis aborted.

for membrane function. In plants the situation may be different. Therefore, in terms of biological response, the preference for one or another sterol can be viewed as a case of better and best (relative function) or all or none (absolute function). In cells that synthesize only one end product, that sterol plays dual roles, for example, ergosterol in yeast.[56,69]

The structural difference between cholesterol and sitosterol is an additional C-24 alkyl group in the latter. In collaboration with the X-ray crystallographers Drs. J. Griffin and W. Duax, we have determined that the 3-dimensional shape of the nucleus for the two sterols is essentially the same.[41,84,85] Similarly, the conformation of the side chain in both cases is preferentially oriented to the right because the configuration at C-20 fixes, through 1,3-diaxial interactions between the 20-H and 18-CH_3, the stereochemistry about the C-20 tetrahedral system.[85] As shown by X-ray crystallographic analysis (Fig. 14), there is much conformational freedom in the terminal portion of the side chain. We believe the varying mobilities of side chains to rotate in the manner described are important in sterol-controlled mediation of membrane fluidity. It seems unlikely, however, that rotation about the 17(20)-bond is also responsible for changes in the mobility of the side chain and sterol-lipid interactions in the bilayer, as speculated by others (ref.[59] cited in ref.[51]). Our evidence that the sitosterol side chain is more mobile in the tail region than the cholesterol side chain is based on what I should like to refer to as a "VT foot-print 1HNMR analysis" of several sterols examined in solution over the variable temperature (VT) range -40 to +80 or in the higher temperature range in tetrachloroethane (+80 - +140^0) (Nes and Benson, unpublished data). Plots were constructed on each compound. The log 1/T versus chemical shift in ppm of the signals corresponding to C-18, C-19, C-21, C-26, C-27, in sitosterol and stigmasterol, C-29 were graphed. From the Arrenhius plots, we could determine that a conformational stability in the nucleus (as flat) exists and the side chain orientation remains fixed to the right over the physiological temperature range. Assuming the tail portion of the side chain is highly mobile in the bilayer, as it is in the two extremes of pure solution and solid state, then a 24-alkylated sterol would be expected to sweep out a bigger cone in the bilayer than 24-desalkylated sterols, such as, cholesterol (Fig. 15). The increased van der Waals radii resulting from the bigger cone may be advantageous to cells where greater changes in membrane fluidity are warranted.

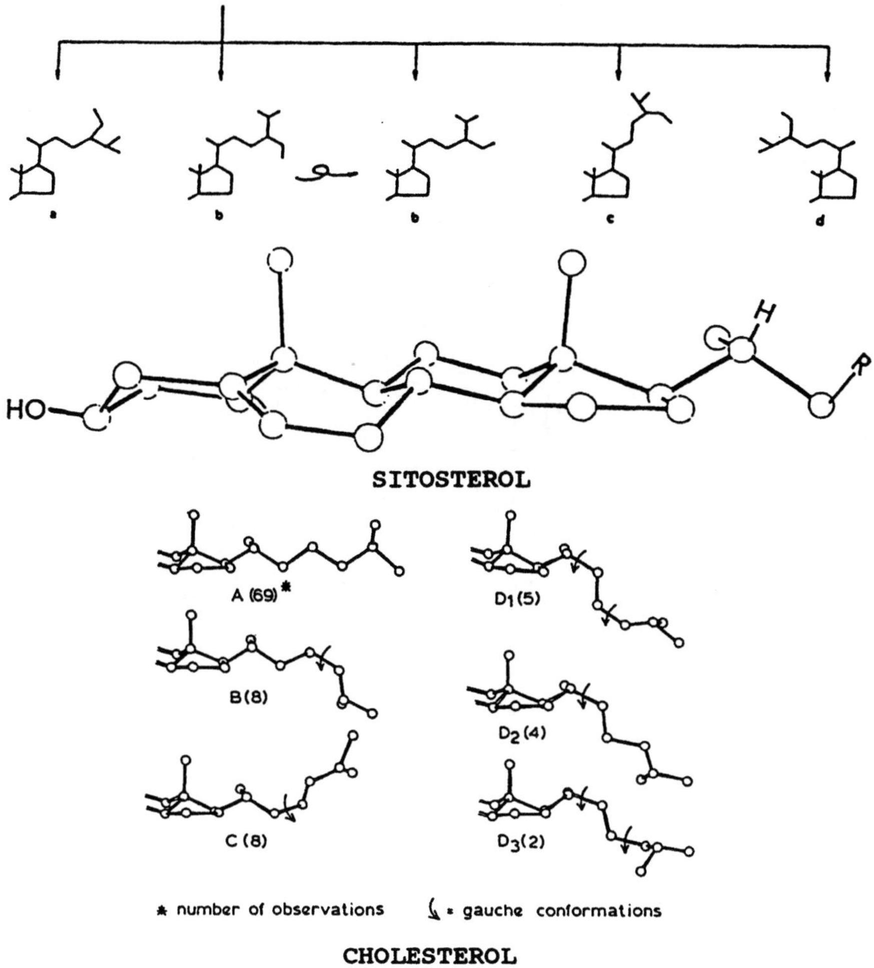

Fig. 14. Different conformations of sterol side chain. X-ray analysis was performed by Ms. R. Wong (USDA) and Drs. J.F. Girffen and W. Duax (Medical Foundation of Buffalo), cf.[41,97]).

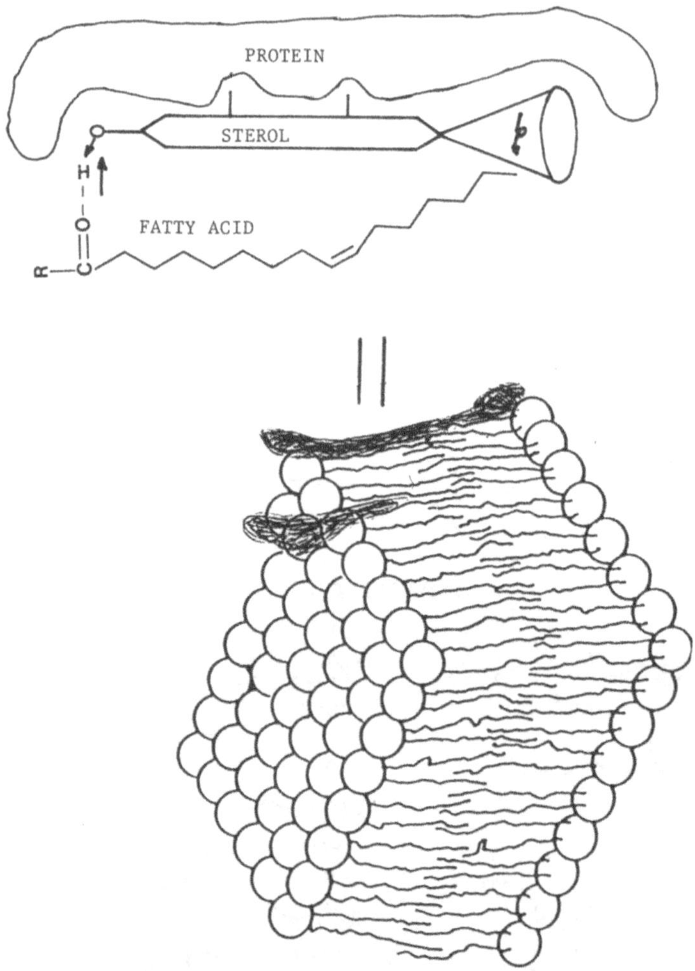

Fig. 15. Bridge model of sterol-protein-lipid interactions. After Nes and Heftman, 1981 (cf.[2]).

The increased fluidity brought about by 24-alkylated sterols can either be an absolute or relative function depending on the ontogeny of the cell. In most cells which synthesize them it is a relative function since cholesterol may replace 24-alkylated sterols in the bulk membrane role to affect cell proliferation (cf. Table 3). However, 24-

Table 3. Effect of sterol structure on biological activity in *S. cerevisiae* and *P. cactorum*.[1]

Growth response Added Sterol	Anaerobic yeast	Anaerobic yeast+Isq.	Semi-anaerobic yeast	Aerobic yeast+IL	GL-7 Growth-arrest	Gl-7 Resting cells	P. cactorum Growth	Oospores
Cholesterol	23	<1	100	-	100	100	100	37
Campesterol	32	<1	100	-	100	100	100	60
Sitosterol	-	<1	100	-	100	100	100	100
Stigmasterol	-	100	100	-	100	100	100	100
Ergosterol	100	100	100	<1[b]	100	100	100	40
Lanosterol	<1	<1	100	-	4	40	80	0 (10)
Cycloartenol	<1	<1	100	-	3	34	100(80)	0 (10)
Euphol	-	-	-	-	0	0	-	-
None	<1	<1	50	<1[b]	2	2	80(80)	0 (10)
α-amyrin	<1	-	-	-	0	0	100	0 (10)
Study	UD	56	72	28, UD	UD	UD	77-82	77-82

[a]Values are given as per cent total cells/ml, radial diameter, or oospores/transect relative to ergosterol grown cells (yeast) or sitosterol grown mycelia (*P. cactorum*). Values in parentesis are of cultures containing trace amounts of cholesterol in the media preparation (Agar).

[b]Compared with growth of wild-type yeast without added Iminolanosterol (IL.). Unpublished data (UD) obtained by Norton and Nes at the Russell Center or by the group at Drexel University.[27,28,56,72,77-82]

alkylation is also known to be essential as a developmental trigger and to influence cell growth and expansion, presumably as a result of sterol-protein associations.[2] As pointed out earlier, there are four approaches to demonstrate essentiality for 24-alkylated sterols: 1) feeding sterol biosynthesis inhibitors; 2) cells growing under anaerobiosis; 3) mutagenesis and 4) molecular cloning techniques. At the Russell Center we have been interested in the first two approaches and recently through collaboration with Dr. Martin Bard, from Purdue University, in approaches three and four. In each of the approaches we have attempted to eliminate an enzyme in the pathway that catalyzes the introduction or removal of a select grouping from the sterol molecule. Approaches one and two interfere with the catalysis of the preformed enzyme, while approaches three and four interfere with DNA governing enzyme production. Feeding sterol biosynthesis inhibitors (SBI) prepared in this laboratory[86] to cultured plant cells demonstrated that C-24 alkylation was important in growth support of sunflower.[87] Similar observations were noted with a commercial SBI fed to celery.[88] In collaboration with G. Popjak[89,90] we have demonstrated the physiological essentiality for reduction of the 24,25-double bond in animals using SBIs. Thus metabolism of the Δ^{24}-bond is important to cellular growth and expansion. In yeast all four approaches have been used to assess the functional significance for the biosynthetic inclusion of the 24-alkyl group into the sterol side chain.[28,56,91,92] Even though data are given in wild-type[27,56] and mutant yeast,[93] Parks, Gaber, Bard and their coworkers continue to assume 24-methylation is not essential in their yeast mutants and, by implication, in wild-type yeast.[91,92] We have reexamined the sterol composition of several of these mutants. They are all leaky, producing hormonal levels of ergosterol (93 and Nes, Norton and Bard, unpublished). The latter findings coupled with the growth-inhibitory effects of SBIs fed to supposed ergosterol-less yeast (erg 6 and GL 7), in the presence of dietary ergosterol (Nes and Norton, unpublished data), or SBI-treated wild-type yeast[28] confirms the importance for a biosynthetic renewal of natural sterols during aerobic growth.[89,90]

The use of transition state inhibitors that interfere with cyclization of squalene-oxide and C-24 methylation (Fig. 16) are particularly useful in structure-growth studies. Benveniste and coworkers have prepared a novel transition state inhibitor that selectively interferes with squalene-

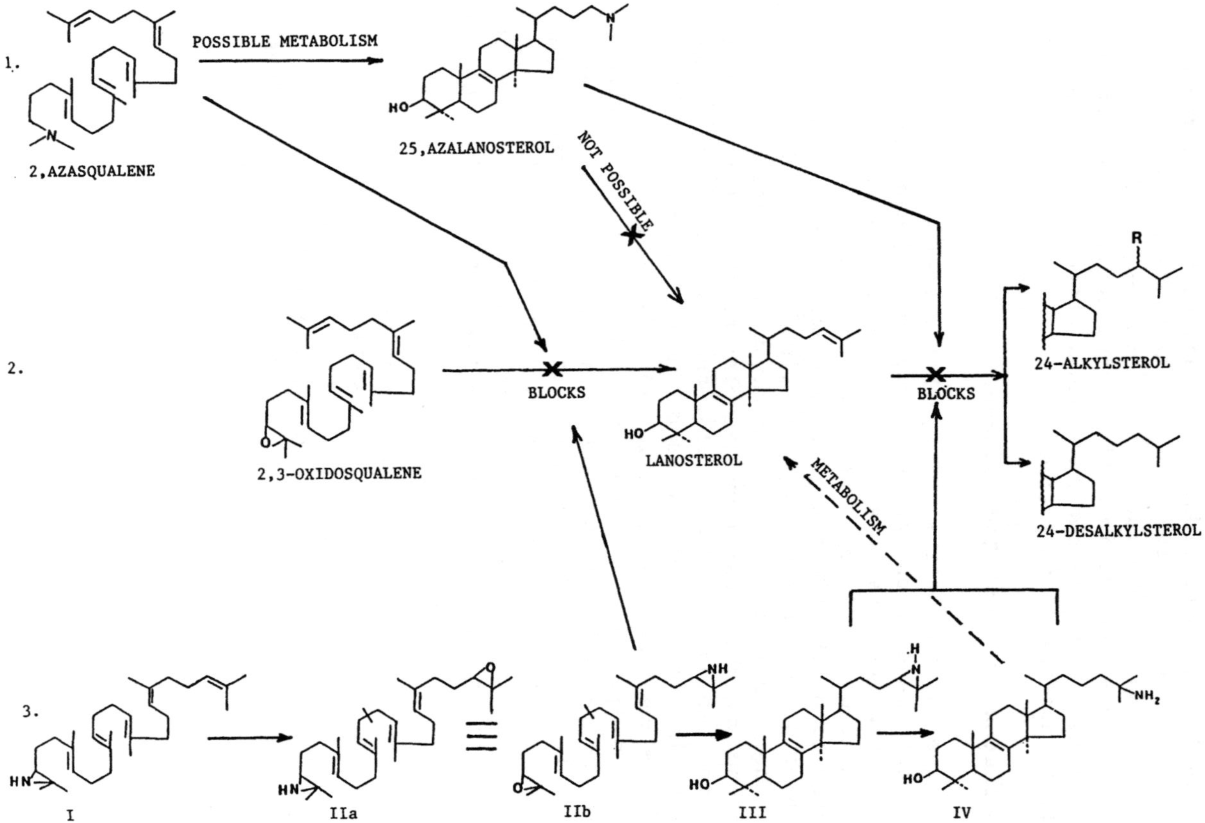

Fig. 16. Metabolism and demonstrable blocks in the sterol pathway induced by N-steroids. Data obtained from ref. 29 and references cited therein.

oxide cyclization to steroids, while allowing for biosynthesis of pentacyclic triterpenoids.[94] The use of this or similar inhibitors to probe the function of the triterpenoids is expected. We have found that aziridinyl N-steroids are potentially advantageous in medicine and agriculture compared with azasteroids because the aziridinyl steroids are transformable to non-toxic end products.[57]

Intermediates Versus End Product

On the assumption that biosynthesis operates in a non-random linear manner, end products may be produced having structural features that are superior to the intermediates from which it was derived. The compounds that may be considered to be the first sterol-like molecules in the cholesterol pathway are the 4,4,14-trimethyl steroids - cycloartenol and lanosterol.[15] In the transformation of cycloartenol to cholesterol by vascular plants, three methyl groups are removed at C-4 and C-14, the cyclopropyl group is isomerized into a $\Delta^{8(9)}$-double bond (producing lanosterol) and the $\Delta^{8(9)}$ is rearranged to Δ^5 (Fig. 6). Furthermore, in sitosterol biogenesis the Δ^{24}-bond is alkylated. Otherwise the steps in the pathway are the same as in cholesterol synthesis. In terms of cellular economy, why perform any of these energy expensive transformations? It is inconceivable that there would not exist some selective advantage in the removal of the geminal methyl groups at C-4 and the addition of methyl groups at C-24.

Bloch[16] and Benveniste[95] have advocated conformational reasoning for the ability of 4,4-dimethyl containing cyclopropyl sterols such as cycloartenol, to replace the natural 4,4,14-trisdesmethyl 24-alkylated Δ^5-sterols (campesterol, sitosterol, etc.) as membrane components. These investigators assume that the geminal methyls at C-4, and the angular 14α-methyl group are not debilitating features in cyclopropyl sterols as they are in lanosterol because of the overriding effect of the bent conformation. However, cycloartenol and lanosterol are equally pseudoplanar in solution and the solid state[96,97] so that, cycloartenol will possess the same pertubating features present on the lanosterol molecule. In terms of differences in van der Waals' dispersion forces compared with cholesterol, there is only about a 10 percent effect in having the 14 α-methyl group. While Bloch has recognized that the α-face planarity is destroyed by the angular methyl group,[16] this vew may have little bearing on sterol-lipid interactions in the bilayer. In fact, the presence of the 14α-methyl group is not deleterious for growth-support as Bloch, Parks and Linnane have shown. Thus, depending on the culture conditions, yeast anxotrophic for sterol can grow on lanosterol or dihydrolanosterol.[69,98,99] Consequently, 14α-methyl sterols may play the bulk role, so long as, the cells are leaky for hormonal levels of ergosterol (our unpublished findings). When the cells are prevented from synthesizing ergosterol, then lanosterol will not play the bulk function.[27,93] Recently we have speculated that the biosynthetic inclusion of the $\Delta^{8(9)}$-bond, compared with forming a saturated B-ring lanostenol ($\Delta^{0,24}$-lanostenol), during the cyclization process of squalene-oxide to a tetracycle product is an evolutionary adaptation selected for because the olefinic linkage is structurally important in the subsequent conversion of lanosterol and its stereoisomers eg., cycloartenol, to Δ^5-sterols.[100] In terms of physical chemistry, there is a significant difference in sterol-lipid interactions based on the H-bonding strength of 4,4-dimethyl steroids relative to cholesterol[101] and perhpas that is why the C-4 methyl groups are removed from cyclartenol and lanosterol. The difference in H-bonding capabilities could explain liposome data in which different sterols are incubated in the systems and the physical biochemistry is altered.[16] The Δ^5-grouping is superior to the Δ^8-grouping in terms of H-bonding strength as a result of its conformational contribution to the A and B rings. As Dreiding models show, when compared to the Δ^8-system, the Δ^5-system tends to flatten out the nucleus, thereby affecting the tilt and hydrogen bonding vector of the OH-group.

CONCLUSION

Cell size and the physical chemistry of the membrane are factors which, during evolution, may determine the suitability of particular sterols. Fine-tuning of biosynthesis is developmentally regulated so that only certain end products serve essential roles. The biosynthesis of 24-alkylated end products, in plants and fungi, was selected for because the end products conferred some advantage to the completion of the life cycle. Finally, the complicated patterns noted in sterol composition that appear during ontogeny and throughout phylogeny may have evolved as a result of structural features of the sterol molecule that are related to developmentally regulated functions.

REFERENCES

1. NES, W.R. 1974. Role of sterols in membranes. Lipids 9: 596-612.
2. NES, W.D., E. HEFTMANN. 1981. A comparison of triterpenoids with steroids as membrane components. J. Nat Prod. 44: 377-400.
3. ROHMER, M., P. BOUVIER, G. OURISSON. 1979. Molecular evolution of biomembranes: structural equivalents and phylogenetic precursors of sterols. Proc Natl. Acad. Sci. USA. 76: 847-851.
4. NES, W.D., R.C. HEUPEL, P.H. LE. 1984. A comparison of sterol biosynthesis in fungi and tracheophytes and its phylogenetic and functional implications. In: Structure, function and metablism of plant lipids. (P.A. Siegenthaler and W. Eichenberger, eds.), Elsevier Publ., Amsterdam, pp. 207-216.
5. PORALLA, K., E. KANNENBERG. 1987. Hopanoids: sterol equivalents in bacteria. ACS. Symp. Ser. 325: 239-251.
6. NES, W.D. 1987. Biosynthesis and requirement for sterols in the growth and reproduction of Oomycetes. ACS Symp. Ser. 325: 304-328.
7. SIPERSTEIN, M.D. 1984. Role of cholesterogenesis and isoprenoid synthesis in DNA replication and cell growth. J. Lipid Res. 25: 1462-1468.
8. HAUGHAN, P.A., J.R. LENTON, L.J. GOAD. 1987. Paclobutrazol inhibition of sterol biosynthesis in a cell suspension culture and evidence of an essential role for 24-ethyl sterol in plant cell division. Biochem. Biophys. Res. Commun. 146: 510-516.
9. GROSSMAN, K., E.W. WEILER, J. JUNG. 1985. Effects of different sterols on the inhibition of cell culture growth caused by the growth retardant tetracyclacis. Planta 164: 370-375.
10. DAHL, D., H. BIEMANN, J. DAHL. 1987. A protein kinase antigenically related to pp60v-src possibly involved in yeast cell cycle control: positive *in vivo* regulation by sterol. Proc. Natl. Acad. Sci. USA. 84: 4012-4016.
11. NES, W.R. 1977. The biochemistry of plant sterols. Adv. Lipid Res. 15: 233-324.
12. GOAD, L.J. 1983. How is sterol synthesis regulated in higher plants? Biochem. Soc. Trans. 11: 548-552.
13. GRAY, J.C. 1987. Control of isoprenoid biosynthesis in higher plants. Adv. Bot. Res. 14: 25-91.
14. BACH, T.J., H.K. LICHTENTHALER. 1987. Plant growth

regulation by mevinolin and other sterol biosynthesis inhibitors. ACS Symp. Ser. 325: 109-139.
15. NES, W.R., M.L. MCKEAN. 1977. Biochemistry of steroids and other isopentenoids. University Park Press, Baltimore, p.690.
16. BLOCH, K.E. 1983. Sterol structure and membrane function. CRC Crit. Rev. Biochem. 14: 47-92.
17. OURISSON, G., M. ROHMER, K. PORALLA. 1987. Prokaryotic hopanoids and other polyterpenoid sterol surrogates. Ann. Rev. Microbiol. 41: 301-333.
18. SANDOR, T., S. SONEA. 1975. Are steroids universal biomolecules? Biochem. Soc. Trans. 3: 1157-1159.
19. CALVIN, M. 1969. Chemical evolution. Clarendon Press, Oxford, p.200.
20. VAN TAMELEN, E.E. 1968. Bioorganic chemistry: sterols and cyclic terpene terminal epoxides. Acc. Chem. Res. 1: 111-120.
21. JOHNSON, W.S. 1976. Biomimetic polyene cyclizations: a review. Bioorgan. Chem. 5: 51-98.
22. RUZICKA, L. 1959. History of the iosprene rule. Proc. Chem Soc. 341-360.
23. KERWIN, J.L., N.D. DUDDLES. 1989 Reassessment of the role of phospholipids in sexual reproduction by sterol-auxotrophic fungi. J. Bacteriol. 171: 3831-3839.
24. NES, W.D. , P.K. HANNERS, E.J. PARISH. 1986. Control of fungal sterol C-24 transalkylation: importance to developmental regulation. Biochem. Biophys. Res. Commmun. 139: 410-415.
25. NES, W.D., G.A. SAUNDERS, E. HEFTMANN. 1982. Role of steroids and triterpenoids in the growth and reproduction of *Phytophthora cactorum.* Lipids 17: 178-183.
26. NES, W.D., P.H. LE. 1988. Regulation of sterol biosynthesis in *Saprolegnia ferax* by 25-azacholesterol. Pest. Biochem. Physiol. 30: 87-94.
27. NES, W.R., B.C. SEKULA, W.D. NES, J.H. ADLER. 1978. The functional importance of structural features of ergosterol in yeast. J. Biol. Chem. 253: 6218-6225.
28. NES, W.D., S. XU, W.F. HADDON. 1989. Evidence for similarities and differences in the biosynthesis of fungal sterols. Steroids 53: 533-558.
29. NES, W.D., S. XU, E.J. PARISH. 1989. Metabolism of 24(R,S),25-epiminolanosterol to 25-aminolanosterol and lanosterol by *Gibberella fujikuroi*. Arch. Biochem. Biophys. 272: 323-331.
30. CARTER, P.W., I.M. HEILBRON, B. LYTHGOE. 1939. The

lipochromes and sterols of the algal classes. Proc. Royal Soc. Lond. B. 128: 82-109.

31. LEVIN, E.Y., K. BLOCH. 1964. Absence of sterols in blue-green algae. Nature Lond. 202: 90-91.
32. DESOUZA, N.J., W.R. NES. 1968. Sterols: isolation from a blue-green alga. Science 162: 363.
33. REITZ, R.C., J.G. HAMILTON. 1968. The isolation and identification of two sterols from two species of blue-green algae. Comp. Biochem. Physiol. 25: 401-416.
34. KAHLHASE, M., P. POHL. 1988. Saturated and unsaturated sterols of nitrogen-fixing blue-green algae (cyanobacteria). Phytochemistry 27: 1735-1740.
35. BERGMANN, W. 1953. The plant sterols. Ann. Rev. Plant Physiol. 4: 383-426.
36. KOHL, W., A. GLOE, H. REICHENBACH. 1983. Steroids from the nyxobacterium *Nannocystis exedens*. J. Gen. Microbiol. 129: 1629-1635.
37. BIRD, C.W., J.M. LYNCH, F.J. Pirt, W.W. REID, C.J.W. BROOKS, B.S. MIDDLEDITCH. 1971. Steroids and squalene in *Methylococcus capsulatus* grown on methane. Nature Lond. 230: 473.
38. PATT, T.E., R.S. HANSON. 1978. Intracytoplasmic membrane, phospholipid and sterol content of *Methylbacterium organophilum* cells grown under different conditions. J. Bacteriol. 134: 36-643.
39. NES, W.D., P.H. LE, L.R. BERG, G.W. PATTERSON, J.L. KERWIN. 1986. A comparison of cycloartenol and lanosterol biosynthesis and metabolism by the Oomycetes. Experientia 42: 556-558.
40. XU, S., W.D. NES. 1988. Biosynthesis of cholesterol in the yeast mutant erg6. Biochem. Biophys. Res. Commun. 155: 509-517.
41. HEUPEL, R.C., Y. SAUVAIRE, P.H. LE, E.J. PARISH, W.D. NES. 1986. Sterol composition and biosynthesis in sorghum: importance to developmental regulation. Lipids 21: 69-75.
42. POPJAK, G., A. MEENAN. 1987. Regulation of 3-hydroxyl-3-methyl glutaryl coenzyme A reductase: search for the enzyme repressor derived from mevalonate. Proc. R. Soc. Lond. B. 231: 391-414.
43. NES, W.R., W.D. NES. 1980. Lipids in evolution. Plenum Press, New York, 240 pp.
44. VAN TAMELEN, E.E.. 1982. Bioorganic characterization and mechanism of the 2,3-oxidosqualene→Lanosterol conversion. J. Amer. Chem. Soc. 104: 6480-6481.
45. CORNFORTH, J.W. 1968. Olefin alkylation in

biosynthesis. Angew. Chem. 7: 903-964.
46. JOHNSON, W.S., S.J. TELFER, S. CHENG, U. SCHUBERT. 1987. Cation-stabilization auxiliaries: a new concept in biomimetic polyene cyclization. J. Amer. Chem. Soc. 109: 2517-2518.
47. GOULD, S.J. 1977. Ontogeny and Phylogeny. Belknap Press, Cambridge, 501 pp.
48. DOBZHANSHY, T., F.J. AYALA, G.L. STEBBINS, J.W. VALENTINE. 1977. Evolution. Freeman and Co., San Francisco, 572 pp.
49. GOAD, L.J., T.W. GOODWIN. 1973. The biosynthesis of plant sterols. Prog. Phytochemistry 1: 113-198.
50. CAVALIER-SMITH, T. 1987. The origin of eukaryote and archaebacterial cells. Ann. New York Acad. Sci. 503: 17-54.
51. BURDEN, R.S., D.T. COOKE, G.A. CARTER. 1989. Inhibitors of sterol biosynthesis and growth in plants and fungi. Phytochemistry 28: 1791-1804.
52. PATTERSON, G.W. 1987. Sterol synthesis and distribution and algal phylogeny. In: The metabolism, structure and function of plant lipids. (P.K. Stumpf, J.B. Mudd and W.D. Nes eds.), Plenum Press, New York, pp. 631-636.
53. KOKKE, W.C.M., J.N. SCHOOLERY, W. FENICAL, C. DJERASSI. 1984. Biosynthetic studies of marine lipids. 4. Mechanism of side chain alkylation in (E)-24-propylidenecholesterol by a chrysophyte alga. J. Org. Chem. 49: 3742-3752.
54. AKIHISA, T., T. TAMURA, T. MATSUMOTO, W.D.M.C. KOKKE, T. YOKOTA. 1989 Isolation of acetylenic sterols from a higher plant. Further evidence that marine sterols are not unique. J. Org. Chem. 54: 606-610.
55. NES, W.D., P.H. LE. 1989. Evidence for separate intermediates in the biosynthesis of 24β-methyl sterol end products in *Gibberella fujikuroi*. Biochem. Biophys. Acta (in press).
56. PINTO, W.J., W.R. NES. 1983. Stereochemical specificity for sterols in *Saccharomyces cerevisiae*. J. Biol. Chem. 258: 4472-4476.
57. NES, W.D., R.C. HEUPEL. 1986. Physiological requirement for biosynthesis of multiple 24β-methylsterols in *Gibberella fujikuroi*. Arch. Biochem. Biophys. 244: 211-217.
58. NES, W.R. 1971. Regulation of the sequencing in sterol biosynthesis. Lipids 6: 219-224.
59. BAISTED, D.F., E. CAPSTACK, W.R. NES. 1962. The biosynthesis of β-amyrin and β-sitosterol in

germinating seeds of *Pisum sativum*. Biochemistry 1: 537-541.
60. BAISTED, D.J. 1971. Sterol and triterpene synthesis in the developing and germinating pea seed. Biochem J. 124: 375-383.
61. HEUPEL, R.C., W.D. NES, J.A. VERBEKE. 1987. Developmental regulation of sterol and pentacyclic triterpene biosynthesis and composition: A correlation with sorghum floral initiation. In: The metabolism, structure and function of plant lipids. (P.K. Stumpf, J.B. Mudd, W.D. Nes, eds.), Plenum Press, New York, pp. 53-56.
62. SAUVAIRE, Y., B. TAL, R.C. HEUPEL, R. ENGLAND, P.K. HANNERS, W.D. NES, J.B. MUDD. 1987. The metabolism, structure and function of plant lipids. (P.K. Stumpf, J.B. Mudd, W.D. Nes, eds.), Plenum Press, New York, pp. 107-109.
63. GARG, V.K., W.R. NES. 1985. Changes in Δ^5- and Δ^7-sterols during germination and seedling development of *Cucurbita maxima*. Lipids 20: 876-883.
64. AKIHISA, T., S. THAKUR, F.U. FOSENSTEIN, T. MATSUMOTO. 1985. Sterols of cucurbitaceae: the configuration of 24-alkyl-Δ^5-, Δ^7- and Δ^8-sterols. Lipids 21: 39-47.
65. KARUNEN, P., K. HAKALA, S. HEINONEN. 1984. Occurrence of esterified triterpenoids alcohols in the leaves of *Pilosella officinarm*. Physiol. Plant. 61: 243-250.
66. SEKULA, B.C., W.R. NES. 1980. The identification of cholesterol and other steroids in *Euphorbia pulcherimma*. Phytochemistry 19: 1509-1512.
67. NODA, M., M. TANAKA, Y. SETO, T. AIBA, C. OKU. 1988. Occurrence of cholesterol as a major sterol component in leaf surface lipids. Lipids 23: 439-444.
68. NES, W.D., T.J. BACH. 1985. Evidence for a mevalonate shunt in a tracheophyte. Proc R. Soc. Lond. B. 225: 425-444.
69. RODRIQUEZ, R.J., C. LOW, C.D.K. BOTTEMA, L.W. PARKS. 1985. Multiple functions for sterols in *Saccharomyces cerevisiae*. Biochem. Biophys. Acta 837: 336-343.
70. NES, W.R., J.H. ADLER, J.T. BILLHEIMER, K.A. ERICKSON, J. JOSEPH, J.R. LANDREY, R. MARCACCIO-JOSEPH, K.S. RITTER, R.L. CONNER. 1982. A comparison of the biological properties of androst-5-en-3β-ols and 21-isopentyl-cholesterol with those of cholesterol. Lipids 17: 257-262.
71. NES, W.R. 1987. Multiple roles for plant sterols. In: The metabolism, structure and function of plant lipids.

(P.K. Stumpf, J.B. Mudd, W.D. Nes, eds.), Plenum Press, New York, pp 3-9.
72. NES, W.R. 1987. Structure-function relationships for sterols in *Saccharomyces cerevisiae*. ACS Sump. Ser. 325: 254-267.
73. HAIGH, G., H.J. FORSTER, K. BIEMANN, N.H. TATTRIE, J.R. COLVIN. 1973. Induction and orientation of bacterial cellulose microfibrils by a novel terpenoid from *Acetobacter xylinum*. Biochem. J. 135: 145-149.
74. CLEJAN, S., R. BITTMAN, S. ROTTEM. 1981. Effects of sterol structure and exogenous lipids on the transbilayer distribution of sterols in the membrane of *Mycoplasma capricolum*. Biochemistry 20: 2200-2204.
75. KRIEF, A.J., SCHAUDER, E. GUITTET, C.G. DUPEHOAT, J. LALLEMAND. 1987. About the mechanism of sterol biosunthesis. J. Amer. Chem Soc. 109: 7910-7911.
76. VAN TAMELEN, E.E., J.R. HEYS. 1975. Enzymic epoxidation of squalene variants. J. Amer. Chem. Soc. 97: 1252-1253.
77. NES, W.D., A.E. STAFORD. 1984. Side-chain structural requirements for sterol-induced regulation of *Phytophthora cactorum* physiology. Lipids. 19: 544-549.
78. NES, W.D., A.E. STAFFORD. 1983. Evidence for metabolic and functional discrimination of sterol by *Phytophthora cactorum*. Proc. Natl. Acac. Sci. USA. 80: 3227-3231.
79. NES, W.D., G.W. PATTERSON, G.A. BEAN. 1980. Effect of steric and nuclear changes in steroids and triterpenoids on sexual reproduction of *Phytophthora cactorum*. Plant Physiol. 66: 1008-1011.
80. NES, W.D., P.K. HANNERS, G.A. BEAN, G.W. PATTERSON. 1982. Inhibition of growth and sitosterol-induced sexual reproduction in *Phytophthora cactorum* by steroidal alkaloids. Phytopathology 72: 447-450.
81. NES, W.D., W.R. NES. 1983. Disassociation of oogonia formation from oospore production by sterols with varying side chain lengths. Experientia 39: 276-278.
82. NES, W.D.. 1984. Control of fungal physiology by polycyclic isopentenoids and its phylogenetic implications. In: Isopentenoids in plants: biochemiostry and function. (W.D. Nes, G. Fuller, L. Tsai, eds.), Dekker, New York, pp 267-290.
83. RUJANAVECH, C., D.F. SILBERT. 1986. LM cell growth and membrane lipid adaptations to sterol structure. J. Biol Chem. 261: 7196-7203.
84. DUAX, W.L., J.F. GRIFFIN, C. CHEER. 1989. Steroid conformational analyses based on X-ray crystal

structure determination. In: Analysis of sterols and other biologically significant steroids, Academic Press, New York, pp. 203-221.
85. NES, W.D., R.Y. WONG, M. BENSON, J.R. LANDREY, W.R. NES. 1984. Rotational isomerism about the 17(20)-bond of steroids and euphoids as shown by the crystal structures of euphol and tirucallol. Proc. Natl. Acad. Sci. USA. 81: 5896-5900.
86. PARISH, E.J., W.D. NES. 1988. Synthesis of new eqimi-noisopenoids. Synth. Commun. 18: 221-226.
87. TAL, B., W.D. NES. 1987. Regulation of sterol biosynthesis: Importance of the C-24 alkyl group to growth of sunflower suspension cultures. Plant Physiol. (suppl.) 83: 161.
88. HAUGHN, P.A., J.R. LENTON, L.J. GOAD. 1988. Sterol requirements and paclobutrazol inhibition of a celery cell culture. Phytochemistry 27: 2491-2500.
89. POPJAK, G., A. MEENAN, W.D. NES. 1987. Effects of 2,3-iminosqualene on cultured cells. Proc. R. Soc. Lond. B. 232: 273-387.
90. POPJAK, G., A. MEENAN, E.J. PARISH, W.D. NES. 1989. Inhibition of cholesterol synthesis and cell growth by 24(R,S),25-iminolanosterol and triparinol in cultured rat hepatoma cells. J. Biol. Chem. 264: 6230-6238.
91. GABER, R.F., D.H. COPPLE, B.K. KENNEDY, M. VIDAL, M. BARD. 1989. Erg6 is essential for normal membrane function in *Saccharomyces cerevisiae* but is not required for the formation of the cell cycle. Mol. Cell. Biol. 9: (in press).
92. MCCAMMON, M.T., M. HARTMANN, C.D.K. BOTTEMA, L.W. PARKS. 1984. Sterol methylation in *Saccharomyces cerevisiae*. J. Bacteriol. 157: 475-483.
93. NES, W.R., I.C. DHANUKA. 1988. Inhibition of sterol synthesis by Δ^5-sterols in a sterol auxotroph of yeast defective in oxidosqualene cyclase and cytochrome P-450. J. Biol. Chem. 263: 11844-11850.
94. TATON, M., P. BENVENISTE, A. RAHIER. 1986. N-[C1,5,8]-trimethyldecyl]-4α, 10-dimethyl-8-aZA-trans-decal-3β-ol, a novel potent inhibitor of 2,3-oxidosqualene cycloartenol and lanosterol cyclases. Biochem. Biophys. Res. Commun. 138: 764-770.
95. BENVENISTE, P., M. BLADOCHE, M.F. COSTET, A. EHRHADT. 1984. Use of inhibitors of biosynthesis to study plasmalemma structure and function. In: Membranes and compartmentalization in the regulation of plant functions. (A. Boudet, G. Alibert, G. Marigo, P. Lea,

eds.), Ann. Proc. Phytochem. Soc., Europe, 24: 283-300.
96. NES, W.D., M. BENSON, R.E. LUNDIN, P.H. LE. 1988. Conformational analysis of 9β, 19-cyclopropyl sterols: Detection of the pseudoplanar conformer by nuclear Overhauser effects and its functional implications. Proc. Natl. Acad. Sci., USA. 85: 5759-5763.
97. DUAX, W.L., Z. WAWRZAK, J.F. GRIFFIN, C. CHEER. 1988. Sterol conformation and molecular properties. In: Biology of cholesterol. (P.Y. Yeagle, ed.), CRC. Press, Boca Raton, pp 1-18.
98. PROUDLOCK, J.W., L.W. WHEELDON, D.J. JOLLOW, A.W. LINNANE. 1968. Role of sterols in *Saccharomyces cerevisiae*. Biochim. Biophys. Acta 152: 434-437.
99. BUTTKE, T.M., K. BLOCH. 1981. Utilization and metabloism of methyl-sterol derivatives in the yeast mutant GL7. Biochemistry 20: 3267-3272.
100. NES, W.D., R.A. NORTON, E.J. PARISH, A. MEENAN, G. POPJAK. 1989. Concerning the role of 24,25-dihydrolanosterol and lanosterol in sterol biosynthesis by cultured cells. Steroids 53: 461-475.
101. XU, S., R.A. NORTON, F.G. CRUMLEY, W.D. NES. 1988. Comparison of the chromatographic properties of sterols, select additional steroids and triterpenoids: gravity flow colum liquid chromatography, thin-layer chromatography and high-performance liquid chromatography. J. Chromatogr. 452: 377-398.

INDEX

GPSR Compliance

The European Union's (EU) General Product Safety Regulation (GPSR) is a set of rules that requires consumer products to be safe and our obligations to ensure this.

If you have any concerns about our products, you can contact us on ProductSafety@springernature.com

In case Publisher is established outside the EU, the EU authorized representative is:

Springer Nature Customer Service Center GmbH
Europaplatz 3
69115 Heidelberg, Germany

Batch number: 10620163

Printed by Printforce, the Netherlands